# Kant: Studies on Mathematics in the Critical Philosophy

There is a long tradition, in the history and philosophy of science, of studying Kant's philosophy of mathematics, but recently philosophers have begun to examine the way in which Kant's reflections on mathematics play a role in his philosophy more generally, and in its development. For example, in the *Critique of Pure Reason*, Kant outlines the method of philosophy in general by contrasting it with the method of mathematics; in the *Critique of Practical Reason*, Kant compares the Formula of Universal Law, central to his theory of moral judgement, to a mathematical postulate; in the *Critique of Judgement*, where he considers aesthetic judgement, Kant distinguishes the mathematical sublime from the dynamical sublime. This last point rests on the distinction that shapes the Transcendental Analytic of Concepts at the heart of Kant's Critical philosophy, between the mathematical and the dynamical categories. These examples make it clear that Kant's transcendental philosophy is strongly influenced by the importance and special status of mathematics. The contributions to this book explore this theme of the centrality of mathematics to Kant's philosophy as a whole.

This book was originally published as a special issue of the *Canadian Journal of Philosophy*.

**Emily Carson** is Associate Professor of Philosophy at McGill University, Montreal, Canada. She works in early modern philosophy, with a focus on Kant and the philosophy of mathematics, and is a member of the Editorial Board of the *Canadian Journal of Philosophy*.

**Lisa Shabel** is Associate Professor in the Department of Philosophy at Ohio State University, Columbus, USA. Her primary interests are in the history and philosophy of mathematics and Kant's Critical philosophy.

**Previous titles published from the *Canadian Journal of Philosophy*:**

**New Essays on Thomas Reid**
*Edited by Patrick Rysiew*

**New Essays on the Nature of Propositions**
*Edited by David Hunter and Gurpreet Rattan*

# Kant: Studies on Mathematics in the Critical Philosophy

*Edited by*

**Emily Carson and Lisa Shabel**

Routledge
Taylor & Francis Group

LONDON AND NEW YORK

First published 2016 by Routledge

2 Park Square, Milton Park, Abingdon, Oxon OX14 4RN
711 Third Avenue, New York, NY 10017, USA

*Routledge is an imprint of the Taylor & Francis Group, an informa business*

First issued in paperback 2017

*British Library Cataloguing in Publication Data*
A catalogue record for this book is available from the British Library

ISBN 13: 978-1-138-92502-1 (hbk)
ISBN 13: 978-1-138-09481-9 (pbk)

Typeset in Times
by RefineCatch Limited, Bungay, Suffolk

**Publisher's Note**
The publisher accepts responsibility for any inconsistencies that may have arisen during the conversion of this book from journal articles to book chapters, namely the possible inclusion of journal terminology.

**Disclaimer**
Every effort has been made to contact copyright holders for their permission to reprint material in this book. The publishers would be grateful to hear from any copyright holder who is not here acknowledged and will undertake to rectify any errors or omissions in future editions of this book.

# Contents

*Citation Information*     vii
*Notes on Contributors*     ix

Introduction     1
*Emily Carson and Lisa Shabel*

1. Spatial representation, magnitude and the two stems of cognition     6
   *Thomas Land*

2. Infinity and givenness: Kant on the intuitive origin of
   spatial representation     33
   *Daniel Smyth*

3. Kant on the Acquisition of Geometrical Concepts     62
   *John J. Callanan*

4. Kant (vs. Leibniz, Wolff and Lambert) on real definitions
   in geometry     87
   *Jeremy Heis*

5. Definitions of Kant's categories     113
   *Tyke Nunez*

6. Arbitrary combination and the use of signs in mathematics:
   Kant's 1763 Prize Essay and its Wolffian background     140
   *Katherine Dunlop*

7. Kant on the construction and composition of motion in
   the Phoronomy     168
   *Daniel Sutherland*

# CONTENTS

8. Kant on conic sections           201
   *Alison Laywine*

9. 'With a Philosophical Eye': the role of mathematical beauty
   in Kant's intellectual development       241
   *Courtney David Fugate*

   *Index*           271

# Citation Information

The chapters in this book were originally published in the *Canadian Journal of Philosophy*, volume 44, issues 5–6 (December 2014). When citing this material, please use the original page numbering for each article, as follows:

**Introduction**
*Introduction*
Emily Carson and Lisa Shabel
*Canadian Journal of Philosophy*, volume 44, issues 5–6 (December 2014)
pp. 519–523

**Chapter 1**
*Spatial representation, magnitude and the two stems of cognition*
Thomas Land
*Canadian Journal of Philosophy*, volume 44, issues 5–6 (December 2014)
pp. 524–550

**Chapter 2**
*Infinity and givenness: Kant on the intuitive origin of spatial representation*
Daniel Smyth
*Canadian Journal of Philosophy*, volume 44, issues 5–6 (December 2014)
pp. 551–579

**Chapter 3**
*Kant on the Acquisition of Geometrical Concepts*
John J. Callanan
*Canadian Journal of Philosophy*, volume 44, issues 5–6 (December 2014)
pp. 580–604

**Chapter 4**
*Kant (vs. Leibniz, Wolff and Lambert) on real definitions in geometry*
Jeremy Heis
*Canadian Journal of Philosophy*, volume 44, issues 5–6 (December 2014)
pp. 605–630

## Chapter 5

*Definitions of Kant's categories*
Tyke Nunez
*Canadian Journal of Philosophy*, volume 44, issues 5–6 (December 2014)
pp. 631–657

## Chapter 6

*Arbitrary combination and the use of signs in mathematics: Kant's 1763*
*Prize Essay and its Wolffian background*
Katherine Dunlop
*Canadian Journal of Philosophy*, volume 44, issues 5–6 (December 2014)
pp. 658–685

## Chapter 7

*Kant on the construction and composition of motion in the Phoronomy*
Daniel Sutherland
*Canadian Journal of Philosophy*, volume 44, issues 5–6 (December 2014)
pp. 686–718

## Chapter 8

*Kant on conic sections*
Alison Laywine
*Canadian Journal of Philosophy*, volume 44, issues 5–6 (December 2014)
pp. 719–758

## Chapter 9

*'With a Philosophical Eye': the role of mathematical beauty in Kant's*
*intellectual development*
Courtney David Fugate
*Canadian Journal of Philosophy*, volume 44, issues 5–6 (December 2014)
pp. 759–788

For any permission-related enquiries please visit:
http://www.tandfonline.com/page/help/permissions

# Notes on Contributors

**John J. Callanan** is Lecturer in Philosophy at King's College London, UK. His current research interests include history of philosophy (particularly early modern philosophy), and Kant's Critical philosophy, epistemology and metaphysics.

**Emily Carson** is Associate Professor of Philosophy at McGill University, Montreal, Canada. She works in early modern philosophy, with a focus on Kant and the philosophy of mathematics, and is a member of the Editorial Board of the *Canadian Journal of Philosophy*.

**Katherine Dunlop** is Assistant Professor in the Department of Philosophy at the University of Texas at Austin, USA.

**Courtney David Fugate** is Assistant Professor in the Civilization Sequence Program at the American University of Beirut, Lebanon. His research interests include Immanuel Kant, early modern philosophy, Wilfrid Sellars and post-Sellarsian philosophy, and epistemology.

**Jeremy Heis** is Associate Professor in the Department of Logic and Philosophy of Science at the University of California, Irvine, USA. His research interests include the history and philosophy of mathematics and logic, Kant, and early analytic philosophy.

**Thomas Land** is the Donnelley Research Fellow at Corpus Christi College, University of Cambridge, UK. He is primarily interested in the philosophy of Kant.

**Alison Laywine** is Associate Professor in the Department of Philosophy at McGill University, Montreal, Canada. Her specialist areas are Kant's metaphysics and natural philosophy, the history of early modern philosophy, and the history of philosophical reflection on music theory.

**Tyke Nunez** is currently a PhD student in the Philosophy Department at the University of Pittsburgh, USA.

**Lisa Shabel** is Associate Professor in the Department of Philosophy at Ohio State University, Columbus, USA. Her primary interests are in the history and philosophy of mathematics and Kant's Critical philosophy.

**Daniel Smyth** is currently a PhD student in the Department of Philosophy at the University of Chicago, USA.

**Daniel Sutherland** is Associate Professor in the Department of Philosophy at the University of Illinois at Chicago, USA. His primary interests include the relationship between philosophy, mathematics and science in the work of Immanuel Kant, and mathematical cognition more generally.

# Introduction

Despite the fact that what Kant has to say about mathematics is scattered throughout many texts, and is sometimes very cryptic, Kant's philosophy of mathematics has long been a fertile research area. It has emerged from work in this area that the development of Kant's Critical philosophy, his metaphysical and epistemological doctrines, and his conception of the method and systematicity of philosophy were clearly informed by his serious reflection on mathematics, logic and the exact sciences. Indeed, Kant uses mathematical truths – most famously that $7 + 5 = 12$ – as his primary examples of synthetic *a priori* judgments, the question of the possibility of which frames his entire project in the *Critique of Pure Reason*. Accordingly, the task of explaining Kant's philosophy of mathematics has largely centred on understanding his conception of mathematics as a domain of demonstrable synthetic *a priori* truths. More recently, however, Kant scholars have begun to consider the broader implications of this work for the Critical philosophy itself by examining the role of Kant's reflections on mathematics in his philosophy more generally.

While Kant's explanation of the synthetic apriority of mathematical truth hinges on his theory of mathematical concept construction, these potentially narrow observations about mathematics are deeply connected to many more far-reaching philosophical commitments. Ultimately, then, Kant's account of the philosophy of mathematics must be understood relative to a variety of aspects of the Critical philosophy. Accordingly, classic debates about the role of Kantian intuition in mathematical reasoning will be augmented by research into Kant's understanding of mathematical practice and the applicability of mathematics in scientific understanding and perceptual experience; the general definability of mathematical concepts; the relation between mathematical thought and our concepts of space, time, magnitude and purposiveness; and the specific role that mathematics plays in motivating Kant's unique account of our representational capacities and his defense of the transcendental ideality of the objects of experience. We hope that these papers will show how Kant's appeal to mathematics as a model of synthetic *a priori* cognition sheds light on our cognitive faculties in general.

Thomas Land's contribution to the volume explores precisely this question. In his paper, "Spatial Representation, Magnitude and Two Stems of Cognition", Land uses Kant's philosophy of mathematics to shed light on the distinction between sensibility and understanding, and in particular on how the two faculties cooperate to produce sensible representations. Land shows how Kant's account of mathematical construction can serve as a model for understanding the notion of the productive synthesis of imagination, which is supposed to involve both a sensible and a spontaneous aspect, and which underlies the empirical apprehension of appearances. More specifically, he argues that the representation of determinate spaces in intuition depends on an empirical version of the kind of pure synthesis involved in geometrical construction. This gives precise sense to the idea that both the concepts of mathematics and the categories are concepts of the form of appearances. Land's result has significant implications for the debate over conceptualist vs. non-conceptualist readings of Kant: it suggests that intuitions, although distinctively sensible, are also dependent on spontaneity, and thus on concepts – in particular, the concept of magnitude, which is the subject matter of mathematics.

Like Land, Daniel Smyth shows how mathematical considerations – in this case, about infinity – underlie Kant's distinction between the faculties of sensibility and understanding, and therefore also underlie the related distinction between intuition and concept. Against the common view that Kant simply presupposes his distinction among kinds of representations, Smyth contends in his "Infinity and Givenness" that the Metaphysical Exposition of the Concept of Space contains the first stage of an argument in support of Kant's distinction between conceptual and intuitive representation. According to Smyth, the argument begins with a functional conception of sensibility as object-giving, and turns on the infinitary structure of space – its continuity and open-endedness – which cannot be accounted for by the understanding. While the connection between intuition and infinity is familiar, Smyth turns the standard view on its head, claiming that the singularity of intuition is a *consequence* of the argument of the Exposition, rather than a presupposition: the Exposition fills out the minimal functional conception of intuition with which Kant begins the first *Critique*. In this way, Smyth argues that mathematical considerations underlie the fundamental distinction between the faculties of sensibility and understanding that Kant takes to be one of his most significant innovations.

An examination of mathematical concepts and definitions, and, in particular, of the precise role that such representations play in justifying mathematical judgments, is crucial for understanding Kant's philosophy of mathematics. The papers by Callanan and Heis provide careful analyses of these fundamental notions. Even in his pre-Critical work, Kant recognized that the method of mathematics gave it, its concepts, and its definitions, a special status. In his paper "Kant on the Acquisition of Geometrical Concepts", John Callanan argues for an as yet underappreciated distinction between mathematical and other knowledge claims. Kant importantly distinguishes between the acquisition conditions of concepts and the justification conditions for the use of those concepts. Callanan argues, though, that geometrical construction is both a form of concept acquisition *and* is sufficient to justify the use of those concepts. This peculiarity of mathematical judgments, according to Callanan, explains Kant's claim that mathematical judgments are 'combined with consciousness of their necessity'. In this way, the paper points to a new way of accounting for the modal phenomenology of mathematical judgments.

In his contribution to the volume, "Kant (vs. Leibniz, Wolff and Lambert) on Definitions in Geometry," Jeremy Heis spells out Kant's theory of real definitions by contrasting it to those of three of his predecessors. While Leibniz, Wolff, Lambert and Kant all share the view that geometrical definitions are real definitions, they nevertheless disagree about what constitutes a real definition, and so about what accounts for the status and reality of geometrical definitions of, for example, circle and parallel lines. Heis argues that Kant rejects definitions that Leibniz, Wolff and Lambert each accept because of the uniquely stringent requirements his philosophy of geometry imposes on real definitions. In particular, Heis shows that Kant's conception of mathematics as rational cognition from the construction of concepts, and his claim that concepts (including mathematical concepts) "rest on functions", explains his disagreement with his predecessors over real definitions.

The considerations adduced by Heis are standardly thought to show that for Kant, *only* mathematical concepts admit of real definition. But by means of a detailed analysis of Kant's treatment of definitions, Tyke Nunez challenges this widely accepted view to show that some philosophical concepts do, in fact, admit of definition. In his "Definitions of Kant's Categories," Nunez argues that for Kant, the schemata of the categories ensure the reality of their definitions. Because these definitions do not provide a rule for the construction of their objects, they will not be genetic like those of the concepts of mathematics. Nonetheless, in both cases, the reality of the definitions rests on the claim that they are concepts of the form of objects of possible experience. In this respect, mathematics and philosophy turn out to differ less than we might have expected.

The broad themes of construction and definition are brought together in Katherine Dunlop's paper "Arbitrary Combination and the Use of Signs in Mathematics". She uncovers a role for the perceptibility of mathematical symbols in Kant's Prize Essay, and compares it to the role that sensibility plays for Wolff and Leibniz, revealing a continuity between the views of Kant's predecessors and his own pre-Critical views on the role of signs in mathematics. Dunlop argues that this role for sensibility secures the objective reference of mathematical concepts, but it cannot address the question of their universal applicability. That problem is addressed in the first *Critique*, not by appeal to the constructibility of the concepts, as recent commentators have argued, but by the claim that pure intuition is the form of empirical intuition. Again, as in the case of the papers by Land and Nunez, we see how mathematics is implicated in Kant's doctrine of the form of appearances.

Daniel Sutherland's paper reveals the significance of the notion of mathematical construction beyond Kant's philosophy of mathematics by focusing on its centrality to Kant's *Metaphysical Foundations of Natural Science*. He shows how Kant's account of the possibility of mathematical cognition and his explanation of the applicability of mathematics in the first *Critique* provides a grounding for and shapes Kant's account of the possibility of

Newtonian mathematical physics in the *Metaphysical Foundations*. Because mathematical cognition rests on construction, the mathematizability of physical concepts – above all, motion – depends on their constructibility. He explains how motion and composition of motion are constructed in the Phoronomy in a way that accommodates eighteenth-century treatments of instantaneous velocity without running counter to Kant's claims that motion is an intensive magnitude and is apprehended in an instant. At the same time, Kant's argument in the Phoronomy reveals features of construction that are also important for our understanding of the cognition underlying *pure* mathematics.

In her "Kant on Conic Sections," Alison Laywine illustrates the general theme of this volume by arguing that the four passages where Kant discusses conic sections directly "mirror and illuminate Kant's evolving views on central aspects of his philosophy" from 1763 to 1790. Like Land and Smyth, she takes up the question of the relations among the faculties, in this case, of reason, imagination and judgment, and investigates them in the context of her treatment of Kant's discussion of conic sections. Kant cites Apollonius' theory of conic sections as exhibiting a high degree of systematic unity. We might naturally think that this kind of unity is best brought about by an algebraic treatment of conic sections. But this gives rise to a puzzle: the texts in which Kant discusses Apollonius' conic sections reveal his commitment to *geometrical* construction in the theory of conic sections. How does Kant see geometrical construction as yielding systematic unity in any way comparable to algebraic techniques? What seems to be required here is something analogous to the role the Schematism plays in explaining how we arrive at universal conclusions from particular images, but in this case at the level of reason rather than understanding. Laywine suggests that perhaps the notion of reflective judgment from the third *Critique* could fulfill this function. As she points out, it should then come as no surprise that the section on the critique of teleological judgment begins with a discussion of mathematical purposiveness.

This topic of purposiveness is taken up by Courtney Fugate in his paper "'With a Philosophical Eye': The Role of Mathematical Beauty in Kant's Intellectual Development", again while attending to the task of illuminating the relations between the faculties. Fugate traces the historical and metaphysical roots of Kant's notion of mathematical purposiveness with the aim of showing, like Laywine, that it has greater significance for Kant than just serving to illustrate the different types of purposiveness. That significance lies in the role of reflections on mathematical purposiveness in Kant's discovery of the synthetic *a priori*, and, in particular, in what Fugate calls an inexplicable "purposiveness between our own *a priori* faculties of intuition and understanding".

It should be clear from the foregoing that Kant was struck by the special nature of mathematics: its certainty, its evidence, the nature of mathematical proof and the remarkable fit between mathematics and the natural world, all of which informed his more general philosophical views. The papers in the current volume examine these issues in new ways. It is our hope that these papers inspire still more work on the broad implications of Kant's philosophy of mathematics.

Emily Carson and Lisa Shabel
*McGill University, Montréal, QC, Canada H3A 2T7*
*The Ohio State University, Columbus, OH, 43210 USA*
http://dx.doi.org/10.1080/00455091.2014.977379

# Spatial representation, magnitude and the two stems of cognition

Thomas Land

*Corpus Christi College, Cambridge, UK*

The aim of this paper is to show that attention to Kant's philosophy of mathematics sheds light on the doctrine that there are two stems of the cognitive capacity, which are distinct, but equally necessary for cognition. Specifically, I argue for the following four claims: (i) The distinctive structure of outer sensible intuitions must be understood in terms of the concept of magnitude. (ii) The act of sensibly representing a magnitude involves a special act of spontaneity Kant ascribes to a capacity he calls the productive imagination. (iii) Contrary to what is assumed by many commentators, it is not the case that the Two Stems Doctrine implies that a representation is either sensible or spontaneity-dependent, but not both. (iv) Outer sensible intuitions are both sensible and spontaneity-dependent – they are sensible because they exhibit the kind of structure Kant takes to be distinctive of outer sensible intuitions, and they depend on spontaneity because they are cases of sensibly representing a magnitude.

## 1. Introduction

The doctrine that there are two 'stems' of the cognitive capacity, sensibility and understanding, whose contributions to cognition are distinct, yet equally necessary, is of central importance for Kant's position in the *First Critique*.[1] Yet exactly what this doctrine says is less clear. If we think of it as the conjunction of two claims, the Heterogeneity Thesis and the Cooperation Requirement, there is controversy among commentators with regard to each of these. The Heterogeneity Thesis says that the respective contributions of sensibility and understanding – intuitions and concepts – are distinct in kind. The Cooperation Requirement says that both kinds of representation are essential to cognition.

The recent debate concerning so-called conceptualist readings of Kant's theory of empirical intuition is a case in point.[2] Participants to this debate disagree whether sensible intuitions, which are characterized by Kant as representations of objects, possess objective purport independent of concepts or only in connection with them.[3] This issue is often framed in terms of the 'unity' that representations which purport to be about objects are said to possess. Advocates of conceptualism argue that the relevant unity is due to the application of concepts by the understanding. They take this to be entailed by the correct interpretation of the Cooperation Requirement. Proponents of nonconceptualism disagree. They hold that this cannot be the correct interpretation of the Cooperation Requirement because according to them it falls afoul of the Heterogeneity Thesis. Rather, nonconceptualists argue that the unity on account of which sensible intuitions are representation of objects is a distinctively sensible kind of unity, which is independent of concepts.

The aim of this paper is to show that attention to Kant's philosophy of mathematics sheds light on the Two Stems Doctrine. Specifically, it helps clarify Kant's conception of the Heterogeneity Thesis. As a consequence, attention to Kant's philosophy of mathematics promises not only to advance the debate about Kant's alleged conceptualism but also to further our understanding of Kant's conception of the fundamental cognitive capacities generally. Specifically, I will argue for the following four claims: (i) The kind of unity distinctive of outer sensible intuitions must be understood in terms of the concept of magnitude. (ii) The act of sensibly representing a magnitude involves a special act of spontaneity, which Kant ascribes to a capacity he calls the productive imagination. (iii) Contrary to what is assumed by commentators on both sides of the debate about Kant's conceptualism, it is not the case that the Two Stems Doctrine implies that a representation is either sensible or spontaneity-dependent, but not both. (iv) Outer sensible intuitions are both sensible and spontaneity-dependent – they are sensible because they exhibit the kind of structure Kant takes to be distinctive of outer sensible intuitions, and they depend on spontaneity because they are cases of sensibly representing a magnitude. My aim, then, is to clarify the Two Stems Doctrine by focusing on what consideration of Kant's philosophy of mathematics can show us regarding the correct interpretation of the Heterogeneity Thesis and, specifically, of Kant's conception of intuition.[4]

The concept of magnitude and its relation to the notion of intuition plays a central role in Kant's philosophy of mathematics because Kant holds both that mathematical knowledge is knowledge of magnitudes and that such knowledge is intuition-dependent. Recent scholarship in this area has done much to clarify this relation, and I draw on this work in making my case for these four claims.

The doctrine of the productive imagination is notoriously difficult. Kant claims that this capacity combines both a spontaneous and a sensible aspect, and it is not clear whether the Two Stems Doctrine is compatible with this. I will argue that we can make progress here if we consider this doctrine against the background of Kant's conception of what it is to construct a mathematical concept in intuition. This topic, too, has been the subject of important recent scholarship, whose implications for our understanding of the Two Stems Doctrine have not been sufficiently explored. I will argue, then, that reflection on the notions of magnitude and mathematical construction sheds light on Kant's theory of intuition and, more specifically, puts us in a position to see in what way intuitions are both distinctively sensible and dependent on acts of spontaneity.

I proceed as follows: I begin by presenting a recent interpretation of the Heterogeneity Thesis. According to this interpretation, intuitions and concepts are distinguished by the kind of part–whole structure they each possess: While intuitions for Kant are holistic, so that the whole is prior to the part, concepts are conceived atomistically, in the sense that the part is prior to the whole (Section II). Next, I raise two objections to this interpretation, which I take to be sufficiently strong to motivate an alternative approach (Section III). I go on to present an alternative account of the Heterogeneity Thesis, according to which it is a distinctive structural feature of outer intuitions that they are magnitudes (Section IV). Next, I argue that the representation of a magnitude depends on the act of the productive imagination and therefore involves a distinctive kind of cooperation of sensibility and understanding (Section V). Finally, I further elucidate the notion of the productive imagination by considering Kant's conception of mathematical construction (Section VI).

## 2.   The simple mereological account

Any account of the Two Stems Doctrine must provide an interpretation of the Heterogeneity Thesis. Perhaps the most obvious way of doing so is to focus on the well-known characterizations Kant gives of intuitions as singular representations, which relate immediately to objects, and concepts as general representations, which relate to objects only mediately.[5] However, in this paper, I will pursue a different strategy. This is not to suggest that these characterizations are irrelevant. A comprehensive interpretation of the Heterogeneity Thesis must certainly address them. But the following two considerations suggest that it would be desirable to have an account of the Heterogeneity Thesis which possesses some measure of independence from these characterizations. First, it is not fully clear exactly how these characterizations are themselves to be understood.[6] Second, some commentators have argued that the distinction between singularity and generality is not exclusive, in the sense that a representation may exhibit both. Thus, these commentators claim that intuitions must be understood as themselves having both a singular and a general aspect.[7] We might hope to avoid these complications by looking for an account of the Heterogeneity Thesis that centres on the characteristic *structure* of intuitions and concepts, respectively. Given that the difference between intuitions and concepts is one in kind, not in degree, it seems plausible to assume that these representations differ not just in logical quantity and in the manner of relating to objects, but also in structure.

A promising strategy for developing such an account is suggested by the hylomorphic terminology Kant frequently employs: The form of a representation of type $R$ is, presumably, that which makes it a representation of this type. Attending to Kant's respective characterizations of the forms of sensibility and understanding therefore promises to yield insight into his conception of the distinction between intuitions and concepts.

Because my focus in this paper is on intuitions, I will mostly be concerned with what Kant calls the forms of intuition. More specifically, I will be concerned with what he calls the form of outer intuitions, viz. the pure intuition of space. The hope is that attention to Kant's account of the pure intuition of space allows us to identify structural features of intuitions generally, by means of which intuitions can be distinguished from concepts. In this section, I will present a recent interpretation that follows this strategy.

Colin McLear has argued that intuitions and concepts are heterogeneous for Kant because they each exhibit a distinctive kind of part–whole structure (see McLear forthcoming-b). According to McLear's view, which I call the Simple Mereological Account, intuitions possess a structure in which the whole is prior to the part, while concepts possess a structure in which the part is prior to the whole.

McLear supports the claim about the structure of intuitions by appeal to Kant's arguments for the thesis that the original representation of space is an intuition rather than a concept, which are presented in the Metaphysical Exposition of Space and known as the Third and Fourth Space Arguments. A crucial premise of the Third Space Argument is the claim that space is represented in such a way that any part of space is represented as part of the single whole of space. More precisely, any part of space is represented as a 'limitation' of the single whole of space. But this, Kant argues, implies that the representation of space is an intuition rather than a concept.

The Fourth Space Argument turns on the idea that space has infinitely many parts. Kant argues that if the original representation of space was a concept, this would have to be a concept that contains infinitely many representations in it because space is represented as containing infinitely many parts. But according to Kant there can be no concepts that meet this condition (at least for finite minds like ours). Therefore, the original representation of space is not a concept. It is, rather, an intuition because according to Kant, intuitions and concepts are the only two kinds of elementary representations there are.

McLear extracts from these characterizations of the representation of space an account of the structure constitutive of sensible intuitions in general. Thus, he claims that intuitions in general exhibit the kind of part–whole structure Kant attributes to the representation of space. Accordingly, for McLear a representation is a sensible intuition just in case its parts depend on the whole.

Contrast the representations of the understanding, which according to McLear exhibit a dependence in the opposite direction. So in McLear's view representations of the understanding are such that the whole of the representation depends on its parts. For instance, a concept is a whole which depends on the marks that make up its intension as its parts. A judgement is a whole which depends on the concepts that constitute what Kant calls its matter as its parts. An inference is a whole which depends on the judgements that constitute its premises and conclusion as its parts. Generally, for McLear, intellectual activity proceeds from part to whole, while sensible representation is essentially holistic.[8]

I wish to make two comments about the Simple Mereological Account. First, according to this account the distinction between part–whole dependence and whole–part dependence is exclusive. As a consequence, a representation is either the product of sensibility or the product of the understanding, but not both. Second, the priority of part over whole is regarded as both necessary and sufficient for acts of the understanding. So any representation that exhibits a dependence of whole on part is, according to this account, an intellectual as opposed to sensible representation.[9]

### 3.  Problems for the Simple Mereological Account

I wish to raise two objections to the Simple Mereological Account. My first objection concerns the construal of discursive unity, the second pertains to the claim that the distinction between these is exclusive.

According to Kant, the synthetic unity of apperception is 'the highest point,' on which 'all use of the understanding' depends (B134). It is presupposed even by the subject's consciousness of her own identity, which is itself a presupposition of any act of the understanding.[10] The synthetic unity of apperception is, moreover, 'original' (B131, B137, B143), that is, not derived from anything else, but something that is constitutive of the capacity for discursive thought and thus present wherever the latter is exercised.

Although the content of these claims is far from clear, and giving an account of them would go beyond the scope of this paper, I think we can extract the claim that, for Kant, acts of the understanding depend on the availability of a representation of unity. But if this is right, then it is not true that in discursive activity the part is prior to the whole. Let me explain why this is so.

By way of introducing the doctrine of the synthetic unity of apperception, Kant argues that the understanding must be conceived, fundamentally, as the capacity for the representation of synthetic unity (B130f). The paradigmatic exercise of this capacity is in judgement, which, in Kant's view, involves the deployment of the categories or pure concepts of the understanding. Of these concepts, Kant says that 'they are grounded on logical functions in judgment;' 'in these, however, combination, hence unity of given concepts, is already thought' (B131).

The implication is that the act of judgement itself depends on a representation of the unity of 'given concepts'. If the latter are regarded as the parts that jointly make up the judgement as a whole, then the claim is that the act of combining these parts into a whole itself depends on a representation of the whole; viz. of the kind of unity that is 'already thought' in what Kant calls the logical functions of judgement.

Again, I do not have the space fully to develop the idea at issue here, but the following may serve as an illustration of what I take to be the salient point for our purposes: Kant famously holds that the categories are pure concepts of the understanding, where this entails that anyone possessing the power of understanding possesses these concepts. Because, moreover, the categories are closely related to what Kant regards as the elementary logical forms of judgement, it seems to be Kant's view that possession of the power of understanding requires a grasp of the elementary logical forms of judgement. But the forms of judgement are forms of the kind of whole that, according to the Simple Mereological Account, is dependent on, rather than presupposed by, its parts. And if a grasp of these forms is required for possession of the capacity for judgement, and thus presupposed by any determinate act of judgement, then there is a sense in which on Kant's conception of judgement the whole is prior to the part rather than posterior to it. So the Simple Mereological Account cannot be accurate.[11]

To be sure, proponents of the Simple Mereological Account could respond that we need to distinguish between the kind of whole that a determinate judgement is and the kind of whole that a form of judgement is, and insist that the priority-claim pertains only to the former. But even if we grant that it is true to say of determinate judgements that the part is prior to the whole, this response will not work. For this kind of priority of part to whole does not distinguish intellectual representation from sensible representation. Thus, an associative unity of representations is aptly characterized by saying that the whole depends on a prior availability of the parts. But Kant insists that association is not an act of the understanding and that the unity due to association is different in kind from the unity characteristic of acts of the understanding.[12] So dependence of representational whole on representational part is not sufficient for distinguishing intellectual activity from merely sensible processes.

My second objection concerns the claim that the distinction between sensibility and understanding is interpreted as exclusive, so that it is not possible for a representation to be both sensible and dependent on acts of the understanding. This claim stands in tension with the doctrine of the productive imagination, which Kant puts forth in the course of the Transcendental Deduction and which plays a role also in the Axioms of Intuition. According to the characterizations Kant gives of this capacity, the representations it is responsible for are sensible, yet at the same time dependent on acts of the understanding because the productive imagination is in fact the understanding acting in a certain way. Consider the following passage:

> Now since all of our intuition is sensible, the imagination, on account of the subjective condition under which alone it can give a corresponding intuition to the concepts of the understanding, belongs to sensibility; but insofar as its synthesis is still an exercise of spontaneity, which is determining and not, like sense, merely determinable, [...] the imagination is to this extent a faculty for determining sensibility a priori and its synthesis of intuitions, in accordance with the categories [...], is an effect of the understanding on sensibility [...]. Insofar as the imagination is spontaneity, I sometimes call it the *productive* imagination and thereby distinguish it from the *reproductive* imagination, whose synthesis is subject merely to empirical laws, viz. those of association [...]. (B151f)

Kant says here that the productive imagination 'gives' intuitions and is therefore sensible. At the same time, he says that the act of the productive imagination is an exercise of spontaneity and an 'effect of the understanding' on sensibility, which I take to imply that the act of the productive imagination *is* an act of the understanding.[13] It seems to be Kant's view, then, that there are acts of the understanding which generate intuitions; that is, representations exhibiting characteristically sensible features.

Although both objections I have raised would warrant further discussion, we are I think entitled to conclude that the Simple Mereological Account faces serious problems. If this is right, we have reason to look for an alternative account of the Heterogeneity Thesis, which avoids these objections. My aim in the remainder of this paper is to propose such an alternative. In so doing, I will focus on the side of sensibility and consider the understanding only under the aspect of the productive imagination. So my alternative will not be a complete account of the Heterogeneity Thesis.

Before I present this alternative, I wish to emphasize that I am sympathetic to the overall strategy pursued by the Simple Mereological Account, of focusing on structural features of the relevant representations and more specifically on their part–whole structure. Kant's arguments in the Metaphysical Exposition of Space suggest that he thinks that the distinction between concepts and intuitions can be drawn in these terms. But the upshot of the objections I have raised is that we need a more complex account. For one thing, such an account should make room for what seems to be the implication of the doctrine of the productive imagination, viz. that being a sensible representation is compatible with being dependent on acts of spontaneity. I will suggest below that to do so, we should distinguish between the characteristic *structure* of a representation and its *unity*, though I will not be in a position to introduce this distinction until Section 5.

## 4.   Magnitude and strict logical homogeneity

The aim of this section is to show that consideration of the concept of magnitude, which plays a central role in Kant's philosophy of mathematics, helps us obtain a better understanding of the Heterogeneity Thesis. More precisely, I will argue that consideration of this concept shows that it is a distinguishing feature of outer intuitions that an outer intuition admits of being regarded as a manifold of qualitatively identical parts. Although in making this argument, I will focus on the case of space and outer intuition, parallel considerations apply to time as the form of inner intuition; the feature of intuitions I will be concerned with is sufficiently generic to be neutral with regard to the distinction between time and space. But I will not argue for this here.

The strategy I am pursuing can be motivated as follows: In Kant's view, all objects of mathematical cognition are magnitudes.[14] Because the pure intuition of space is an object of mathematical cognition, it is a magnitude. Because, furthermore, the pure intuition of space constitutes the form of intuition, consideration of the former should provide insight into the nature of outer intuitions as such, on the assumption that the notion of form Kant has in mind is a broadly Aristotelian one.[15]

I will first introduce the notion of magnitude, as Kant understands it, then give evidence that space is a magnitude, and finally argue that this property of space provides the basis for an account of the characteristic structure of outer intuitions. This last argument will depend on the contention that the representation of an essential property of magnitudes is in a fundamental sense intuitional (or, equivalent for present purposes, sensible), as opposed to conceptual.

## 4.1   The concept of magnitude

In a series of recent papers, Daniel Sutherland has argued that the notion of magnitude is fundamental to Kant's conception of mathematics and that his understanding of this notion is informed by the Greek mathematical tradition, specifically the Eudoxian theory of proportions, as it figures in Euclidean geometry.[16] In explaining this notion, we should first note that Kant draws a distinction between two different species of magnitude. These are called, by their Latin names, *quantum* and *quantitas*. A *quantum* is a concrete magnitude, for example a line or a plane figure. The term 'quantum' thus denotes an ontological kind: Magnitude in this sense is not something that an object *has*, but rather something that an object *is*. By contrast, 'quantitas' denotes the magnitude that an object has, its size. Thus, the *quantitas* of something is that which is given in response to the question 'How big?' or 'How many?'.[17]

Two central features of the concept of magnitude, which play an important role in Kant's conception of mathematics are, first, that magnitudes can be composed. For instance, two *quanta* can be added to one another so as to yield more of the same. Second, magnitudes are capable of standing in comparative size-relations. That is, one magnitude can be larger, equal to or smaller than another. Both of these features require that magnitudes exhibit a part–whole structure. Thus, for the composition of *quanta* to be possible it must be possible to say, e.g., that two *quanta* are parts of a whole that is larger than either of them. Likewise, to say that one magnitude is greater than another is to say, on this view, that the second is equal to a proper part of the first.[18]

Because for Kant, geometry serves as the paradigmatic sub-field of mathematics, this theory of magnitudes applies in the first instance to spatial magnitudes. Against this background, we can begin to see why it is of crucial importance to Kant that, as he asserts in the Metaphysical Exposition, space as a whole is a magnitude (see A25/B39). We can also begin to see why his discussion there emphasizes the part–whole structure of the representation of space. For if space is a magnitude, then it exhibits a certain kind of part–whole structure, and this must be part of the content of its representation.

However, if space is a magnitude, then its part–whole structure is of a very specific character. We shall see that contrary to the Simple Mereological Account the central feature here is not that the whole is prior to the part. It is, rather, that all the parts of space are homogeneous in a demanding sense.[19]

To see this, consider the requirement that magnitudes can be composed, where this means that, e.g., two *quanta* can be added to one another so as to yield more of the same. Evidently, *quanta* that are composable in this manner must be logically homogeneous in the sense that they belong to the same kind. Intuitively, I can sum three apples and two oranges only if there is a sortal concept under which I can bring all of them, e.g., 'pieces of fruit'. As this example also shows, however, *insofar as* objects are regarded as magnitudes, hence as composable and as standing in comparative size-relations, they are considered in abstraction from *any* qualitative differences they may exhibit. That is, insofar as objects are regarded as magnitudes, they are regarded as exhibiting what Sutherland calls strict logical homogeneity, which is defined as follows:

> *Strict Logical Homogeneity*: A whole $w$ is *strictly logically homogeneous* iff, for every $x$ and every $y$, if $x$ and $y$ are parts of $w$ and $x \neq y$, then there is no monadic concept $F$ such that $F$ is instantiated by $x$ but not by $y$.

A whole is strictly logically homogeneous, then, just in case its parts are qualitatively identical, yet numerically distinct. Magnitudes, for Kant, are wholes that exhibit strict logical homogeneity of their parts.[20]

Now, Kant regards space as a magnitude in precisely this sense. So space for Kant is strictly logically homogeneous. This is shown by his discussion of the Leibnizian doctrine of the Identity of Indiscernibles in the Amphiboly-chapter of the *First Critique*.[21] In objecting to this doctrine, Kant asks the reader to suppose that two drops of water have exactly the same intrinsic properties, so that we may say that they are indiscernible, or qualitatively identical. But, he argues, because the two drops occupy different locations in space, there is not just one drop (as the Identity of Indiscernibles would entail), but two. So the two drops are numerically distinct despite being qualitatively identical.

Consider also the following passage, in which the distinction between qualitative and numerical identity is explicitly mentioned by Kant in connection with space:

> The concept of a cubic foot of space, wherever and however often I think it, is in itself *completely identical*. But two cubic feet are nevertheless distinguished in space by the mere difference of their locations (*numero diversa*); these locations are conditions of the intuition wherein the object of this concept is given; they do not, however, belong to the concept but entirely to sensibility. (A282/B338, my emphasis)

What Kant says here is that every instance of the concept 'cubic foot of space' has exactly the same intrinsic properties and yet these instances are numerically distinct from one another, because they are different parts of space. This shows that space, for Kant, has a part–whole structure that exhibits strict logical homogeneity.

However, saying this is not meant to exclude the possibility of recognizing qualitative differences between different parts of space. Thus, Kant regards differences in shape – say, the difference between a triangle and a circle – as qualitative differences (see 720/B748). So the point is not to deny that there are such differences. The point is rather that it is *always* possible to abstract from such qualitative differences and to consider a region of space purely as a magnitude. So to say that space is a magnitude is to say that, for any region of space, it is always possible to regard it as a magnitude, hence as a manifold of qualitatively identical parts. But doing so may involve abstraction from various qualitative differences. Thus, each of the lines making up the figures in the example can be regarded as a manifold of strictly logically homogeneous parts, in this case of line segments of equal length.[22]

## 4.2  Magnitude and concepts

To make the case that Kant's characterization of space as a magnitude furnishes the materials for an account of the Heterogeneity Thesis that avoids the problems faced by the Simple Mereological Account, I now wish to argue that strict logical homogeneity cannot be represented by concepts. Once we see why this is so, it will emerge that strict logical homogeneity is an essential characteristic of intuitions, which can serve as the basis of such an account.

A concept, for Kant, is an essentially *general* representation. As he puts it, a concept represents objects 'by means of a mark, which can be common to several things' (A320/B377).[23] However, Kant thinks of generality, not on the model of a mathematical function, as Frege would, but rather on the model of a biological taxonomy.[24] This conception is based on the Aristotelian theory of definition in terms of nearest genus and specific difference, as illustrated by Porphyrian genus–species trees. What matters for our purposes is that on this theory the intension of a concept is represented as a combination of other concepts, which here play the role of what Kant calls marks. The theory then assigns to a conceptual mark one of two roles: either the mark denotes a genus, or it indicates a specific difference.[25] Thus, if I take the concept ⟨animal⟩ and treat it as a genus, I may distinguish two species by introducing ⟨rational⟩ as a mark representing a specific difference. So the genus ⟨animal⟩ is divided into the two species ⟨rational animal⟩ and ⟨non-rational animal⟩. In each case, a mark is added to the intension of my original concept to form the new species-concept. I can continue this process and distinguish further species by, say, dividing my concept ⟨rational animal⟩ into ⟨scholar⟩ and ⟨non-scholar⟩, and so on.[26] In each case, I effect the division of a genus into species by adding to its intension a mark which sorts the members of the genus into those that exhibit the relevant characteristic and those that do not.

Let us now ask how, in such a system, one would represent bare numerical differences. That is, how would one represent a species that contains under it a plurality of members which do not differ from one another by belonging to different species? It appears that numerical difference in general can be represented only by way of adding marks. Moreover, this means that one could represent numerical difference only by means of introducing further qualitative differences. So the answer is that in such a system it is impossible to represent numerical difference without specific difference.[27]

This shows that it is not possible to represent magnitudes, as Kant conceives of them, by purely conceptual means.[28] For it is essential to the notion of a magnitude that magnitudes exhibit strict logical homogeneity. Now, the representation of space exhibits strict logical homogeneity, according to Kant, and it also functions as a form of intuition. This suggests that the fundamental characteristic in virtue of which the representation of space is an intuition, as opposed to a concept, is its strict logical homogeneity. If this is right, it follows that outer intuitions generally are distinguished from concepts by the fact that the former, but not the latter, exhibit strict logical homogeneity.

Again, however, this claim needs qualification.[29] I said above that space exhibits strict logical homogeneity in the sense that it is always possible to abstract from qualitative differences that may be present in a region of space (e.g., differences of shape) and consider this region as a magnitude. A similar qualification is needed to account for empirical intuitions. Empirical intuitions involve sensation, and this suggests that empirical intuitions often exhibit qualitative differences among their parts; for instance, differences in colour. Here, too, Kant's view appears to be that it is always possible to bracket these qualitative differences and attend only to the spatial characteristics an empricial outer intuition invariably has. To do that, however, is to consider the intuition as a magnitude, that is, as a manifold of strictly logically homogeneous parts.[30]

By way of illustration, let me give two examples. First, consider the case of having a visual intuition of a chessboard. Its surface exhibits differences in colour, so there are qualitative differences among different parts of this intuition. However, insofar as this surface is regarded as being extended, as having a determinate area, it is conceived as a magnitude. This means that it is conceived as having a part–whole structure which makes it possible to think of this area as composed of parts. But this requires that the area is conceived as consisting of parts that are strictly logically homogeneous.

Second, consider a modified version of Kant's raindrop example. Suppose two drops of water do exhibit qualitative differences, in respect of both shape and colour. Still, insofar as the two drops are represented as being in different locations, they are represented as occupying distinct parts of space. This means that they are represented as occupying different parts of a strictly logically homogeneous manifold. This is implied by the point the original raindrop example is intended to support: viz. that a difference in the location that two objects are represented as occupying is sufficient for representing them as numerically distinct regardless of their qualitative determinations.

Bearing in mind this qualification, the thesis that intuitions are magnitudes yields a building block for an account of the Heterogeneity Thesis that provides an alternative to the Simple Mereological Account. On this account, intuitions are heterogeneous to concepts because the former exhibit strict logical homogeneity, while the latter do not.

## 5. Strict logical homogeneity and the productive imagination

I have argued that consideration of the concept of magnitude yields a partial account of the Heterogeneity Thesis, in that it allows us to see that sensible representations, for Kant, are characterized by strict logical homogeneity. I have also argued that strict logical homogeneity cannot be represented by purely conceptual means and for this reason serves to distinguish the sensible from the conceptual. I now wish to argue that for Kant outer intuitions nonetheless depend on acts of spontaneity and thus on the understanding. That is, although outer intuitions exhibit a structure that is distinct in kind from the structure of conceptual representations, the actualization of the capacity for having outer intuitions nonetheless depends on acts of the intellect, the capacity for concepts and judgement.

Before I support this contention, let me make its content a little more precise. To do so, I wish to introduce a distinction between the *structure* of a complex representation and its *unity*. To give the structure of a complex representation is to specify the kinds of parts it is made up of as well as the manner in which these parts hang together. Thus, to characterize intuitions as strictly logically homogeneous is to give an account of their structure. For it is to say that the parts of such a representation are qualitatively identical and stand in relations of coordination (as opposed to, say, subordination). By the unity of a complex representation I mean the fact that it is represented *as* complex. Thus, to say that a complex representation has unity is to say that the subject entertaining the representation has a consciousness of its character as complex. This consciousness need not be explicit or articulated. It is sufficient that the subject entertaining the representation has some kind of grasp, or appreciation, of the complex nature of the representation.

By means of this distinction, the claim I wish to defend can now be characterized more precisely as saying that the *structure* of sensible representations is independent of spontaneity, but their *unity* is not. Put differently, Kant holds that in the absence of the sensible stem of the cognitive capacity, the representation of strict logical homogeneity would be impossible for us. For this reason, strict logical homogeneity serves to distinguish sensible representations from conceptual ones. At the same time, the sensible stem is not self-standing in the sense that its actualization in determinate representations, which exhibit strict logical homogeneity, depends not just on itself, but also on acts of spontaneity.

If this is right, the implication of the Simple Mereological Account, that the Heterogeneity Thesis implies that a representation is either sensible or depends on spontaneity, but not both, is false. I already argued in Section III above that Kant's doctrine of the productive imagination puts pressure on this implication. The argument of the present section is intended to lend additional support to this claim. I will first present some textual evidence for the claim that the unity of sensible representations depends on acts of spontaneity and then argue that Kant has good philosophical reasons for holding this. These reasons derive from the conception of sensibility as a receptive capacity in conjunction with the claim just considered, that space is a magnitude. In the following section, I will expand on these reasons by considering Kant's theory of mathematical construction.

Textual evidence for the thesis that the unity of outer intuitions depends on acts of spontaneity comes from passages such as the following:

> We cannot think a line without *drawing* it in thought, cannot think a circle without *describing* it [ ... ]. (B154)

Lines and circles are paradigm examples of magnitudes for Kant. Drawing a line and describing a circle are acts of sensible spatial representing, that is, of outer intuitions. Yet these acts are Kant's usual examples of an act of the productive imagination, which he further characterizes as a synthesis of the understanding by means of which the latter 'is capable of itself determining sensibility internally with regard to the manifold which may be given to it in accordance with its form of intuition' (B153).

Next, consider the following passage from Section 26 of the B-Deduction:

> [ ... ] if, e.g., I make the empirical intuition of a house into perception through apprehension of its manifold, my ground is the necessary unity of space and of outer sensible intuition in general, and I as it were draw its shape in agreement with this synthetic unity of the manifold in space. This very same synthetic unity, however, if I abstract from the form of space, has its seat in the understanding, and is the category of the synthesis of the homogeneous in an intuition in general, i.e. the category of quantity, with which that synthesis of apprehension, i.e. the perception, must therefore be in thoroughgoing agreement. (B162)

The category of quantity is identical with the concept of magnitude in general. To say that the synthetic unity of space *is* the category of magnitude, as Kant does here, is thus to say, first, that space has the structure characteristic of magnitudes, i.e. strict logical homogeneity. But it is also to say that the unity of a spatial representation – in the sense in which I use that term here, that is, the consciousness that the homogeneous manifold constitutes a whole – depends on an act of synthesis and thereby, as the context makes clear, on an act of spontaneity (which is once again described as 'drawing').

Finally, consider this claim from the Axioms of Intuition:

> All appearances contain, as regards their form, an intuition in space and time, which grounds all of them a priori. They cannot be apprehended, therefore, i.e., taken up into empirical consciousness, except through the synthesis of the manifold through which the representations of a determinate space or time are generated, i.e., through the composition of that which is homogeneous and the consciousness of the synthetic unity of this manifold (homogeneous). (B202f)

In this passage, Kant speaks of the representation of determinate spaces and times. These are sensible representations, that is, intuitions. Yet these representations are 'generated' through acts of synthesis; specifically, acts of composition of a homogeneous manifold. Such acts are characterized by Kant, in the same section, as cases of the 'synthesis of the productive imagination in the generation of shapes' (A163/B204).

All three passages concern the representation of magnitudes in intuition, yet in all three cases these representations are said to depend on acts that, as the second and third passage make clear, are acts of synthesis involving the category of magnitude, which Kant attributes to the productive imagination. On the picture painted by these passages, then, spatial representation in intuition is not an actualization merely of sensibility (conceived as a self-standing capacity) but involves other capacities besides. In particular, it involves exercises of spontaneity in the guise of the productive imagination.[31]

Although the textual evidence just considered lends support to the claim that spatial representation for Kant involves more than actualizations of sensibility, we should ask what philosophical reasons Kant has to uphold this claim. I will give one such reason, which I take to be comparatively strong. It derives from Kant's characterization of sensibility as a receptive capacity. The basic idea is that spatial representation exhibits a kind of unity that is such that a receptive capacity on its own could not account for it.

To make the case for this, I will employ the notion of a merely receptive capacity for representation.[32] According to Kant, as I read him, the kind of sensibility human beings possess is not a capacity of this kind because on my reading human sensibility depends essentially on spontaneity. By contrast, the notion of a merely receptive capacity is defined so as to exclude this kind of dependence. In considering the idea of a merely receptive capacity, then, we should be careful not to assume that the various characterizations Kant gives of human sensibility find application. If any of these characterizations do hold of a merely receptive capacity, then this must be shown by deriving them from the concept of such a capacity.[33]

A merely receptive capacity for representation is one whose actualization depends entirely on being affected by something outside it. As Kant puts it, such a capacity represents by means of 'the receptivity of impressions' (A50/B74). This means that such a capacity represents current affection. It represents, as we might put it, the here and now. But this is not a sufficient characterization, for it is ambiguous. We need to distinguish between merely representing the here and now, on the one hand, and representing the here and now *as* the here and now, on the other. By the latter I mean a representation such that the here and now is (perhaps implicitly) understood to contrast with the there and then: what is represented as being here and now is thereby understood not to be at a different spatio-temporal location.

Our ordinary grasp of spatial and temporal indexicals is of the latter kind. It includes an understanding of spatial and temporal relations such that what is understood to be here and now is thereby understood as not being at any number of other spatio-temporal locations, to which it is, however, related in determinate ways.[34] By contrast, a merely receptive capacity lacks such an understanding. It *merely* represents the here and now. It is, we might say, 'locked into' its current location and the present moment. It does not grasp that location or this moment as one in a system of locations or a continuous series of moments. For such a grasp, a representation of the relations that obtain between the content of current affection and the contents of other, possible affections would be required; for instance, what Gareth Evans has called a primitive theory of space.[35]

The idea is, then, that if sensibility were a merely receptive capacity, it would represent the here and now, but not *as* the here and now. Kant makes a closely related point, I think, in a passage near the opening of the A-Deduction:

> Every intuition contains a manifold within itself, which however would not be represented as such if the mind did not distinguish the time in the succession of impressions on one another; for *as contained in one moment* no representation can ever be anything other than absolute unity. (A99, emphasis in original)

By 'absolute unity' I take Kant to mean the absence of internal complexity. The contrast is with the unity of a manifold, that is, something that is a one, yet internally complex.[36] Kant's point in the passage is that a succession of impressions is represented *as* a succession of impressions only if there is an awareness of time, where this must include an awareness of the present moment as one in a continuous series of moments. Absent such awareness, there is no representation of manifoldness, no awareness of complexity, but merely a succession of impressions. And this is what a merely receptive consciousness would be able to represent. Again, the image of such a conciousness's being 'locked into' each moment suggests itself.

It is clear that Kant thinks spatial representation, at least of the kind he is concerned with in the *First Critique*, is not like that.[37] He says so, for instance, in the continuation of the passage from A99:

> Now for unity of intuition to come from this manifold (*as, say, in the representation of space*), it is necessary first to run through and then to take together this manifoldness, which action I call the synthesis of apprehension, since it is aimed directly at intuition, which to be sure provides a manifold but can never effect this *as such*, and indeed *as contained in one representation*, without the occurrence of such a synthesis. (A99, my emphases)

Kant says here that the representation of space involves not just manifoldness, but unity of the manifold, where this must mean that the representation of space is the representation of a manifold as manifold. In the temporal idiom employed in the preceding paragraph, the contrast is between a succession of representations and a representation of succession. For Kant, then, spatial representation involves unity in at least the sense that the present location is represented as one in a network of possible locations. The here and now is (at least implicitly) understood to contrast with the there and then.[38]

If spatial representation for Kant exhibits this kind of unity, it cannot be the product of a merely receptive capacity of representation. Something over and above a capacity to represent present affection is needed. As the passages considered earlier in this section show, this is the act Kant calls synthesis. However, because spatial representation exhibits strict logical homogeneity, a particular kind of synthesis is required, viz. one that is compatible with this structural feature of spatial representation. In the passage from B202 quoted earlier Kant calls this kind of synthesis 'the composition of that which is homogeneous', and in the passage from B162 he speaks of the 'synthesis of the homogeneous in an intuition in general'.

As the context of these passages, and other similar ones, makes clear, the synthesis of the homogeneous in intuition is an act of spontaneity. It follows that the kind of spatial representation Kant is concerned with in the *Critique* is dependent for its unity, though not for its structure, on the spontaneous stem of the cognitive capacity.

Now the notion of spontaneity is closely linked to the idea of a capacity for judgement. The spontaneous stem of the cognitive capacity appears to be the power of applying concepts in judgement. But this presents a problem. For judgements are different in structure from spatial intuitions. The latter exhibit strict logical homogeneity, the former do not. So how can the power of judgement account for the unity of a kind of representation whose structure consists of strict logical homogeneity?

The solution to this problem lies in the fact that spontaneity for Kant is not exhausted by the capacity for judgement. Rather, spontaneity can also be exercised in the guise of the productive imagination. This allows us to draw the following distinction: The exercise of spontaneity in judgement accounts for the unity of representations whose structure is specifically conceptual. The exercise of spontaneity in acts of the productive imagination accounts for the unity of representations whose structure consists of strict logical homogeniety. In short, spontaneity in the guise of the productive imagination accounts for the unity of the representation of magnitudes.

A full defence of this solution would require discussion of Kant's theory of judgement and its relation to the notion of spontaneity, and that lies beyond the confines of this paper.[39] What I wish to do instead is to argue, in the next section, that consideration of Kant's doctrine of mathematical construction lends support to the suggestion that he recognizes an exercise of spontaneity that is distinct from judgement in that it accounts for the unity of representations that exhibit strict logical homogeneity.

## 6. Construction and spontaneity

Kant famously holds that, in contrast to philosophical knowledge, mathematical knowledge is based on the construction of concepts in pure intuition.[40] He holds, further, that such construction depends on acts of the productive imagination. As he puts it in the section entitled Axioms of Intuition, '[on the] successive synthesis of the productive imagination, in the generation of shapes, is grounded the mathematics of extension (geometry)' (A163/B204).[41] We can expect, then, that consideration of the doctrine of mathematical construction sheds light on Kant's conception of the productive imagination. I will argue that this in turn provides at least the basis for an account on which the act of the productive imagination is spontaneous, yet specifically distinct from judgement and thus for an account of the Heterogeneity Thesis that, unlike the Simple Mereological Account, does not entail that a representation's being sensible is incompatible with its being dependent on acts of spontaneity.

Kant's doctrine of mathematical construction and the role it plays in mathematical proof are subject to controversy among commentators.[42] However, because my discussion of construction will appeal only to very general features of it, which are granted by all sides, there is no need to enter into the details of this debate.[43]

As is well known, the doctrine of construction is closely connected to the status of mathematical knowledge as synthetic a priori. Because mathematics does not proceed by conceptual analysis, but rather by construction, it can synthetically determine the properties of the objects it considers because it can rely, for cognition of these properties, on pure intuition. Thus, when the geometer considers the concept of a triangle and constructs this concept in pure intuition, she can 'read off' this construction properties of triangles which, according to Kant, are not analytically contained in the concept of a triangle (such as, for instance, the property that the internal angle-sum of any triangle is equal to 180°).[44]

As Shabel has emphasized, of crucial importance here are certain kinds of spatial properties.[45] Thus, in constructing a concept in intuition, the geometer can appeal to containment-relations among angles or to topological properties such as one part of a figure's lying outside another figure (e.g., part of a line lying outside a circle). Crucially, these are properties that, according to Kant, cannot be represented by purely conceptual means, for they involve the strict logical homogeneity of space. So they are not properties that could be ascribed on the basis of conceptual analysis. They *can* be ascribed on the basis of the construction of a concept because the construction, as it were, draws on those characteristics of space that make the original representation of space an intuition, as opposed to a concept. The construction, we might say, gives the geometer access to a content that is not part of the concept independently of its construction.

Let me elaborate the point slightly. In constructing a concept, the geometer brings about a particular spatial representation; or, as I will say from now on, using Kant's terminology, she brings about a determinate spatial representation.[46] Unlike the *concept* ⟨triangle⟩ taken by itself, the determinate spatial representation that is the construction of this concept exhibits concrete spatial properties and relations such as, e.g., one circle intersecting another circle. The content of the concept ⟨triangle⟩ is constituted by certain conceptual marks, viz. ⟨figure in the plane enclosed by three straight lines⟩. But this is not an actual figure. By contrast, the construction of this concept *is* an actual figure. In advancing from the bare concept to its construction, the geometer gains epistemic access to the spatial properties and relations that characterize an actual figure.[47]

An important difficulty confronted by Kant's theory of construction concerns the question how an actual figure, which is a particular, can express a concept in its full generality – as it must if it is to underwrite mathematical knowledge. Thus, how can the construction of the concept ⟨triangle⟩, which is a particular triangular figure, underwrite claims about the properties of all triangles? My purposes in this paper do not require me to address this question head-on. But I wish to draw attention to an aspect of Kant's attempt to address it, viz. the rule-governed nature of construction.

Kant holds that a particular constructed figure, such as a triangle, expresses the required generality insofar as we

consider only the act whereby we construct the concept and abstract from the many determinations (for instance, the magnitude of the sides and of the angles), which are quite indifferent, as not altering the concept 'triangle'. (A713f/B741f)

By 'the act whereby we construct the concept', Kant means an act of following certain clearly-defined constructive operations. The most basic operations of this kind are set out in the postulates of Euclid's *Elements*. So, for instance, 'to draw a straight line between any two points on the plane' is an operation of construction, as is 'to describe a circle with any center and distance'. Further operations of this kind can be generated by combining these basic operations with one another and with the definitions of geometrical concepts. Thus, 'to draw straight lines between any three non-collinear points on the plane' is the operation of constructing a triangle.

Again, what it is to 'consider only the act whereby we construct the concept' and how this is supposed to yield the required generality is not something we need to enter into here. The important point for my purposes is that carrying out a constructive operation is an act of spontaneity, which yields a determinate sensible representation. Although I cannot offer an account of spontaneity here, I take it to be intuitively plausible that the following characteristics of construction are sufficient for ascribing it to a spontaneous capacity: (i) Construction is an act of following a rule; of applying a general concept of an act-type so as to bring about an action (perhaps a mental action) that is an instance of this type.[48] (ii) This act is performed in such a way that the subject applying the concept understands what she is doing. For otherwise, she could not 'consider only the act whereby we construct the concept and abstract from [various irrelevant] determinations [of the figure]'. It seems obvious that an act of this type could not be due to a receptive capacity.[49] But within the context of the Two Stems Doctrine, this entails that it must be an act of spontaneity.

Textual evidence for this contention comes from Kant's repeated character-izations of construction as an act of the productive imagination, which, as we saw above in Section 3, he conceives as spontaneous, yet at the same time as yielding sensible representations.[50] Kant's theory of mathematical construction, then, lends support to the claim that he recognizes a kind of exercise of spontaneity whose distinctive character consists in the fact that it *generates* sensible representations. If this is right, then the Heterogeneity Thesis must not be interpreted as entailing that a representation's being sensible is incompatible with its being dependent on an act of spontaneity. Kant allows for such dependence, and in terms of the distinction between struture and unity introduced in the preceding section we can express his view by saying that a representation can have distinctively sensible structure, while depending for its unity on acts of spontaneity.

Indeed, as I wish to argue in conclusion, it seems to be Kant's view that this is true of human outer intuitions generally. Consider once more the passage from B202f quoted in Section 5:

> All appearances contain, as regards their form, an intuition in space and time, which grounds all of them a priori. They cannot be apprehended, therefore, i.e., taken up into empirical consciousness, except through the synthesis of the manifold through which the representations of a determinate space or time are generated, i.e., through the composition of that which is homogeneous and the consciousness of the synthetic unity of this manifold (homogeneous). (B202f)

Kant says here that the representation of determinate spaces depends on a synthesis of that which is homogeneous; that is, a synthesis in accordance with the category of magnitude. Now, of this synthesis he says that it is of the same kind as the act of synthesis involved in mathematical construction:

> The synthesis of spaces and times, as the essential form of all intuition, is that which at the same time makes possible the apprehension of appearance, thus every outer experience, consequently also all cognition of its objects, and what mathematics in its pure use proves about the former is also necessarily valid for the latter. (A166f/B206)

The claim here appears to be that the representation of determinate spaces in empirical intuition depends on the empirical version of a kind of synthesis which is employed purely in the construction of a geometrical figure. If this is right, then some of the characteristic features of mathematical construction – viz. those that are indifferent to the distinction between pure and empricial intuition – also hold of the kind of synthesis involved in the apprehension of appearances. Central among these are the features that pertain to the spatial character of the representation thus generated. As we have just seen, it is essential to Kant's conception of construction that a constructed figure exhibits spatial properties – properties that are essentially properties of magnitudes. If the representation of determinate spaces in the empirical apprehension of appearances shares this feature, then it, too, must be regarded as exhibiting the *structure* characteristic of sensible representations – strict logical homogeneity – while depending for its *unity* – the fact that a particular strictly logically homogeneous manifold constitutes a one, viz. a determinate space, and is represented as such – on an act of spontaneity.

## 7. Conclusion

Although Kant introduces the doctrine that there are two stems of the cognitive capacity in a somewhat casual manner, by way of a 'reminder', making out the precise content of this doctrine is not at all straightforward. What the Heterogeneity Thesis and the Cooperation Requirement are regarded as saying depends in large part on one's interpretation of a number of other important doctrines Kant puts forth in the *Critique*. I have argued that two such doctrines, both of which are closely connected to Kant's philosophy of mathematics, are more salient here than is commonly supposed. These are, first, the doctrine that space is a magnitude and, second, the doctrine of the productive imagination, which informs Kant's theory of mathematical construction.[51]

Consideration of these doctrines has yielded a partial account of the Heterogeneity Thesis and, at the same time, a constraint on any adequate account of the Cooperation Requirement. The partial account of the Heterogeneity Thesis says that it is a distinctive feature of outer intuitions that they exhibit strict logical homogeneity. This does not yet give us a positive characterization of the distinctive features of concepts, discussion of which must await another occasion. What has emerged, however, is that the spontaneous stem of the cognitive capacity must be conceived in such a way as to allow for a kind of exercise that is responsible for the unity of outer intuitions. This result of our discussion constitutes a criterion of adequacy on any account of spontaneity in Kant and, by implication, on any account of the Cooperation Requirement. Any such account must make room for the idea that sensibility, as Kant conceives it, is not a self-standing cognitive capacity. At least one important class of sensible representations, viz. outer intuitions, depend not just on this capacity, but on acts of spontaneity as well. Absent the exercise of spontaneity, no determinate outer intuition is generated. As Kant puts it: a 'determinate intuition [ . . . ] is possible only through the consciousness of the determination of [sensibility] through the transcendental act of imagination' (B154).

The Simple Mereological Account considered in the first half of this paper does not meet this constraint and should, partly for this reason, be rejected. This is not to say, however, that there is nothing of value in this account. On the contrary, the alternative I have presented shares the Simple Mereological Account's strategy of focusing on the structure of intuitions and of turning to Kant's discussion of the forms of intuition for information concerning this structure. What is more, I also agree that the heterogeneity of intuitions and concepts can be brought out by attending specifically to the part–whole structure Kant ascribes to these two kinds of representations. I disagree only (though significantly) with the claim that the heterogeneity of intuitions and concepts consists in the fact that they exhibit opposite directions of dependence between representational parts and representational whole. In my view, what makes intuitions distinct in kind from concepts is that they exhibit the particular kind of part–whole structure that Kant takes to be the distinguishing characteristic of magnitudes.[52]

## Notes

1. '[ . . . ] there are two stems of human cognition which may perhaps arise from a common but to us unknown root, namely sensibility and understanding, through the first of which objects are given to us, but through the second of which they are thought' (A15/B29). 'Intuition and concepts therefore constitute the elements of all

our cognition so that neither concepts without intuition corresponding to them in some way nor intuition without concepts can constitute a cognition.' (A50/B74) – References to the *Critique of Pure Reason* use the A- and B-edition pagination; translations are from Kant (1998), tacitly modified where appropriate. References to other works of Kant's are by volume- and page-number of the Academy Edition (Ak. = Kant 1902ff) using the following abbreviations: Logik = Logik, ed. Benjamin Jaesche. Prolegomena = Prolegomena zu einer jeden künftigen Metaphysik die als Wissenschaft wird auftreten künnen [Prolegomena to any Future Metaphysics that will be able to come forward as Science].

2.  Conceptualist readings of Kant are advocated by, among others, Allison (2004), Ginsborg (2006, 2008), Griffith (2012), McDowell (1998, 2009a, 2009b), Pippin (1982), Sellars (1978, 1992) and Strawson (1966). Nonconceptualist readings have been proposed by Allais (2009), Hanna (2005, 2011), McLear (forthcoming-b), Rohs (2001) and Tolley (2013). See McLear (forthcoming-a) for an overview of the debate.

3.  See A68f/B93f, A320/B376, *Logik* §1 (Ak. IX: 91).

4.  Throughout this paper, I will be concerned with sensible, as opposed to intellectual, intuition and indeed with sensible intuitions of outer, as opposed to inner, sense, but I will not always make this explicit. When I speak of intuition *simpliciter,* this should be taken as shorthand for 'outer sensible intuition,' except where indicated. The same goes for 'understanding' and 'discursive understanding'.

5.  See A68f/B93f, A320/B376, *Logik* §1 (Ak. IX: 91).

6.  See, for instance, the well-known dispute between Hintikka (1969a, 1969b) and Parsons (1982) on whether the singularity of intuition entails its immediacy. Furthermore, Wolff (1995) makes the case that not only intuitions but concepts, too, can relate to objects immediately, so that we need to distinguish between conceptual and intuitive immediacy.

7.  Perhaps the most prominent advocate of this position is Wilfrid Sellars, who argues that although having an intuition is distinct from making a judgement, an intuition must be understood on the model of a complex demonstrative referring expression (a 'this-such') comprising a demonstrative and a sortal term. See Sellars (1967, 1978, 1992).

8.  'The unity of aesthetic representation – characterized by the forms of space and time – has a structure in which the representational parts depend on the whole. The unity of discursive representation – representation where the activity of the understanding is involved – has a structure in which the representational whole depends on its parts' (McLear forthcoming-b, 19f.).

9.  'Thus, if a representation has a structure in which the parts depend on the whole rather than a structure in which the whole is dependent on its parts, that representation cannot be a product of intellectual activity, but must rather be given in sensibility independently of any such activity' (McLear forthcoming-b, 18f.).

10. 'Synthetic unity of the manifold of intuitions, as a priori given, is therefore the ground of the identity of apperception itself, which precedes a priori all *my* determinate thinking' (B134).

11. Kant's conception of analytic judgements highlights another respect in which discursive activity can be said to move from whole to part rather than part to whole. In Kant's view, analytic judgements 'dissect [the subject concept of the judgment] into its component concepts, which were already thought in it (albeit obscurely)' (A7/B11). Notice also that, according to the account offered in §77 of the *Critique of Judgment* (Ak. V: 407), the intuitive intellect proceeds from an intuition of the whole to cognition of its parts. This might be thought to support the contention that part-on-whole dependence is characteristic of intuition in general, whether finite and sensible

or infinite and intellectual. By parity of reasoning, however, it would also show that whole-on-part dependence cannot be a characteristic of intellectual activity in general. This at least suggests (though it does not establish) that such dependence does not constitute the distinguishing characteristic of finite intellectual activity, either. Thanks to Johannes Haag for drawing my attention to this passage.

12. That relation of representations which constitutes a judgement is 'sufficiently distinguished' from a merely associative unity by the fact that the former, unlike the latter, depends on 'principles of the objective determination of all representations, insofar as cognition can arise from them, which principles are all derived from the principle of the transcendental unity of apperception' (B142). Again, we can note the intimate connection between judgement and the doctrine of apperception, while leaving open the precise meaning of the passage. See B141 for the claim that associative unity is due to the reproductive imagination and B151f for the claim that this capacity is sensible rather than intellectual.

13. See also B153, where Kant says that the transcendental synthesis of imagination is an act of the understanding, and B162, where he says that understanding and productive imagination are 'one and the same spontaneity' considered under different aspects.

14. See e.g., A714f/B742f. Note that for Kant mathematical cognition is not defined as being cognition of magnitudes. It is defined, rather, in terms of its method, as rational cognition based on the construction of concepts in pure intuition. It is a consequence of this definition that mathematics cognizes magnitudes. For discussion see Sutherland (2004b).

15. Compare the following passage from the Axioms of Intuition: 'Empirical intuition is possible only through pure intuition (of space and time); what geometry says about the latter is therefore undeniably valid of the former [ . . . ]' (A165/B206). Consider also Kant's claim in the Aesthetic, that the representation of space 'is the ground of all outer intuitions' (A24/B38).

16. See, in particular, Sutherland (2004a, 2004b), to which the discussion that follows is indebted.

17. Cf. the following explanation from the Lectures on Metaphysics: 'That determination of a thing through which one cognizes something as a quantum is quantity or magnitude' (Metaphysik K3, Ak. XXIX: 991). As Shabel (2005) puts the point: For Kant, as for the moderns generally, magnitudes have magnitude.

18. Compare the fourth and fifth of Euclid's Common Notions.

19. Kant appears to hold that the holistic nature of outer intuitions is a consequence of their character as magnitudes and my point here is that the converse implication does not hold (see Sutherland [2004a] for discussion). To see why Kant might hold this, consider that if space exhibits strict logical homogeneity, then the identity conditions of any part of space can be given only in terms of its relation to other parts of space. There is nothing intrinsic about a part that could differentiate it from other parts. So necessarily any talk of a part of space presupposes reference to other parts of space, indeed to the whole of space. Kant seems to say as much at *Prolegomena*, §13 (Ak. IV:286): '[ . . . ] the inner determination of any space is possible only through the determination of the outer relation to the whole of space [ . . . ], i.e. the part is possible only through the whole'. This holistic character of space should then also characterize the representation of space. Thanks to Daniel Smyth for bringing this passage to my attention.

20. This is not to deny that objects which are considered as magnitudes can differ from one another qualitatively. The point is that *insofar as* their character as magnitudes is concerned, they are regarded as strictly logically homogeneous; see Sutherland (2004a, 199f.). I say more about this at the end of this sub-section and in the following sub-section.

21. See A272/B328. See also A725/B753, where Kant speaks of space and time as the 'only original *quanta*'.
22. I am grateful to Tobias Rosefeldt for helping me clarify this point.
23. The notion of a mark is discussed in *Logik*, Introduction, §VIII (Ak. IX, 58ff).
24. For discussion see Anderson (2005) and Schulthess (1981).
25. These roles are, of course, relative. If A is contained under B, and B under C, then B is a species relative to C, which is its genus. But B is a genus relative to A, which is a species of it.
26. The division need not be dichotomous, but can comprise more members, as long as there is no overlap between them and they jointly exhaust the extension of the divided concept. The dichotomous case is simply the easiest for illustration purposes. For discussion see Anderson (2005) and Wolff (1995, 160–170).
27. Consider an example: I begin with the concept ⟨scholar⟩, and the task is to represent a plurality of scholars. Because the only means at my disposal is the introduction of conceptual marks, I can represent, say, two scholars by introducing a mark that sorts scholars into short and tall. But this means that I have succeeded in representing numerical difference only at the cost of introducing further qualitative differences. Clearly, this would be the case no matter how many additional marks I introduce.
28. This is not to deny that there are concepts of magnitudes. Mathematical concepts are of this kind. But they are not purely conceptual in the sense that their content depends essentially on the possibility of being constructed in intuition and so depends on intuition. Kant characterizes such concepts as 'containing' pure intuitions (A719/B747). This distinguishes such concepts from what I am calling purely conceptual representations.
29. Thanks to Stefanie Grüne and Marcus Willaschek for pointing out the need for clarification here.
30. Although this is a separate point, we should also note that Kant holds that sensations themselves, considered in abstraction from their spatial or temporal extent, are magnitudes; see A166/B207–A176/B218. Although sensations are intensive rather than (like space and time) extensive magnitudes, if the account of the general concept of magnitude presupposed here is correct, intensive magnitudes, too, are conceived as manifolds of strictly logically homogeneous parts. For discussion see Sutherland (2004a, 196f.).
31. Because Kant recognizes spatial concepts (see Note 28 above), we should distinguish between a direct and an indirect dependence of spatial representations on acts of the productive imagination. Sensible spatial representations (i.e. outer intuitions) depend directly on exercises of the productive imagination. Conceptual spatial representations depend indirectly on exercises of the productive imagination of spatial magnitudes inasmuch as they depend for their content on being constructible in intuition and thus on sensible spatial representations. I say more about construction and the productive imagination in Section 6.
32. See Land (2006) for discussion.
33. Note that if I am right and spatial representation for Kant partly depends on acts of spontaneity, yet is still in a robust sense sensible rather than discursive, then there is a danger of using the term 'sensibility' ambiguously. For we might mean by this either a capacity that is merely receptive and so on my reading of Kant would not be a capacity for spatial representation of the kind under discussion in the *Critique*. Or we might mean by it a capacity for the latter kind of representation, hence, a capacity for representations that exhibit the characteristic structure of outer intuitions, yet depend also on spontaneity. So we should distinguish between a merely receptive kind of sensibility and a kind of sensibility that is not merely receptive but, as it were, conditioned by spontaneity. Part of the point of the doctrine of the productive

imagination is, I think, to show that the kind of sensibility Kant is concerned with in the *Critique* is the second kind. Perhaps no-one has made the case for this more eloquently than Sellars (see e.g., Sellars 1967, 1978, 1992; see also the lectures published as Sellars 2002). For a defence of the idea see also McDowell (2009a, 2009b). We should also note that according to Sellars (1992) Kant was himself not entirely clear about the commitments entailed by his theory of sensibility and, in a sense, fell victim to the ambiguity just outlined. A related ambiguity is noted by Beck (2002) and Allison (2004).

34. For discussion see, e.g., Evans (1985) and Rödl (2012); see also Sellars (1967).
35. See Evans (1985).
36. The passage is sometimes read differently. Thus, Tolley (2013) reads the claim that a representation contained in a single moment can only have absolute unity as marking one side of a distinction between a manifold's *having* a unity and a manifold's *being represented as having* a unity. However, because a moment is a temporal boundary in the same way that a line or a point is a spatial boundary (see A169/B211) and because, as is explicitly pointed out by Kant in the paragraph preceding the passage, the kind of manifoldness being considered is temporal manifoldness, an absolute unity cannot be the unity of a manifold, whether represented as such or not. For helpful discussion of this point see Henrich (1953).
37. The qualifier is intended to allow for the possibility of kinds of spatial representation that do not depend on spontaneity. The kind of spatial representation enjoyed by non-rational animals is presumably of this kind, for Kant. In the *Critique*, Kant is concerned with finite rational knowers. What he says about spatial representation applies, in the first instance, to those finite rational knowers whose forms of sensibility include space. Whether, and to what extent, it also applies to non-rational creatures is not obvious.
38. For further support of this point, consider that it is a premise of the First Space Argument that in outer experience I represent things as being in different locations in space from my own: 'For in order for certain sensations to be related to something outside me (i.e. to something in another place in space from that in which I find myself) [ . . . ], the representation of space must already be their ground' (A23/B38). For discussion see Warren (1998).
39. I say more about it in Land (2006, forthcoming).
40. See A712/B740ff.
41. Although this passage focuses on geometry, Kant holds that all of mathematics depends on construction. For discussion see Shabel (1998).
42. For the recent debate see Carson (1997), Friedman (1992, 2000, 2012), and Shabel (1998, 2003).
43. Kant's most extensive (though still rather abbreviated) discussion of mathematical construction is found at A713f/B741f.
44. This idea of 'reading off' properties of the figure is most naturally understood in terms of the Euclidean diagram. The diagram allows the geometer literally to see that certain spatial relations obtain between different parts of the constructed figure. For discussion see Shabel (2003).
45. See Shabel (2003).
46. It might be thought that according to Friedman's (2012) account of construction, this is not the case. For Friedman rejects the view, propounded e.g., by Shabel (2003), that the Euclidean diagram plays an essential role in Kant's theory of construction. However, despite his reservations concerning diagrams, Friedman is still committed to the idea that construction involves the generation of determinate spatial representations. On his account, Kant's notion of construction is understood in terms of the notion of a function. For example the construction of the concept 'triangle'

must be understood, according to Friedman, in terms of a 'function or constructive operation which takes three arbitrary lines (such that two together are greater than the third) as input and yields the triangle constructed out of these three lines as output' (2012, 237). Crucially, the outputs of such functions 'are indeed singular or individual representations' (2012), that is, what I am calling determinate spatial representations.

47. As Kant puts it, the geometer proceeds by 'determining [her] object in accordance with the conditions of [ ... ] pure intuition' (A718/B746).
48. This suggests that there is a sense in which the concept at issue is a practical concept; that is, a concept of a way of acting. Note that Kant characterizes a mathematical postulate as 'the *practical* proposition that contains nothing except the synthesis through which we first give ourselves an object and generate its concept' and gives one of Euclid's postulates as an example (A234/B287, my emphasis).
49. See the discussion above, in Section 5.
50. For example, B154, A163/B204, A164f/B205.
51. This is not to say, of course, that these are the only salient doctrines. A full account would have to give due consideration to, for instance, the arguments put forth in the Transcendental Deduction of the Categories.
52. For comments and suggestions I am grateful to audiences in Chicago, Frankfurt, Oxford and Potsdam, and in particular to James Conant, Stefanie Grüne, Johannes Haag, Till Hoeppner, Colin McLear, Tobias Rosefeldt, Daniel Smyth, Andrew Stephenson, Daniel Sutherland and Marcus Willaschek. Work on the paper was supported by a research fellowship at Corpus Christi College, Cambridge.

## References

Allais, Lucy. 2009. "Kant, Non-Conceptual Content, and the Representation of Space." *Journal of the History of Philosophy* 47: 383–413.

Allison, Henry. 2004. *Kant's Transcendental Idealism: An Interpretation and Defense.* 2nd ed. New Haven, CT: Yale University Press.

Anderson, Lanier. 2005. "The Wolffian Paradigm and Its Discontents: Kant's Containment Definition of Analyticity in Historical Context." *Archiv für Geschichte der Philosophie* 87: 22–74.

Beck, Lewis White. 2002. "Did the Sage of Königsberg Have No Dreams?" In *Selected Essays on Kant*, edited by Hoke Robinson, 85–101. Rochester, NY: University of Rochester Press.

Carson, Emily. 1997. "Kant on Intuition in Geometry." *Canadian Journal of Philosophy* 27: 489–512.

Evans, Gareth. 1985. "Things Without the Mind." In *Collected Papers*, edited by Antonia Phillips, 249–290. Oxford: Clarendon Press.

Friedman, Michael. 1992. *Kant and the Exact Sciences*. Cambridge, MA: Harvard University Press.

Friedman, Michael. 2000. "Geometry, Construction, and Intuition in Kant and His Successors." In *Between Logic and Intuition*, edited by Gila Sher, and Richard Tieszen, 186–218. Cambridge: Cambridge University Press.

Friedman, Michael. 2012. "Kant on Geometry and Spatial Intuition." *Synthese* 186: 231–255.

Ginsborg, Hannah. 2006. "Kant and the Problem of Experience." *Philosophical Topics* 34: 59–106.

Ginsborg, Hannah. 2008. "Was Kant a Nonconceptualist?" *Philosophical Studies* 137: 65–77.

Griffith, Aaron. 2012. "Perception and the Categories." *European Journal of Philosophy* 20: 193–222.

Hanna, Robert. 2005. "Kant and Nonconceptual Content." *European Journal of Philosophy* 13: 247–290.

Hanna, Robert. 2011. "Kant's Non-Conceptualism, Rogue Objects, and the Gap in the B Deduction." *International Journal of Philosophical Studies* 19: 399–415.

Henrich, Dieter. 1953. "Zur theoretischen Philosophie Kants" [On Kant's Theoretical Philosophy]. *Philosophische Rundschau* 1: 124–149.

Hintikka, Jaako. 1969a. "Kant on the Mathematical Method." In *Kant Studies Today*, edited by Lewis White Beck, 117–140. La Salle, IL: Open Court.

Hintikka, Jaako. 1969b. "On Kant's Notion of Intuition (Anschauung)." In *The First Critique: Reflections on Kant's Critique of Pure Reason*, edited by T. Penelhum, and J. J. MacIntosh, 38–53. Belmont: Wadsworth.

Kant, Immanuel. 1902ff. *Kants gesammelte Schriften*. Edited by the Königlich Preußische Akademie der Wissenschaften. Berlin: de Gruyter and predecessors.

Kant, Immanuel. 1998. *Critique of Pure Reason*. Translated and edited by Paul Guyer and Allen Wood. Cambridge: Cambridge University Press.

Land, Thomas. 2006. "Kant's Spontaneity Thesis." *Philosophical Topics* 34: 189–220.

Land, Thomas. Forthcoming. "No Other Use than in Judgment? Kant on Concepts and Sensible Synthesis." *Journal of the History of Philosophy*.

McDowell, John. 1998. "Having the World in View: Sellars, Kant, and Intentionality." *Journal of Philosophy* 95: 431–491.

McDowell, John. 2009a. "Avoiding the Myth of the Given." Chap. 14 in *Having the World in View: Essays on Kant, Hegel, and Sellars*. Cambridge, MA: Harvard University Press.

McDowell, John. 2009b. "Sensory Consciousness in Kant and Sellars." Chap. 6 in *Having the World in View: Essays on Kant, Hegel, and Sellars*. Cambridge, MA: Harvard University Press.

McLear, Colin. Forthcoming-a. "The Kantian (Non)-Conceptualism Debate." *Philosophy Compass*.

McLear, Colin. Forthcoming-b. "Two Kinds of Unity in the Critique of Pure Reason." *Journal of the History of Philosophy*.

Parsons, Charles. 1982. "Kant's Philosophy of Arithmetic." In *Kant on Pure Reason*, edited by Ralph Walker, 13–40. Oxford: Oxford University Press.

Pippin, Robert. 1982. *Kant's Theory of Form*. New Haven, CT: Yale University Press.

Rödl, Sebastian. 2012. *Categories of the Temporal*. Cambridge, MA: Harvard University Press.

Rohs, Peter. 2001. "Bezieht sich nach Kant die Anschauungunmittelbar auf Gegenstände?" [Does Intuition relate immediately to Objects, according to Kant?]. In *Kant und die Berliner Aufklärung*, edited by Volker Gerhardt, Rolf-Peter Horstmann and Ralph Schumacher, vol. 2, 214–228. Berlin: de Gruyter.

Schulthess, Peter. 1981. *Relation und Funktion*. Berlin: de Gruyter.

Sellars, Wilfrid. 1967. "Some Remarks on Kant's Theory of Experience." *Journal of Philosophy* 64: 633–647.

Sellars, Wilfrid. 1978. "The Role of Imagination in Kant's Theory of Experience." In *Categories: A Colloquium*, edited by Henry Johnstone, 231–245. University Park, PA: Penn State University Press.

Sellars, Wilfrid. 1992. "Sensibility and Understanding." Chap. 1 in *Science and Metaphysics: Variations on Kantian Themes*. Atascadero, CA: Ridgeview.

Sellars, Wilfrid. 2002. *Kant and Pre-Kantian Themes: Lectures by Wilfrid Sellars*, edited by Pedro Amaral. Atascadero, CA: Ridgeview.

Shabel, Lisa. 1998. "Kant on the 'Symbolic Construction' of Mathematical Concepts." *Studies in History and Philosophy of Science* 29A: 589–621.

Shabel, Lisa. 2003. *Mathematics in Kant's Critical Philosophy: Reflections on Mathematical Practice*. New York: Routledge.

Shabel, Lisa. 2005. "Apriority and Application: Philosophy of Mathematics in the Modern Period." In *Oxford Handbook of Philosophy of Mathematics and Logic*, edited by Stewart Shapiro, 29–49. Oxford: Oxford University Press.

Strawson, Peter. 1966. *The Bounds of Sense*. London: Methuen.

Sutherland, Daniel. 2004a. "Kant's Philosophy of Mathematics and the Greek Mathematical Tradition." *Philosophical Review* 113: 157–201.

Sutherland, Daniel. 2004b. "The Role of Magnitude in Kant's Critical Philosophy." *Canadian Journal of Philosophy* 34: 411–442.

Tolley, Clinton. 2013. "The Non-Conceptuality of the Content of Intuitions." *Kantian Review* 18: 107–136.

Warren, Daniel. 1998. "Kant and the Apriority of Space." *Philosophical Review* 107 (2): 179–224.

Wolff, Michael. 1995. *Die Vollständigkeit der kantischen Urteilstafel. Miteinem Essay über Freges 'Begriffsschrift'* [The Completeness of Kant's Table of Judgments. Including an Essay on Frege's 'Begriffsschrift']. Frankfurt a.M. Klostermann.

# Infinity and givenness: Kant on the intuitive origin of spatial representation

Daniel Smyth

*Department of Philosophy, University of Chicago, Chicago, IL, USA*

I advance a novel interpretation of Kant's argument that our original representation of space must be intuitive, according to which the intuitive status of spatial representation is secured by its infinitary structure. I defend a conception of intuitive representation as what must be given to the mind in order to be thought at all. Discursive representation, as modelled on the specific division of a highest genus into species, cannot account for infinite complexity. Because we represent space as infinitely complex, the spatial manifold cannot be generated discursively and must therefore be given to the mind, i.e. represented in intuition.

Space is represented as an infinite **given** magnitude. (*Critique of Pure Reason*, B39)[1]

Kant's distinction between sensibility and understanding is arguably the keystone of his critical enterprise. His detailed account of this dichotomy ought therefore to be one of the most controversial aspects of his system. Of course, the history of philosophy abounds with versions of some broad contrast between 'intellect' and 'sense' or between 'higher' and 'lower' cognitive faculties. This historical prevalence can mask the heterodoxy of Kant's account, but its strangeness becomes glaring in light of his unprecedented claim that our knowledge of the infinite (and, indeed, all our knowledge of pure mathematics) is ultimately grounded in *sensibility*. Kant himself was eager to emphasize the originality and importance of his critical distinction between conceptual and intuitive representation (cf. A271/B327). Yet one is hard-pressed to find direct and explicit arguments for the details of these distinctions in Kant's published critical writings. Indeed, it can seem that the *Critique* and the *Prolegomena* begin by presupposing, stipulating or otherwise hypothesizing certain robust conceptions of judgement, intuition, conceptual representation, mathematical cognition, etc. and then proceed to demonstrate (with more or less success) the fruitfulness of these conceptions indirectly, by showing how they (alone?) serve to resolve various philosophical difficulties.[2] In what follows, I will resist this impression and suggest that Kant does, in fact, provide the materials for an extended argument in favour of his nuanced conceptions of conceptual and intuitive representation over the course of the Aesthetic and Analytic. I will confine myself to one brief but crucial stage of that argument, which bears on his account of sensible, intuitive representation. My focal point will be the penultimate section of the Metaphysical Exposition of the Concept of Space,[3] which argues that our original representation of space is intuitive. The reason for this focus is simple. Any argument to the effect that a given representation is intuitive must trade on a particular conception of what intuitive representation consists in. Accordingly, a satisfactory interpretation of the latter MEs must spell out Kant's conception of intuition at that point in the *Critique*. And the adequacy of any such interpretation will depend, in part, on how plausible it is to suppose that Kant is entitled to that conception at the relevant stage in his argument. I will propose that Kant articulates a functional conception of sensibility as the ability to be given objects that exist independently of our spontaneous acts of thought. The penultimate ME, then, seeks to show that our concept of space must derive from such an object-giving representation. It shows this, I will argue, by observing that discursive acts of thought cannot account for the infinitary structure of space – in particular, its continuity and open-endedness. Consequently, such an infinitary manifold must (originally) be given to us in order to be thought at all and is, *ipso facto*, represented in sensible intuition (insofar as it is represented at all). One advantage of this reading is that it enables us to identify a stage in Kant's 'synthetic' or progressive argument for his distinctive account of intuitive representation. If the argument that spatial representation is intuitive principally turns on the infinitary structure of space and a conception of intuition as object-giving, and if the arguments for the apriority of space are sound, it follows that all outer intuitions are singular representations, because they necessarily represent unique portions of a single, essentially unitary space. Thus, assuming parallel arguments can be made for time, the MEs collectively establish at least one crucial feature of Kant's conception of sensibility: namely the singularity of sensible representation. This is a step in a 'synthetic' argument in the sense that it enriches the conception of intuition with which we began – from *object-giving representation* to *singular, object-giving representation*.

Section 1 surveys the criteria of conceptual and intuitive representation that are typically invoked in reconstructions of the penultimate ME. I argue that none of these criteria can fund a textually and philosophically satisfying interpretation. Section 2 sketches the functional account of intuition Kant articulates and defends in the Introduction to the *Critique*. Section 3 deploys this functional conception in a novel interpretation of Kant's argument, on which the intuitive status of spatial representation follows from its infinitary structure. Section 4 concludes by highlighting some consequences of my interpretation for an overall reading of the argumentative structure of the *Critique*.

## 1. The interpretive challenge posed by the penultimate ME

Space is not a discursive, or, as one says, a general [allgemeiner] concept of relations of things generally [überhaupt], but rather a pure intuition. For, first, one can only represent a unitary [einigen] space, and when one speaks of many spaces, one understands by that only parts of one and the same solitary [alleinigen] space. Nor can these parts precede the unitary, all-encompassing space as, so to speak, components [Bestandteile] (from which it might be composed [daraus eine Zusammensetzung möglich sei]); rather, they can only be thought **in it**. Space is essentially unitary [einig]; the manifold in it, and thus even the general concept of spaces as such [überhaupt] rests solely on limitations [Einschränkungen]. It follows from this that, with regard to space, all concepts of it are grounded upon an intuition a priori (one that is not empirical). (A24f./B39)

This passage clearly presupposes some criterion of intuitive and/or conceptual representation. For any argument to the effect that a certain representation is an intuition and not a concept must inevitably trade on some account of what intuitions and concepts are. Our interpretive challenge is to determine what the relevant differentiating criteria are and how they bear on the features of space (or spatial representation) Kant highlights.[4] To meet this challenge an interpretation must (a) provide ample and clear textual evidence that Kant adopted the criteria in question, (b) show that Kant might plausibly have taken himself to be entitled to invoke these accounts as premises at this point in the *Critique* and (c) by combining these accounts with what Kant explicitly says, yield an argument of sufficient cogency to have been endorsed by someone of Kant's philosophical acumen. The chart below outlines the criteria typically invoked by interpreters, along with citations that are often provided as textual evidence for them.[5]

| *Criteria of intuitive representation* | *Criteria of conceptual representation* |
| --- | --- |
| **Singularity** | **Generality** |
| – Intuitions are essentially singular representations, i.e. they intrinsically refer to exactly one object as such. | – Concepts are essentially general representations, i.e. they are intrinsically able to refer to indefinitely many objects. |
| (*De Mundi* 2:402; *Logic* 9:91; B137; A320/B376f.) | (*Logic* 9:91f.; A320/B376f.) |
| **Holistic containment** | **Atomic containment** |
| – The representations contained in an intuition are posterior to the intuition that contains them; i.e. the 'whole' is the ground of the possibility of the 'parts'. | – The representations contained in a concept are prior to the concept that contains them; i.e. a concept's constituent 'parts' (or marks) ground the possibility of the 'whole'. |
| (A169f./B211f.) | (*Logic* 9:35, 58; or 'traditional logic') |
| **Immediacy** | **Mediacy** (i.e. discursivity) |
| – An intuition relates immediately to the object(s) it represents. | – A concept relates to the object(s) to which it refers through mediating marks or features common to those objects. |
| (A19/B33; A320/B376f.) | (*Logic* 9:58ff.; B33; A320/B376f.) |

Prevailing interpretations differ only in which of these criteria they invoke and whether they treat them as necessary and/or sufficient in tracing a valid argument to the desired conclusion.[6] Yet I will argue that no combination of these criteria provides a textually and philosophically compelling reconstruction of Kant's argument – i.e. one that meets our threefold interpretive challenge. Let's consider them in descending order.

It's quite natural to suppose that the above passage turns on the criteria of singularity and generality. First, the opening sentence seems to gloss 'discursive' as 'general', which suggests that the aim of the passage is to show that our representation of space is not a general representation. Moreover, the passage is peppered with phrases that seem to affirm the singularity (or at least particularity[7]) of space, as we represent it. For example, Kant claims that 'one can only represent a unitary [einen einigen] space', since any plurality of spaces represents only 'parts of one and the same solitary [alleinigen] space', which he goes on to describe as 'the unitary [einigen], all-encompassing space' and as 'essentially unitary [wesentlich einig]'.[8] These features have suggested to many commentators that Kant's argument is some version of the following:

| | |
|---|---|
| (1) Only intuitions are singular. | (1′) All concepts are general. |
| (2) Our representation of space is singular. | (2′) Our representation of space is singular. |
| (3) Our representation of space is intuitive. | (3′) Our representation of space is not conceptual. |

This had better not be Kant's argument, however. For, as several commentators have observed, Kant seems to admit a variety of singular representations that are not intuitive, but conceptual.[9] He admits ideas of reason such as the $<ens$ $realissimum>$ [10] (A605/B633; cf. A568/B596), mathematical concepts such as $<$ sum of 7 and 5 $>$ (B15; cf. A164/B205), pure concepts of the understanding such as $<$ substance $>$ (or $<$ matter $>$ cf. A182, B224), and empirical concepts such as $<$ coldest known temperature $>$ (cf. *Prolegomena*, 4:273). It is important here to distinguish various senses in which these concepts might be deemed 'singular'. Clearly, each purports to refer to exactly one thing. But it is also necessary, in one sense or another, that each refers to exactly one thing (provided it has reference at all). It is arguably a logical necessity that there can be no more than one most real being or coldest known temperature; it is a mathematical necessity that $<$ sum of 7 and 5 $>$ refers to exactly one thing (namely 12); and it is, at least for Kant, a metaphysical necessity that there is but one (conglomeration of) matter or substance, which can neither increase nor decrease.

Now Kant famously declares that

it is a mere tautology to speak of general or common concepts – an error that arises from an incorrect categorization [Eintheilung] of concepts into general, particular, and singular. Not concepts themselves, but only their use [Gebrauch] can be so categorized. (*Logic* 9:91)

This might suggest that Kant would (or should) treat the concepts listed above, not as singular representations proper, but as general representations put to a singular use. Certainly the concept < coldest known temperature > invites such an analysis. And it is at least plausible to hold that the concept < substance > intrinsically admits of a plural and is therefore a general representation, even if philosophical considerations subsequently demonstrate that there can be but one substance proper (namely matter), so that the only legitimate use of < substance > is a singular one.

But other concepts do not yield so easily to such treatment. They seem not merely to have been put to a singular use, as when one accompanies a concept with the definite article or a demonstrative expression – as in '*the* sage of Königsberg' or '*this* lame example'. They appear to satisfy a more ambitious notion of singularity, for they necessarily represent exactly one thing *as such*. The concept < sum of 7 and 5 > not only refers to exactly one object as a matter of mathematical necessity; it is also intrinsic to the content of the concept, on Kant's view, that it 'contains nothing more than the unification of both numbers *into a single one* [in eine einzige]' (B15, my emphasis). Now, it is crucial to Kant's theory of mathematical cognition that < sum of 7 and 5 > does not represent, as part of its content, *which* particular individual it refers to: < 12 > is not part of the content of < sum of 7 and 5 > nor *vice versa*.[11] But such an arithmetical concept is nevertheless a discursive representation of an individual as such, for it is part of the content of the concept that it refers to exactly one object (i.e. one number), even though we rely on intuition to determine *which* object that is. This is not a case of a general concept being put to a singular use, for the definite article in 'the sum of 7 and 5' is otiose: one can just as well say (as Kant often does) 'a sum of 7 and 5'.[12] The representation, < sum of 7 and 5 >, owes its singularity to the content of the mathematical function it invokes (namely addition), not to the fact that it is (invariably?) put to a singular use.[13]

The essential singularity of < *entis realissimi* > is equally difficult to dispute. Kant is emphatic that the transcendental ideal is 'the concept of a singular entity [einzelnen Wesens]' (A576/B604; cf. A574/B602, A576/B604).[14] But he is also keen to emphasize the oddity of this concept, which is the only ideal human reason can conceive 'because only in this unique [einzigen] case is an intrinsically [an sich] general concept thoroughly [durchgängig] determined through itself and cognized [erkannt] as the representation of an individual' (A576/B604). Despite its 'intrinsic generality', the ideal is essentially singular not because it is put to a singular use, but because it contains the idea of thoroughgoing determination.[15] In particular, it is singular because it invokes a totality of possible predicates and picks out a determinate subset of them – namely those that express a 'reality [*Realität*]' rather than a 'privation [*Mangel*]'. And this ought to alert us to an extensive family of essentially singular discursive representations: namely those thought through the pure category of the understanding, < totality >.[16] To argue that the singular purport of every totality concept is merely the result of an essentially general concept's having been put to a singular use is tantamount to banishing < totality > from the table of pure concepts of the understanding. Kant is adamant that, although '**allness** (totality) is nothing other than multiplicity considered as unity', the former cannot be reduced to the latter:

> For the combination of the first and second [categories in each heading] in order to bring forth the third requires a special act of the understanding, which is not identical to the act performed in the first and second [cases]. (B111)

That is to say, the category < totality > makes a distinctive cognitive contribution to the content of any concept thought through it – a cognitive contribution Kant links to the form of singular judgement in the Metaphysical Deduction.[17] If we were to suppose that the singularity of totality concepts is due solely to their use (rather than to their content), we could perhaps understand the need for a distinctively singular *form of judgement* (associated with the singular use of general concepts), but it would remain unclear what cognitive contribution < totality > might make to the *content* of such a judgement, thus meriting it a place among the pure categories of quantity.

In view of this proliferation of essentially singular discursive representations, it should be clear that no simple appeal to the singularity of intuition can secure the sensible, non-conceptual status of spatial representation.[18] Even if one is tempted to explain away the apparent singularity of the sorts of concepts I have discussed, such explanations will prove extremely complicated and will tend to draw on argumentative resources outside the Aesthetic. The more complicated and extraneous these explanations become, the less plausible it is to invoke them in reconstructing Kant's argument in the MEs. A natural move, at this point, would be to augment the appeal to singularity with a further differentiating factor. And to many commentators, this is precisely the function of Kant's remarks about the priority of the whole of space over its parts.

Let us turn, then, to the second pair of criteria I identified: atomic versus holistic containment structure. The idea is to supplement the foregoing argument with this one:

---

(i) The parts (or marks) of a concept are prior to the concept that contains them.
(ii) The parts of our representation of space are posterior to that representation as a whole.
(iii) Therefore, our representation of space is not a concept, but an intuition.

---

Though there are weighty reasons for thinking,[19] as the majority of commentators do, that Kant accepts some version of (i),[20] I would like to explore a line of thought that suggests it is actually incompatible with fundamental features of his views about definition.

In the Discipline of Pure Reason, Kant argues that it is impossible to define either empirical or a priori 'given' concepts – that is to say, concepts we do not consciously invent but simply find ourselves with.[21] A definition, for Kant, is the clear and complete presentation of a concept's marks.[22] It is impossible to define concepts we simply find ourselves with, in Kant's view, because we can never be sure we have identified all their marks. In the case of given, empirical concepts, this is because 'one employs certain marks only so long as they suffice for making distinctions; new observations remove some [marks] and add others' (A728/B756). Similarly, it is impossible to define given, a priori concepts (such as 'substance, cause, right, equity') because we can never be sure we have not 'passed over [certain obscure representations] in our analysis, though we constantly depend on them in the application [of the concept to be defined]' (A728/B756). In both cases, whether a certain mark belongs to a concept is a function of our use of that concept in judgements (cf. A68/B93).[23] Use, for Kant, is the touchstone of analysis, the criterion of markhood.[24] This suggests that the marks of a concept that we simply find ourselves with (which is, of course, the normal case) are dependent on and hence *posterior* to the concept that contains them.[25] They are posterior to the whole concept in the sense that the identity conditions of the marks depend on the identity of the whole concept, whereas the identity of the whole concept is not determined by its marks, but by something else – namely its competent use in (its cognitive contribution to) potentially knowledgeable judgements and cogent inferences.[26]

So it is not obvious that Kant would accept that the parts of a discursive representation are necessarily prior to the whole that contains them. Moreover, it is not clear how the introductory sections of the *Critique* could possibly support such a view. Thus, even if Kant does treat discursive representations as atomically structured (which I've tried to cast some doubt on), it would be a philosophical weakness of his argument and an interpretive weakness of any reconstruction of his argument to invoke that (disputable) theory in establishing the intuitive status of spatial representation.

That leaves the criteria of mediacy and immediacy. Kant clearly articulates and even seems to argue for these criteria in the Aesthetic (A19/B33), but it is not obvious how they might be combined with Kant's remarks (at A25/B39) to form a sound argument for the intuitive status of our original representation of space. Why, one wonders, should the features of spatial representation Kant highlights (namely essential unity and holistic structure) prevent us from representing space mediately? After all, the MEs are expositions of our discursively mediated *concept* of space. Far from demonstrating the immediacy of our representation of space, the features Kant highlights are themselves marks through which that representation is mediated. Indeed, we must be able to mediately represent anything we can think about at all, for all thought and judgement involves discursive representation. Unless one supplements the mediacy criterion with an account of conceptual representation as general or as atomically structured – which we have seen would be problematic – it is unclear how mediacy might play a pivotal role in Kant's argument.

A more plausible (and, I think, compelling) case can be made for the criterion of intuitive immediacy. This might seem surprising if one takes immediacy merely to mean that a representation refers to its object(s) without thereby referring to any other 'intervening' representations.[27] For then the same question recurs – why should the essential unity and holistic structure of our representation of something (such as space) entail that that representation relates to its object immediately? I will instead suggest that we view the immediacy criterion as a further (synthetic) specification of Kant's initial and, I would maintain, fundamental characterization of intuition as that (type of) representation through which objects are *given* to us.

## 2. Kant's functional conception of intuition

To bring into view the conception of intuition that is operative in the MEs, one must look to the introductory discussions preceding them. I will suggest that the Introduction to the *Critique* and the opening section of the Aesthetic articulate and vindicate a distinctive conception of the finitude of the human mind – a conception which should be acceptable to rationalists and empiricists alike and which sets the framework for Kant's critical project.[28] In theoretical philosophy, the crucial aspect of our cognitive finitude is that, without experience, we would have no knowledge at all: 'There is absolutely no doubt [gar kein Zweifel] that all our knowledge begins with experience' (B1).[29] This is because, as Kant puts it, the human mind ('das Erkenntnisvermögen') must 'be awakened into action' (B1). That is, even if our cognitive activity (once begun) is spontaneous and self-sustaining, it must still be triggered, occasioned, or otherwise set into motion by something distinct from it. And our knowledgeable cognitive activity is paradigmatically triggered (at least initially) by the objects of which it is knowledgeable. This is all it means to say, at the outset of the *Critique*, that our faculty of intuition is receptive – namely that our capacity to stand in a potentially knowledgeable relation to objects depends on our being given objects that exist independently of our acts of thinking about them.[30] Such object-giving representation is, for us, sensible intuition (A15/B29). The opening of the Aesthetic then further specifies (synthetically enriches) this conception by arguing that such object-giving representations must be immediate. For a discursively mediated representation of an object – one which refers to its object by representing some feature it may share with other objects – applies to indefinitely many possible objects, some of which may not even exist.[31] Thus, because mediate representations cannot guarantee the existence of the objects they represent, all object-giving representations must relate to their object(s) immediately.

This functional conception of sensibility – as that faculty whose (immediate) representations give us objects that exist independently of our thoughts about them – is the only conception Kant has articulated and vindicated by the beginning of the MEs and is, accordingly, the only conception on which they may legitimately trade.[32] Thus, if we can show that the latter MEs depend on precisely this conception of sensibility (and not on notions of conceptual or intuitive representation justified only later on, or not at all), we will have revealed them to be impressively well grounded. Moreover, if we can also show that the MEs provide the conceptual resources to further enrich this initial conception of intuitive representation, we will have revealed them to be exemplars of the 'synthetic' procedure Kant claims to have pursued in the *Critique* (cf. *Prolegomena* §4, 4:274). Our challenge, then, is to explain why our representing something as essentially unitary and holistically structured entails that what is thus represented must be given to us in order for us to represent it at all. The answer, I think, is to be found in Kant's grounds for attributing essential unity and holistic structure to space in the first place. It is precisely because space is infinite – and, in particular, continuous – that it is essentially unitary and holistically structured. The finitude of the human mind, Kant reasonably maintains, is incapable of accounting for such a holistic, infinitary manifold. To the extent that we can (and do) represent such a manifold, therefore, it must be given to us – i.e. our original representation of it must be intuitive.

Before delving into the details of Kant's argument, it behooves us to reflect for a moment on the sort of argument the MEs are supposed to embody. Kant writes:

> By **exposition** (expositio) I mean the distinct (if not complete) representation of that which belongs to a concept; an exposition is **metaphysical** when what it contains exhibits the concept **as given a priori**. (B38)

The argument we are considering is engaged in a particular kind of conceptual analysis – an investigation of the content of a concept, which reveals that concept to have an a priori origin.[33] It is natural to wonder what sorts of considerations may legitimately be invoked in arguing that a concept contains a particular mark. The answer, I think, is that everything – every thinkable content – is fair game. As I argued above, the criterion of markhood is our competent use of concepts in judgements. In analysing a concept, therefore, we may call upon any potentially knowledgeable judgement which competently uses the concept in question. It may be that, for such purposes, false judgements are relatively uninformative or downright misleading, while knowledgeable judgements are most illuminating. But, in principle, it should not matter whether the judgements guiding our analysis are true, false, analytic, synthetic a priori, or empirical, so long as they perspicuously exhibit aspects of the concept's cognitive contribution to the judgements and inferences in which it figures. This approach undermines a widely held view of Kant's method in the Aesthetic. Some commentators seem to think that Kant's strategy of 'isolating sensibility' (A22/B36) culminates in something like a phenomenological inspection of our bare, unconceptualized intuitions of space and time (as though there were such things). This misconceives Kant's self-avowed task. Kant is not engaged in introspective phenomenology or any other attempt to tap into conceptually unadulterated sensible representations: he is engaged in conceptual analysis. And that means he may freely bring to bear the full range of our conceptual apparatus. The goal, of course, is still to show that the *origin* of our representation of space must be intuitive. But this does not mean we must (or can) enjoy such intuitive representations (or have knowledge of them) independent of our conceptual activity, nor that such originally intuitive representations are themselves unconceptualized. Why should we start by assuming that the MEs run afoul of Kant's famous claim that 'thoughts without content are empty, intuitions without concepts are blind' (A51/B75)?[34]

Kant's strategy in the MEs is thus to exploit our judgements about space in determining what belongs to our concept of it.[35] His argument that our original representation of space must be intuitive will then turn on the claim that some of the features revealed in the course of this investigation − some of the marks of our concept of space − could only have an intuitive source. The marks he highlights are the *essential unity* and *holistic mereological structure* of space, to which we now turn.

## 3. The essential unity and holistic mereological structure of space

Kant offers two considerations in favour of his conclusion, each of which is supposed to capture an essential aspect of our concept of space. First, we can only represent one single space; any plurality of 'spaces' is conceived to constitute only parts of that single space.[36] Second, these spaces cannot be conceived as prior to the whole of all-encompassing space, as though space could be constructed out of them.[37] Rather, any plurality of spaces must be represented as *in* one, essentially unified space. These are related observations. The first says that space is a single whole containing a plurality of parts, while the second clarifies the notions of 'part' and 'whole' at issue. These reflections are then summarized in the remark that '[space] is essentially unitary [einig], the manifold in it [ . . . ] rests solely on limitations' (A25/B39). It is at this point that Kant draws his conclusion: 'it follows from this that [ . . . ] an intuition a priori [ . . . ] must underlie all concepts of [space]' (A25/B39). If, as I have argued, the only conception of intuitive representation available to Kant at this point is the idea of a representation which gives us an object that exists independently of our mental activity, then we can interpret Kant's argument here as claiming that the essential unity and holistic structure of spatial representation shows that space could only be given to us − i.e. that anything we represent as having these characteristics could not be a product of our own discursive mental activity.

There is a fairly straightforward and by no means unprecedented line of thought which might lead Kant to assert this. After all, our faculty of thought is limited in extent and acuity − that is, our thoughts can only ever be finitely complex. That is not to say that we cannot think or know anything about things that are infinitely complex. It is just to say that our thoughts and concepts of those things will not themselves be infinitely complex, i.e. they will not consist of or contain an infinite (much less non-denumerable) manifold of representations: 'no concept, as such, can be thought as containing an infinite collection [Menge] of representations **in itself**' (B39-40).

The products of our spontaneous, discursive, mental activity may be arbitrarily complex, since there is no 'lowest species' and we may accordingly contrive concepts as complex as we please. But there is, for Kant, a 'highest genus' – namely the concept of (merely) *something* [*Etwas*]. So no concept or thought is infinitely complex, for every concept results from a finite number of specific differentiations of the highest genus.[38] Therefore, if Kant can show that we represent space to be infinitely complex, he will have shown that space (as we represent it to be) could not be a product of our power of thought, and must consequently be given to us in order for us to know or represent it as such.[39] And since sensibility is precisely our capacity to be given what is independent of our mental activity, Kant would thereby have shown that our representation of space (as an infinitely complex manifold) can only have its source in our faculty of sensibility: 'when it comes to space, an intuition a priori underlies all concepts of it' (A25/B39).

That this is indeed the argumentative route Kant favours is strongly suggested by the opening line of the final ME: 'Space is represented as an infinite **given** magnitude' (B39). Yet questions remain about how this claim relates to Kant's explicit observations in the penultimate ME. Emily Carson has argued that the essential unity and holistic character of space are meant to establish that space is infinite in expanse and divisibility:

> The boundlessness of space is shown by the fact that any given space, however large, is given as bounded by more of the same. Similarly, particular spaces are given only as limitations of the all-encompassing space. These latter two facts seem to me to underlie Kant's claim that the progression of intuition is limitless . . . in both directions [i.e. outward and inward]. (Carson 1997, 499)

The textual evidence for this aspect of Carson's reading is quite strong.[40] Kant often seems to take the fact that every determinate space is surrounded by still more space to imply that space is infinite in expanse, in the sense that the whole of space is strictly greater than any space that can be determined in it.[41] Similarly, Kant seems to take the fact that all determinate spaces are (or 'rest on') mere limitations of the whole of space to imply its infinite divisibility.[42] However, both these views are philosophically problematic. As Parsons (1998, 53) observes, it can be true that every space is surrounded by still more space even while the whole of space is finite in size – provided that space is dense and that the size difference between each space and the space surrounding it converges to zero as one proceeds further out. Similarly, the idea that determinate spaces rest on limitations of an all-encompassing space only entails the latter's infinite divisibility if one assumes the boundlessness of such phenomenological 'zooming in'. But this assumption is dubious, for we seem to enjoy a finite fineness of phenomenological grain, and this notion of limitation simply begs the question against a sceptic about infinite divisibility. Carson's reading thus saddles Kant (not without textual grounds) with several dubious, unsupported assertions.[43]

One obvious way to support these assertions would be by appeal to certain results of geometry or assumptions of mathematical physics. This would effectively reverse the inference Carson portrays Kant as making. Instead of concluding that space is infinite (in both directions) on account of its essential unity and holistic mereological structure, Kant would instead be relying on precisely this twofold infinity of space to justify those characterizations of it. There is ample historical precedent for the latter inference. Most notably, Leibniz holds that the infinite – and, in particular, the continuum – is prior to its parts and therefore cannot be 'composed' out of parts, since its parts, considered in themselves, are indeterminate. This is because any determination of its parts depends on reference to the whole in the same way that determining a quantity through a fraction (e.g. 'half of my paycheck') depends on reference to the whole quantity.[44] Leibniz reasons that anything with infinitely many parts has more parts than can be expressed in a determinate number. Accordingly, there is no one determinate set of parts out of which one can say that such a thing is composed. In the case of the infinitely large, composition is impossible because it cannot come to an end; in the case of the infinitely small, composition is impossible because it cannot find a place to begin, i.e. a basic unit of composition.[45] To maintain that the parts of the spatial continuum are prior to the whole, one must contend either that its 'parts' are extended, or that they are unextended. If they are extended, then they too are continua and thus depend on *their* parts, and so on *ad infinitum* – a regress Leibniz and Kant quite reasonably take to be vicious.[46] If, on the other hand, the 'parts' of space lack extension, one must explain how points of zero measure can 'sum' or 'aggregate' to a non-zero magnitude. Kant follows his major interlocutors in denying that this is coherent.[47] But even assuming it is coherent, a natural strategy would be to assign the points Cartesian coordinates (or something analogous).[48] Yet the ability to assign these coordinates presupposes the intelligibility of the coordinate system within which they have their sense. A point (or part) is only identified as the part it is (and as spatial at all) by reference to an antecedently intelligible coordinate system (or whole) within which its location (and, consequently, its identity as a point in space) is determined. And this is precisely to treat the whole (system) as prior to its parts, contrary to the hypothesis.

Thus, we might read Kant's remarks about the essential unity and holistic structure of space – the fact that all spaces are located in a single, all-encompassing space of which they are limitations – as reminding the reader that we think of space as an open-ended continuum, i.e. as a reminder that this is part of what one means 'when one speaks of many spaces' (A25/B39). This meaning becomes explicit in judgements that are available to us from geometry, mathematical physics and our ordinary beliefs about the possibility of enduring, continuous motion.[49] And insofar as these judgements competently deploy the concept of space, we may legitimately rely on them in identifying the marks of that concept.

It is not my primary concern here to decide whether Kant argues for the twofold infinitary structure of space on the basis of its essential unity and holistic articulation or the reverse. The point I wish to insist on is simply that the intuitive status of spatial representation – the fact that space must be given to us in order for us to represent it at all – turns on our representation of the spatial manifold as infinitely complex.[50] Because this infinite complexity cannot be a product of our discursive mental activity, whose representations are always finite, our representations of it must originate in intuition.[51]

This line of thought might seem suspect, however. After all, intuitive representations belong to our finite minds, too; so why should an *intuition* be capable of containing an infinite manifold if a *concept* is not? Here, it is helpful to recall the conception of sensible representation with which Kant is operating. To say that a representation is sensible is just to say that it involves features that discursive thought cannot give rise to. It is not to say that intuition can *represent* things thought cannot (something Kant would surely deny!). Anyway, Kant not only acknowledges, but *insists* that the whole of infinitely complex space is not (and cannot be) an object of intuitive representation:

> The mere form of intuition, without substance, is not in itself an object [Gegenstand], but merely the formal condition of one (as appearance), just as pure space and pure time are admittedly something, as forms of intuit*ing* [Formen anzuschauen], but are not themselves objects that *get* intuited [Gegenstände die angeschauet werden] (*ens imaginarium*). (A291/B347, my italics)[52]

Infinitely complex space is a condition of the possibility of experience; it is not (and need not be) an *object* of possible experience, for no intellectual synthesis is capable of comprehending it.[53] Spatial representation is intuitive because its content must be given to us – a fact which is quite independent of whether we can become fully conscious of that content by framing space itself as an object of intuition. We must not hold Kant's critical conception of sensible intuition hostage to notions of perception or phenomenological discrimination. These are surely paradigm cases of intuition for Kant. But that is only because they fit his functional characterization of sensibility: they are cases in which something independent of our discursive thought is given to us. A striking consequence of Kant's functional conception of sensibility is that we can and do represent in intuition contents we could never perceptually discriminate or become explicitly conscious of. Whether something fulfils the criterion of giving us an object is determined not through phenomenological investigation, but conceptual analysis. The MEs attempt to determine the primary marks of our concept of space based on what we think (judge, know) about space. Even without endorsing these (mathematical, physical, ordinary) conceptions of space as knowledgeable, we can ask: how are these contents so much as available to thought? Kant's aim is to identify the conditions under which certain kinds of human knowledge are possible, to spell out the cognitive abilities any finite human knower must possess. He is not concerned (in the *Critique*, anyway) to explain how such capacities are realized *in concreto* in the human animal, nor to characterize what it is like, as it were, 'from the inside' to exercise such capacities. It simply does not matter whether we can become phenomenologically aware of every aspect of what we attribute to intuition. Kant's notion of intuition is purely functional: whatever cannot originate in thought but must be given to it is, *ipso facto*, represented in intuition (if it is represented at all). It is this functional account of our cognitive finitude that underlies Kant's heterodox view that the mathematically infinite both *can* be represented sensibly and *can only* be so represented. And it is this account that enables him to argue (to the consternation of opponents and proponents alike) that the 'coarseness [of the senses] doesn't at all concern [überhaupt nicht angeht] the form of possible experience' (A226/B273).[54] To claim that spatial representation is ultimately sensible is merely to register that we conceive of space as an unbounded continuum and that our discursive mental activity cannot account for such holisitic, infinite complexity. This does not mean we have (or are able to have) some quasi-perceptual acquaintance with infinitely complex space. It only requires us to acknowledge our cognitive finitude in accounting for our concept of an infinite spatial continuum.

## 4. The singularity of intuition – a step in Kant's synthetic argument

I will close by briefly indicating how this interpretation enables us to construct an argument for the singularity of intuitive representation. This serves to partly confirm Kant's claim to have proceeded 'synthetically' in the *Critique*, and it helps to illustrate how protracted and meticulous his elaboration and defence of the critical notions of understanding and sensibility truly are. The initial conception of intuition with which Kant begins is what one might call 'cognitive access' – namely a knowledge enabling cognitive relation to an object. He then argues, fairly trivially, that our mode of intuition is receptive – i.e. that the objects that are given to us, the objects to which we have cognitive access, exist independently of our acts of thinking about them.[55] An initial synthetic step comes in the opening of the Aesthetic (A19/B33), which argues that only immediate representations can give us objects that exist independently of our thought (since mediate representations apply equally well to possible, though non-existent, objects). We thereby enrich the notion of an *object-giving* representation by including the idea of *immediacy*. The two quartets of the MEs then purport to establish that space and time are a priori *forms* of intuition. But since the MEs reveal the intuitive roots of spatiotemporal representation without appealing to the singularity of intuition, Kant can subsequently invoke the singularity (and unique coordination) of space and time to argue that all intuitive representation is *singular*. Since the form of our intuition is essentially (spatio-)temporal, all intuitions represent determinate portions of (space-)time.[56] And since we represent spacetime as essentially unitary and as mereologically articulated, we necessarily represent it and all its 'parts' as particulars.[57] It follows that all our intuitions essentially represent unique parts of a mereologically structured and holistically articulated unity. They are, that is, essentially singular representations. This is a further enlargement of our concept of intuition as *object-giving, immediate, singular representation*. The lynchpin of this argument is our concept of spacetime as a holistic, infinitary structure. For this simultaneously supports the claims that (i) spacetime is represented as essentially unitary and (ii) spatiotemporal representation is ultimately intuitive.

What I have tried to argue here is that it is, in part, this surprising connection between infinity and givenness which distinguishes Kant's critical conception of sensibility as unprecedented and ingenious.

## Acknowledgements

The debts of gratitude I have gladly incurred in writing this paper are too many to list. For valuable comments, questions and encouragement I owe particular thanks to Ian Blecher, Matt Boyle, Jim Conant, Robert Pippin, Anat Schechtman, Daniel Sutherland, Clinton Tolley and an anonymous reviewer.

## Notes

1. All translations are my own, though I have consulted the standard editions. Kant's emphases are in bold, my own are in italics and noted parenthetically. References to the *Critique of Pure Reason* refer to the 1781 (A) and 1787 (B) edition pagination. References to Kant's other writings cite the volume and page number of the Akademieausgabe of *Kants Gesammelte Schriften* and are preceded by an abbreviated title of the cited work. I abbreviate the titles of Kant's works as follows: *Physical Monadology* ( = *The Employment in Natural Philosophy of Metaphysics Combined with Geometry, of which Sample 1 Contains the Physical Monadology*); *Only Argument* ( = *The Only Possible Argument in Support of a Demonstration of the Existence of God*); *Inquiry* ( = *Inquiry Concerning the Distinctness of the Principles of Natural Theology and Morality*); *De Mundi* ( = *On the Form and Principles of the Sensible and the Intelligible World*); *Critique* ( = *Critique of Pure Reason*); *Prolegomena* ( = *Prolegomena to Any Future Metaphysics that Will Be Able to Come Forward as Science*); *Groundwork* ( = *Groundwork of the Metaphysics of Morals*); *Discovery* ( = *On a Discovery According to which Any New Critique of Pure Reason is to be made Superfluous by an Older One*); *Prominent Tone* ( = *On a Recently Prominent Tone of Superiority in Philosophy*); *Logic* ( = *Jäsche Logic*); *DW-Logic* ( = *Dohna-Wundlacken Logic*); *Kästner* ( = *Über Kästners Abhandlungen*). Reflections are preceded by an 'R' and include Adickes's estimate of their date, where applicable.
2. This impression is perhaps encouraged by the fact that the sensibility/understanding distinction is first presented 'by way of introduction' as a 'preliminary reminder [Vorerinnerung]' (A15/B29). The opening claims of the Transcendental Logic are liable to reinforce this impression, because it is hardly any easier to find an explicit, non-question-begging argument for the details of Kant's concept/intuition distinction in the vicinity of his much celebrated declaration that 'thoughts without content are empty, intuitions without concepts are blind' (A51/B75; cf. A271/B327).
3. I will refer to the numbered sections of the Metaphysical Exposition of the Concept of Space in the B edition simply as the 'MEs'. There are significant similarities between these and the corresponding discussions of the concept of time, so much of what I say about the one will apply *mutatis mutandis* to the other. But the two are sufficiently different to warrant separate treatments.
4. My guiding assumption throughout is that Kant's claims pertain to our *representation* of space – i.e. to space *as we represent it to be*. Without this implicit qualification, the question whether 'space is [ ... ] a discursive [ ... ] concept' already presupposes that space itself is a mere representation (and, hence, ideal), which is rather supposed to be one of the 'Conclusions from the Above Concepts' (A26/B42). The MEs are expressly concerned with our *concept* of space –

i.e. with space as we conceive of it. Their task is to provide a 'clear (if not complete) representation of what belongs to [this] concept' (B38; see Section 2).

5. The chart provides only the roughest glosses on these notions, and it would be hasty to attribute them, as stated, either to Kant or to any particular commentator's account of his views. They are meant only to indicate some directions in which precise criteria for intuitive and conceptual representation might be sought. My discussion aims to accommodate differences of opinion about how these criteria are to be spelled out and about how they interrelate. With one exception – namely the purportedly atomic containment structure of conceptual representation, discussed below – I do not dispute that the listed features are criterial for intuitive and conceptual representation. What I dispute is that these criteria may be legitimately invoked to generate a cogent reconstruction of the argument in question.

6. A smattering of examples should suffice to confirm this. Vaihinger treats the concept/intuition distinction as exclusive and invokes singularity as a necessary and sufficient criterion of intuition (1892, 211f., 223; citing *De Mundi* and the *Nachlass*), generality as a sufficient condition of concepts (211f.; citing the *Nachlass*) and atomic containment structure as a sufficient condition of concepts (219; citing the *Logic*). Kemp Smith (1923, 105, 107) treats the distinction as exclusive and exhaustive and invokes singularity and immediacy as severally necessary and sufficient conditions of intuition, and then generality, mediacy and atomic containment structure as severally sufficient conditions of concepts. Paton (1936, I:115) presupposes the exclusivity of the distinction and invokes generality as a sufficient condition of concepts (citing the *Logic*) and declares singularity a sufficient condition of intuitions. Ewing (1950, 37) invokes singularity as a sufficient condition of intuition and atomic containment structure as a necessary condition of concepts. Strawson (1966, 64) claims that the distinction is exclusive and exhaustive and invokes singularity as a necessary and sufficient criterion of intuitions and generality as a necessary condition of concepts. Parsons (1992, 63, 69f., 1998, 46) invokes singularity and immediacy as severally sufficient conditions of intuition (citing the *Stufenleiter* at A320/B376f., the *Logic* and the Aesthetic). Pippin (1982, 64ff.) invokes singularity and immediacy as severally necessary and sufficient conditions of intuition (citing the *Stufenleiter* and B136n.). Allison (1983, 90, 2004, 109) presents the argument as consisting of 'two steps', the first of which presupposes the exhaustiveness of the concept/intuition distinction and invokes singularity as a sufficient condition of intuition. The second step Allison identifies (1983, 91, 2004, 110) presupposes the exclusiveness of the concept/intuition distinction and invokes atomic containment as a sufficient criterion of conceptual representation. Guyer (1987, 346, 348) invokes singularity as a necessary and sufficient condition of intuition. Falkenstein (1995, 218) invokes singularity as part of the 'definition' of intuition and thus, presumably, as at least a necessary condition for intuition (citing *De Mundi*, *Logic* and *Discovery*). Falkenstein (1995, 230, 234f.) also describes atomic containment structure as a necessary condition of concepts (oddly citing Kant's conception of substance (not any theory of concepts) in the *Physical Monadology* and *De Mundi*). Carson (1997, 494, 496, 498) implicitly invokes singularity as a sufficient criterion of intuition and atomic containment structure as a necessary condition of conceptual representation, thereby following Allison in presenting two distinct argumentative routes to the desired conclusion. Gardner (1999, 78) invokes singularity and immediacy as severally sufficient conditions of intuition. Rosenberg (2005, 66) invokes only the mediacy or discursivity of concepts. Buroker (2006, 52) invokes generality as a necessary condition of conceptual representation. Shabel (2010, 100, 102) invokes only singularity as a sufficient condition of intuition, implicitly treating the concept/intuition distinction as exclusive.

7.  Kant is often interpreted as asserting that there is (necessarily) exactly one space – a claim some challenge by arguing that the idea of wholly dissociated spaces is coherent (cf. Quentin 1962; Hollis 1967). The basic worry is that Kant commits a fallacious quantifier inversion. The fact that every space is part of some greater space does not entail that there is an all-encompassing space of which every space is part. But the weaker point may suffice for Kant's purposes because he is ultimately concerned with space as the framework of *our* outer experience. The unity of space, then, is a function of the unity of experience (cf. B136n., B138, B144n, B160, B161n.). The point is that it is part of our concept of experiential space that it cannot be fragmentary and that anything we might possibly experience as outer must be located within it. Falkenstein (1995, 219) accordingly interprets Kant as claiming only that space (and anything in it) is a particular, but not that it is unique. In any case, a possible plurality of spaces only seems to threaten Kant's position if one takes his argument to turn on the singularity of intuition, and it is my present task to expose the flaws of this interpretation. Section 2 gives an account of intuition that does not presuppose singularity. Section 4 then indicates how the singularity of intuition can be viewed as a consequence of Kant's argument.

8.  Guyer and Wood translate 'einig' as 'single'. Now 'einig' can indeed mean *single* or *solitary*, as in Luther's translation of Genesis 22.2: 'nim Isaac deinen einigen son' ['take Isaac, your only son']. But it can also signify unity or undividedness, as in Kant's claim that the necessary being 'is unified [einig] in its essence, simple [einfach] in its substance' (*Only Argument* 2:89). Indeed, in its primary colloquial sense, 'einig' is used (and was used at the time) to signify agreement or solidarity (i.e. unity) among various parties, and this is the sense that most frequently appears in Kant's writings. Kemp Smith preserves the ambiguity between these senses of 'einig' by translating it as 'one' or eliding it altogether. Though I take 'einig' here to mean *unified*, I have opted for 'unitary' in order to preserve the ambiguity of the original and to avoid begging any questions. If Kant had used 'einzig' rather than 'einig', 'single' would be apt. Since he did not, a neutral translation seems preferable. See the entry for 'einig' in the Grimm Wörterbuch, from which the above examples are drawn.

9.  See, for example, Kemp Smith (1923, 107), Allison (1983, 90–91, 2004, 109–110), Parsons (1992, 70) and Falkenstein (1995, 67–69). Vaihinger mentions, but does not evaluate this objection, attributing it to Riehl, while also citing Trendelenburg (1892, 213). Commentators typically cite singular ideas of reason as potential counter-examples – e.g. $<$ the world $>$ , $<$ nature $>$ or $<$ God $>$ . But the full force of this objection only becomes clear when one considers the wide range of ostensibly singular concepts Kant's system allows, as I do below.

10. I follow the common practice of using chevrons around words to mention the concepts they signify: thus, $<$ Pferd $>$ and $<$ horse $>$ denote the same concept, viz. the concept of the natural species *equus ferus*.

11. For a compelling account of why this should be so, see Anderson (2004).

12. E.g. B15 (twice), B16 and *Prolegomena* 4:268.

13. One might think the singularity of such mathematical concepts can be explained by invoking construction in intuition and the singularity of intuition. There is surely something right about this, but it does not advance our understanding of A25/B39. A first difficulty with the proposal is that some mathematical concepts admit of multiple references: e.g. $< \sqrt{4} >$ refers both to 2 and to $-2$. Kant implies that square roots have multiple solutions in *Negative Magnitudes* (2:173) in arguing that combining negative magnitudes is a case of addition rather than subtraction (i.e. that $-2$ and $-2$ yield $-4$). Because the product of $-2$ and 2 is just the sum of $-2$ and $-2$ (i.e. $-4$), the product of $-2$ and $-2$ should have the opposite sign, i.e. $= 4$ (cf.

Euler 1911, *Vollständige Anleitung zur Algebra*, §33). Thus, 2 and $-2$ are referents of $< \sqrt{4} >$. Kästner (1758) also argues that the product of two negative numbers is positive in §105 of his *Anfangsgründe der Arithmetik*, which Kant mentions approvingly in *Negative Magnitudes* (2:170). Moreover, the solution of physical problems involving scalar quantities will require that we allow the products of negative magnitudes to be positive. Multiply referential mathematical concepts show that construction in intuition does not ensure singularity in the sense of reference to exactly one object (unless we construe the referent of $< \sqrt{4} >$ as a set or an unordered pair, which seems unjustifiably anachronistic). Moreover, this plurality of reference seems to result from the relevant mathematical operation, not from its construction in intuition, which supports my contention that singular mathematical concepts owe their singularity to the content of the mathematical functions they invoke (and not to their having been put to a singular use). A second difficulty with the proposal is that the theory of mathematical construction in intuition (along with the singularity of intuition itself, I would argue) is something the Aesthetic is supposed to help establish, not something it can take for granted as a premise. So, one cannot appeal to this theory in order to defuse counterexamples to a proposed reconstruction of Kant's argument in the MEs. It is more natural to suppose that the order of argumentation in the *Critique* is precisely the reverse: geometrical concepts, for example, must be constructed in intuition *precisely because* space is the form of outer sense and geometry is the pure science of space. Finally, even if Kant were entitled to invoke this theory of mathematical construction in establishing the originally intuitive status of spatial representation, the role of construction in mathematics no more undermines the genuinely discursive nature of mathematical concepts than the originally intuitive status of spatial representation undermines the discursive character of the concept $<$ space $>$. So we would still be lacking an account of singularity on which *only* intuitions are singular and/or an account of generality such that *all* concepts are general, which is what the mooted reconstruction of the argument calls for.

14. Recall that Kant explicitly argues that the concept of the most real being is singular in *Only Argument* (2:83f.).

15. It should be obvious that the singular purport of $<entis\ realissimi>$ cannot be borrowed from intuition, because it is 'a concept which we can never exhibit in concreto in accordance with its totality' (A573/B601).

16. The idea that $<$ totality $>$ can itself constitute a mark (*Merkmal*) of more specific concepts not only affords us a procedure for generating essentially singular discursive representations consistent with Kant's theory of concepts, but it also suggests a way of reconciling these peculiar representations with Kant's claims that concepts are intrinsically general (e.g. *Logic*, 9:91). Although many concepts which contain $<$ totality $>$ as a mark are, for that reason, essentially singular, $<$ totality $>$ is not *itself* a singular representation. (There are, after all, innumerably many totalities.) This ought to remind us that individual marks are essentially general, for they are precisely 'that which is common to *many* objects' (*Logic*, 9:58, my italics, cf. A320/B377). Marks very often impart their generality to the discursive representations mediated through them. But as we have observed, one can generate a singular discursive representation either by including a mark like $<$ totality $>$ in its intension, or by combining other marks in such a way that it is (logically, metaphysically, or mathematically) impossible for more than one object to exhibit those marks. Because all concepts (as discursive) are mediated through marks, one can see the sense in insisting that even $<ens\ realissimum>$ is an 'intrinsically [*an sich*] general concept' despite its being 'the representation of an individual' (A576/B604). Moreover, this explains why the *Stufenleiter* does not actually characterize

concepts as 'general' but instead says that they are 'mediated by means of a mark that can be common to many things' (A320/B377). It also sheds light on Kant's remark that '[t]he generality or general validity of the concept does not rest on the fact that it is a partial-concept [Theilbegriff], but on the fact that it is a ground of cognition' (*Logic*, 9:95). If it is impossible for a certain concept to serve as the ground of cognition of more than a single thing, that concept is essentially (though not intrinsically) singular, notwithstanding its discursive mediation through general *Merkmale*. What a certain concept can enable us to cognize is precisely the sort of consideration from which we abstract in pure, general logic (cf. A58/B83; *Logic* 9:13). Within the context of such an investigation, then, it makes sense to treat all discursive representations as (tautologically) general. Once we are concerned with the conditions of the possibility of knowledge of *objects* (e.g. in the *Critique*), such abstractions are illegitimate and we must ask of every discursive representation *whether* and *of how many* objects it can serve as a ground of cognition. Sometimes, we can answer both questions (e.g. in the case of < sum of 7 and 5 > , or < largest integer > ). Other times, we can answer one but not the other. We may demonstrate that, if *anything* can be cognized through the concept in question (e.g. < complete series of causes > ), only *one* thing can be cognized through it, even though the objective validity of the concept as a ground of cognition must remain forever problematic for human reason. The crucial point is that, in the context of a critique of our ability to know objects, the discursivity of a representation is not a matter of its generality, but a matter of its mediation through marks and its consequent inability to *give us* the object(s) it represents. It is, I shall suggest, precisely this feature that fundamentally distinguishes concepts from intuitions.

17. The link between the forms (and functions) of judgement and the categories that is supposed to be established in the Metaphysical Deduction clearly involves some reference to intuition in general: '[The categories] are concepts of an object in general, through which the intuition of that object is regarded as **determined** with respect to one of the **logical functions** to judge' (B128). This might tempt one to yet again (see note 13) attempt to explain away the singular purport of such totality-concepts by invoking their connection to intuition (and the singularity of intuition). There is surely something right about this, but the strategy faces a number of complications. First, the category < totality > remains a discursive representation that originates in the understanding alone (cf. A137/B176, B377; *Prolegomena* 4:330; *Logic* 9:92), so totality-concepts still seem to provide examples of essentially singular discursive content, even if that content bears some intrinsic link to intuition. Second, the strategy does not neatly apply to a number of totality-concepts (such as < *ens realissimum* > ), which cannot, in principle, be exhibited in intuition and which, therefore, cannot derive their singular purport from intuition in any straightforward manner. Finally, it is interpretively suspect to draw on the results of the Metaphysical Deduction (and, one suspects, the notion of figurative synthesis from the Transcendental Deduction) in reconstructing Kant's argument in the MEs. Even if it is possible to explain away the ostensible singularity of totality-concepts (and their kin) by exploiting argumentative resources found elsewhere in Kant's corpus, it strains credulity to claim that these argumentative resources are available at the outset of the Aesthetic. So while it may be possible to resolve the apparent tension between such singular discursive representations and Kant's claims that concepts are intrinsically general (perhaps along the lines I suggest in note 16), it is illegitimate to employ such an account in reconstructing Kant's argument at A25/B39.

18. It is worth noting that Kant nowhere characterizes intuitions as *singular* in the text leading up to the MEs. He says that intuitions alone *give* us objects (A16/B29, A19/B33), he says they relate *immediately* to the objects they give us (A19/B33), and he

says that they contain a *manifold* and have a hylomorphic structure (A20/B34). But *singularity* does not figure in the characterizations of intuitive representation that open the *Critique*. Allison (1983, 2004) points to Kant's parallel discussion of time to justify invoking the singularity of intuition in reconstructing Kant's argument. There, Kant says that 'the representation which can only be given through a single object [einen einzigen Gegenstand] is intuition' (A32/B47). But this does not state that intuitions alone are singular representations. What it literally says is that intuitions are the only representations that can *solely* arise through isolated affections by objects. Whether or not concepts *can* arise through affection by a single object (and it is doubtful that they can), it is clear that they *also* (and paradigmatically) arise when we reflect on what many objects (which we have compared) have in common, and when we abstract from certain of their features to common ones (perhaps while recombining these with others; cf. *Logic*, 9:95). Thus, it is not true that concepts can *only* arise through cognitive contact with a single object. This is at once what enables concepts to represent objects and their properties in the absence of those objects and what disables them from guaranteeing (on their own) the objective validity of what they represent. It is because of this discursivity or mediacy that concepts cannot give us objects. (Cf. note 16 above.) Given the exhaustiveness of the concept/intuition distinction, this means that intuitions are the only representations that can *only* arise through affection, which is just what A32/B47 says. It thus recapitulates the opening sentences of the Aesthetic, which connect the idea that intuitions give us objects with the idea that intuitions arise through affection: 'The latter [sc. intuition] only takes place [findet nur statt], insofar as the object is given to us; and this, in turn, is only possible, for us humans at least, if it [sc. the object] affects the mind in a certain way'. (A19/B33)

In Section 4, I will argue that the singularity of intuition is not mentioned before the MEs precisely because it is a *consequence* of them: since all intuition is spatiotemporal and since spacetime is a unique, unitary structure, all intuition is singular, for it represents unique portions of that structure, as such.

19. For example, in the first edition version of the final MEs concerning time, Kant writes: 'But where the parts themselves and every magnitude of an object can only be determinately represented through limitations, there the entire representation cannot be given through concepts (*for there [sc. in concepts] the partial representations are prior)*' (A32, my italics). However, Kant changes this parenthetical phrase in the second edition to read '(for they [sc. concepts] contain only partial representations)'. This change reflects Kant's more substantial revision of the corresponding argument about space in light of his realization that the concept/intuition distinction does not turn on the *priority* of part versus whole, but on the kind of *relation* that obtains between contained and container. See note 50 for my reading of the final MEs.

20. Readers may also have misgivings about (ii), inasmuch as the text of A25/B39 seems to speak not about our *representation* of space and its parts, but about the parts of space itself. But see note 4 and Section 2.

21. The sense in which concepts may be 'given' differs from the sense in which objects may be given (in intuition). With respect to objects, the contrast is between being *given* and being (merely) *thought*. Here, the contrast is between *conceptus dati*, which are given, and *conceptus factitii*, which are made or fabricated (cf. A730/B758; *Logic* 9:93; Vienna Logic 24:913–918). Fabricated concepts are 'willkürlich' or 'arbitrary' in that their marks are determined by our elective choice (*arbitrium*). They accordingly have an atomic containment structure: their parts (marks) are prior to the whole. But not all concepts are fabricated and concept fabrication presupposes our possession of 'given' concepts, which, I argue above, are not atomically, but holistically articulated. Paradigmatic fabricated concepts are those of mathematics and the technical concepts of natural science. The former can be defined, in Kant's

ambitious sense, for they are generated through the construction of their objects in pure intuition, and we can therefore secure their objective validity a priori (cf. A730-2/B758-60; *Logic* 9:63f.). A non-mathematical, fabricated concept cannot be defined because merely combining given marks does not indicate 'whether [the fabricated concept] has an object' (A729/B757). It will have the logical form of a concept but may, for all that, '[have] no meaning [Sinn] and [be] completely devoid of content' (A239/B298). A would-be definition of such a concept is thus a 'declaration (of my project)' (A729/B757) of demonstrating its application to objects of experience – namely by 'compel[ing] nature to answer [my] questions' (B xiii; cf. *Logic* 9:36f.). This is the project of natural science, which extends our knowledge of the phenomenal world even as it secures the meaningfulness of our fabricated concepts.

22. Cf. A727/B755; *Logic* 9:140–145.

23. For example, by noting the contradiction involved in thinking of a thoroughly permeable material body, we can establish that < impenetrable > is a mark of the concept < body > . For if something is completely permeable, it cannot properly be said to *occupy* a given space, because its 'presence' in no way prevents anything else from occupying that space. But every material body must occupy a space. Thus, by reflecting on competent, non-trivial uses of concepts in judgements and by drawing on our comparatively primitive ability to recognize contradictions, valid inferences, and so forth, we can establish that certain marks are necessary criteria of a given concept's application. Yet we cannot, by this means, establish what marks are *sufficient* criteria for a given concept's application. And that is why given concepts cannot be defined.

24. Kant's emphasis on the standing possibility of human ignorance and error in identifying concepts' marks might prompt the objection that use (in judgements) concerns only the *ratio cognoscendi*, not the *ratio essendi* of concepts' marks – i.e. how we come to *know* a concept's marks but not what it is to *be* a mark of a concept. Yet such fallibility does not suggest that there is some standard for the content of concepts *apart* from their knowledgeable use. Granted, we can be no more certain of the marks of our concepts than we can be certain that our apparently virtuous motives are not surreptitiously corrupted by self-love (cf. *Groundwork* 4:407). Yet, in neither case does pervasive uncertainty vitiate the fundamentally *self-conscious* nature of the representation or the internality of the standard against which it is measured. Just as a good will is an expression of reason's self-knowledge, so too is the full content of a concept an expression of the understanding's self-understanding – i.e. its knowledge of its own activity in applying concepts in potentially knowledgeable judgements. Despite the fallibility of our efforts at self-knowledge (e.g. through analysis of our concepts), our use of concepts in knowledgeable judgements is constitutive of their meaning precisely because their meaning consists in the contribution they are capable of making to our knowledge. Since knowledge is essentially self-conscious (epistemic hiccups notwithstanding), for concepts, *esse* is *concipi*.

25. This comports well with Kant's emphasis that a mark is not only a 'Partialvorstellung' but a 'partial-representation *insofar as* it is considered [a] ground of cognition *of the whole representation*' (*Logic* 9:58, my italics). The cognitive contribution of the (whole) concept is the standard against which a mark's membership in its intension is determined. Cf. also *Logic*, 9:35f., 58f., 64, 95, 96, 145; *Critique* B39f., A69/B94, B133f.n., A728/B756; *Discovery* 8:199; *Prominent Tone* 8:399.

26. That the marks of a concept are posterior to and dependent on the whole of that concept is thus the complement of Kant's view that concepts are essentially 'predicates of possible judgments' (A68/B93).

27. Longuenesse (1998, 24n.13) seems to understand the immediacy criterion in this sense. Parsons (1992) advances an alternative conception of immediacy as

'phenomenological presence to the mind' (66). I prefer to reserve the term 'immediacy' for non-mediation through discursive marks, but I do think Parsons's account gets closer to the *nervus probandi* of Kant's reasoning here. For, as I will argue, Kant's conclusion does turn on a sort of cognitive presence to the mind – albeit one that should not be construed phenomenologically.

28.  I argue for this in detail in Chapter 3 of my dissertation (Smyth, n.d.), which is indebted to Engstrom 2006. I go beyond his account in locating a sound argument for Kant's conception of our cognitive finitude in the opening passages of the *Critique*, as befits Kant's 'synthetic' method (cf. *Prolegomena* 4:274 and note 35).

29.  The A edition Introduction opens with the same thought, though it is already entangled with a particular picture of the dependence of human knowledge on experience, which Kant excises from the B edition: 'Experience is, without doubt, the first product which our understanding produces, in working up the raw material [Stoff] of sense impressions' (A1).

30.  For an intuitive intellect, however, the faculty of thought is also a faculty of intuition. There is an intuitive moment in all theoretical knowledge (finite or infinite), because all knowledge, for Kant, relates to objects. Rosenkoetter (unpublished manuscript) fruitfully suggests that Kant takes knowledge to be of objects (rather than states of affairs, say) in order to explain the meaningfulness of false propositions.

31.  See A19/B33 and notes 16 and 18, above.

32.  This suggests an analogous functional characterization of the understanding as that faculty which secures the aspects of our knowledge which could not possibly be given to us. The Transcendental Analytic initiates a (synthetic) transition from a 'merely negative' conception of the understanding as a 'non-sensible cognitive faculty' (A67/B92) to a richer, positive conception of the understanding as a spontaneous capacity. This synthetic enrichment of our conception of the understanding begins by noting features of our knowledge that our passive sensibility cannot account for, e.g. combination in general (cf. B129).

33.  For a refreshingly detailed account of Kant's conception of exposition as analysis, see Messina (forthcoming).

34.  Falkenstein (1995) rightly emphasizes that the MEs should be read in light of Kant's 'blindness thesis'. This corrects a widespread misconception that the Aesthetic presupposes cognitive access to unconceptualized intuitions, which the Analytic denies is possible. Yet while he insists that unconceptualized intuitions are 'for us as good as nothing' (A111), Falkenstein still thinks Kant implies the existence of such intuitions. Kant does hold that we can 'isolate' (A22/B36) sensibility's distinctive contribution to cognition by abstracting from cognitive features due to discursive thought, once 'extended practice has made us attentive to [them] and skilled in separating [them] out' (B1f.). But such notional separability need not imply that these aspects can exist or be conceived on their own. Nor do I see what explanatory role such nugatory intuitions could play in Kant's transcendental epistemology and hence why we should read him as committed to their existence.

35.  Carson (1997, 495) objects that these judgements cannot include the claims of geometry, for that would '[go] against Kant's explicit assertion in the *Prolegomena* that in the *Critique*, he is pursuing the "synthetic method" which is "based on no data except reason itself"' [4:274]. I agree that Kant cannot legitimately appeal here to geometrical principles as objectively valid cognitions we can endorse as true, but I do think he can appeal to them as merely thinkable contents. Such thinkable content provides a sufficient basis for the 'clear (if incomplete)' analyses that make up the Metaphysical and Transcendental Expositions (B52). These analyses nevertheless contribute to a 'synthetic' argument, i.e. one that progresses from relatively simple principles to relatively complex ones and their consequences (*Logic* 9:149; *DW-*

*Logic* 24:779; R3831 (from 1769) 17:353; cf. also R3343 (c. 1772-5) 16:789, and §422 of Meier's 1752, *Auszug aus der Vernunftlehre*). For example, as I suggest below, the marks identified in the MEs enable us to conclude that intuitions are not only immediate and object-giving, but also *singular* representations. Or we can infer any of the notorious claims Kant makes in the section 'Conclusions from the Above Concepts', e.g. that space does not represent any property of things in themselves (A26/B42). Each of these inferences constitutes a synthetic step in that it enriches the principles we began with (e.g. our concept of intuition, or our concept of space). Yet insofar as they depend solely on analyses of concepts (i.e. contents treated as merely thinkable), they do not 'rest upon any fact' (4:274). This is how I interpret Kant's claim that, in philosophy, 'we can conclude various things from a few marks drawn from an incomplete analysis [i.e. an exposition] before we arrive at the complete exposition, i.e. the definition' (A730/B758; cf. *Logic* 9:145). Even while progressing synthetically with the 'assured gait of science', there are moments of analysis and exposition when one foot is planted, enabling the next step forward.

36.   Hopefully, this mention of singularity no longer tempts us to short-circuit the argument at this point. Recall that intuitions have not yet been characterized as singular and that many discursive representations are essentially singular. See notes 7 and 18.

37.   The priority here is obviously not supposed to be temporal, for all parts of space are simultaneous (B40). Nor is it likely that the priority in question is 'ontological dependence', for Kant emphatically rejects the Newtonians' reification of space: space is not a 'thing', not an 'object' or a 'substance' (see, e.g. A39/B56; A291/B374). Despite his own idealism about space, Leibniz makes the same mistake as the Newtonians in applying the principles of a substance ontology to space – e.g. in claiming that the parts of a continuum depend on the whole of it as the modes of a substance depend on that substance (cf. Leibniz 1923, A6.3:502, 520, and 553). Kant rather seems to hold that the part-whole dependence relation in continua pertains to the sortal-identity of the parts – to *what* they are, not merely *that* they are. It is in virtue of being (represented as) situated within the whole of space that its parts are (represented as) *spatial* in the first place. I think this is why Kant expresses himself by saying that the parts of space 'can only be *thought* as **in it**' (A25/B39, my italics), rather than by saying that they can only *exist* as in it. Nor is his point simply that the numerical identity or difference of distinct spaces is secured by their co-membership in one unitary space (per Melnick 1973, 9). Questions of numerical identity can only arise once the elements in question have been identified as homogeneous – e.g. once the 'parts' of space are identified *as spaces* (as *spatial*). It is a subsidiary point that spatial disjointness (at a given time) suffices for the numerical distinctness of spaces and spatial objects, as such: the question of *what* they are is prior to the question of *which* they are or *whether* they are.

38.   Cf. A290/B346, A659/B687; *Logic* 9:59, 146. See also Kant's notes on Logic 24:755 (cited by Carson 1997) and the Vienna Logic (24:912). For a helpful discussion of Kant's conception of specific differentiation, see Anderson (2004, 507–14).

39.   This line of thought bears some structural similarity to Descartes's so-called 'causal' argument for the existence of an infinite being (namely God) in the third *Meditation* (see Descartes 1964–1974, AT 7:40ff.). Descartes maintains that (i) we have an idea of an infinite being (i.e. God), (ii) we could not have derived this idea from anything but an infinite being and (iii) because we know ourselves to be finite beings, we cannot have derived the idea from ourselves, therefore, (iv) there must be an infinite being (God), which is distinct from us, from which we have derived this idea. Accordingly, Descartes and Kant face structurally similar objections. Yet while it is plausible for Descartes's critics to argue (against (i)) that we do *not* in fact have a

genuine or adequate idea of an infinite being, and (against (ii)) that an idea of an infinite being *can* be derived from our ideas of finite beings, these challenges lose much of their force against Kant. For it is not similarly plausible to maintain (against a Kantian analogue of (i)) that we do not represent space as infinite in geometry, physics or everyday reasoning. Moreover, if I am correct in arguing that the continuity of space secures its holistic structure (see below and notes 37, 44 and 51), then it is demonstrably false to claim (against a Kantian analogue of (ii)) that the idea of space's infinite complexity can be derived from an idea of finitely complex space (s). Thus, Kant's responses to these objections are different from (and arguably stronger than) any responses available to Descartes. If Kant's argument is to be overturned, it must be on the grounds that human spontaneity can indeed account for the infinitary structure of space (time, etc.). This, I take it, is the strategy both Hegel and, in an entirely different manner, Michael Friedman pursue. For a compelling account of Descartes's argument, see Schechtman, 2014.

40.  Despite her well-placed and novel emphasis on the twofold infinity of space, Carson's overall reading still falters because it presupposes the essential singularity of intuition. See note 6.

41.  Kant operates with a number of different conceptions of infinity. He sometimes says that 'the collection [Menge] which is not a part' is infinite [R4764 (from 1770s) 17:721]. Space would meet this criterion, though it clearly does not entail what we would call 'metrical infinity'. More frequently, Kant characterizes infinity as 'great beyond all measure [über alle Maße groß]' or 'greater than any number' (*Critique* A32/B48, A432n./B460n.; *Kästner* 20:419ff.; R4673 (from 1774) 17:637, R5338 (from 1770s) 18:155). This would imply metrical infinity. To establish that space is infinite in this sense, however, it is not enough to show that every space is surrounded by more; one must show that there is no upper limit on the magnitude of spaces that are delimitations of the whole of space. Kant clearly believes this, but his reasons are not entirely apparent.

42.  *De Mundi* 2:403n.; *Critique* A169f./B211f.; R4183 (early 1770s) 17:448; R4424 (c. 1771) 17:541 and R5831 (c.1783/4) 18:365; see also his corollary view that spatial points, which are simple and thus indivisible, are mere limits (cf. A169/B211, B419, A438/B466, A439/B467). It is probable that Kant (like his contemporaries) did not clearly distinguish between infinite divisibility (i.e. denseness) and continuity. Nevertheless, Kant is clearly committed to both the denseness and continuity of space, for he held that the possibility of continuous motion entailed the continuity of the spatiotemporal manifold. See note 49.

43.  There is some evidence that Kant might have granted that these claims were unsupported by argument – not because they were dogmatic, but because they were indemonstrable. In *Inquiry*, Kant calls it 'the most important business of higher philosophy' to adumbrate 'indemonstrable fundamental truths'. The examples of such truths he provides recur throughout the critical corpus: the externality of spaces to one another, the non-substantiality of the spatial manifold, and the three-dimensionality of space, 'etc.': 'Such propositions can very well be elucidated [erläutern] if one examines them **in concreto** so as to cognize them intuitively; but they can never be proved' (*Inquiry* 2:281). The problem Carson's Kant faces, of course, is not just that his assertions are indemonstrable, but that they may be *false*.

44.  In the *New Essays*, with which Kant was familiar, Leibniz offers a *précis* of his solution to the 'labyrinth of the continuum': 'The true infinite [ ... ] precedes all composition and is not formed by the addition of parts' (1981, 2.17.1, A 6.6:157). This non-compositional priority of the whole over its parts is precisely what leads Leibniz to regard continua (and, *a fortiori*, space and time) as ideal (not real): 'It follows from the very fact that a [continuous] mathematical body cannot be resolved

into primary constituents that it is also not real but something mental and designates nothing but the possibility of parts, not anything actual. [ . . . ] [T]he parts are only possible and completely indefinite. [ . . . ] But in real things, that is, bodies, the parts are not indefinite – as they are in space, which is a mental thing – but actually specified in a fixed way'. (Letter to de Volder, June 30, 1704; Leibniz 1978, G 2:268f.; cf. G 2:276f., 2:281f., 3:611f., 4:394, 4:491f., 5:17, 6:394, 7:563; for a helpful discussion, see Levey 1998, 58–68). Leibniz also denies that the infinite can be a 'genuine whole', but this should not be taken to suggest he denies the essential unity of space. Leibniz follows Locke (1959, *Essay* 2.17.7-8, vol. 1, 281) in distinguishing the 'infinity of space' from 'space infinite' – i.e. the idea that, since any determinate space may be extended further, space itself is greater than any assignable quantity, from the idea that there is an all-encompassing space whose measure is actually infinite. To view space as a complete whole is, for Locke and Leibniz, to view it as consisting of a determinate number of parts (given magnitudes). Yet to view it as an *infinite* whole is to view it as having more parts than can be expressed in a determinate number. Thus, 'it would be a mistake to try to suppose an absolute space which is an infinite whole made up of parts. There is no such thing: it is a notion which implies a contradiction' (1981, 2.17.5, A 6.6:158). Leibniz's denial that space is a whole does not imply that space is not a unity (in Kant's sense), but only that the status of its 'parts' (as indeterminate, potential and abstract) ensures that it cannot have the (atomic) compositional structure of a 'real thing'. There is evidence that Kant also took the infinitary structure of space and time to entail their ideality: 'The mathematical properties of matter, e.g. infinite divisibility, proves that space and time belong not to the properties of things but to the representations of things in sensible intuition' (R5876 (c. 1783/4) 18:374). I explore the Leibnizian roots of Kant's idealism in Chapter 1 of my dissertation (Smyth, n.d.).

45. These temporal metaphors are only meant to capture the asymmetric dependence of a composite on the parts that compose it. They neither imply that composition is a temporal process (psychological or otherwise) nor that Kant or Leibniz treat it as one (though Locke may well conceive things this way; cf. Locke 1959, *Essay* 2.17.7, vol. 1, 281).

46. The priority relation in this regress can be interpreted in various ways. If it is a relation of sortal-identity dependence, as I take it to be, then the regress is clearly vicious, since, if it does not terminate, one literally does not know what one is talking about at any given stage in the regress. The regress would arguably also be vicious if such priority expressed ontological dependence, but see note 37.

47. E.g. *Physical Monadology* 1:479; *Critique* A169f./B211, A439/B467.

48. The variables' values could be assigned on analogy with a popularized version of Cantor's diagonal proof:

$(x, y, z)$ 1 . . . 2 . . . 3 . . .   The point is that there is a recursive procedure for picking
1       1/1   1/2   1/3   out spatial points with triplets of rational Cartesian coordinates.
2       2/1   2/2   2/3   We can thus 'construct' an unbounded and infinitely divisible
3       3/1   3/2   3/3   (though not continuous) manifold out of points. Friedman (1992, 66ff.) offers a similar model. Grünbaum 1952 gives a consistent conception of the linear continuum as an aggregate of unextended points, but a discussion of that account and its relation to Kant is out of place here.

49. Both the continuity and infinite extendability of a line segment are secured by Euclid's second postulate, from which the continuity and infinite expanse of space itself follow easily (Euclid 1908, I:154). Kant's favoured geometrical proof of the infinite divisibility of space is drawn from either Jacques Rohault or John Keill (cf. *Physical Monadology* 1:478; Kant 1992, 422n.6) and that proof depends, in turn, on

the infinite extendability of line segments (Kant himself emphasizes this point in R5901 (c.1783/4) 18:379). The physicist's use of infinitesimal calculus in modelling physical motions similarly presupposes the continuity of space (cf. *Metaphysical Foundations*, 4:501, 503, 505, 508, 521f., 551, 557), which Kant explicitly recognizes in noting that the generation of a line (in time) through a fluxion entails the continuity of the spatial manifold (see his notes (R13 and R14 14:53–59) for his 1790 letter to Rehberg (11:207–210), cited by Friedman (1992, 76n.29) and originally highlighted by Parsons 2012). Kant registers the connection between continuity and motion in a figurative but telling manner in the Anticipations of Perception: '[Continuous] magnitudes can also be called **flowing [fließende]** because the synthesis (of the productive imagination) in their generation is a process in time, the continuity of which is paradigmatically designated by the expression 'flowing' ('elapsing')' (A170/B211f.).

50. This may seem to collapse the final two arguments of the MEs, but it does not. The penultimate argument, which we have been considering, turns on the difference in cardinality between the elements we represent space as containing, on the one hand, and the elements conceptual representations are capable of containing, on the other. By contrast, the final argument hinges not on the cardinality of elements contained in each type of representation, but on the different kinds of *relation* that obtain between those elements. Parts of space stand in compositional (viz. mereological) relations: two perfectly homogeneous spaces can compose to form a third space, distinct from the other two. By iterating a line segment, one can produce a new segment twice the size of the original. The sort of relation that obtains between the discursive marks of a concept is entirely different; it is, one might say, 'information-theoretical'. I may 'repeat' a discursive mark as many times as I care to within the intension of a concept, without at all altering the content of the latter: the concept $<$ rational animal $>$ is identical to the concept $<$ rational animal with reason $>$ . Although Kant invokes the infinitary case in order to highlight this difference in the internal articulation of intuitive versus discursive representations, the point he is making does not strictly require that he do so. Mereological relations can also hold among the elements of a discrete and finite manifold (e.g. building blocks). So the final two MEs make complementary, but distinct points. The introduction of compositionality (mereology) further enriches the sense in which an intuition represents an individual: it not only refers to a single manifold, as such, but it also represents all the *parts* of that manifold as themselves individuals. This is what enables Kant to say that 'space and time and *all their parts* are intuitions' (B137n.). For more on this contrast, see Wilson (1975), Sutherland (2004) and Anderson (2004, forthcoming).

51. It is tempting to think we can derive an idea of infinite complexity by reflecting on finitely complex representations – namely by observing that their complexity can be increased without (apparent) limit. If, as Locke argues, 'it be so, that our *Idea* of Infinity be got from the Power, we observe in ourselves, of repeating without end our own *Ideas*', then why should it be impossible for us to arrive at the idea of infinite space by extrapolating, through our own mental activity, on our ideas of *finite* spaces? (Locke 1959, *Essay* 2.17.6, vol. 1, 279; cf. Leibniz 1981, *New Essays* 2.17.3, A6.6:158). Yet this neglects the holistic mereological structure of space and, in particular, the sortal-identity-dependence of the parts of space on the whole (cf. note 37). This strategy for recursively generating (a representation of) infinite space is a non-starter, for one cannot help oneself to the base of the recursion – namely a (representation of any) finite space – without already presupposing the result the recursion is supposed to generate – viz. the (representation of the) whole of space, of which the part is a delimitation. One cannot even *think* of a finite space except as *in* the whole of space (A25/B39).

52.  Cf. also A429n./B457n., A431/B459; *Discovery* 8:222; *Progress* 20:267; R4673 (from 1774), 17:638f.

53.  A412f./B439f., A432/B306, A499f./B527f., A518-20/B546-8, A524/B552, A527/B555.

54.  Hence, Kant is prepared to argue against Eberhard that 'if something is an object of the senses and of sensation, all its simple parts must be as well, even if clarity in their representation is lacking' (*Discovery* 8:205, cf. also *Discovery* 8:209f., 212, 217; A522/B550; *Logic* 9:35). Of course, Kant denies that sensible intuition contains simple parts, but the principle still holds: every part of an object of the senses is represented in sensibility, even if we cannot be conscious of it, since it exceeds our powers of phenomenological discrimination. Thus, the *Critique* holds that 'space and time and *all their parts* are **intuitions**' (B137n., my italics). See also Kant's letter to Reinhold (19 May, 1789) 11:45f. (cited by Domski 2008). To make phenomenological features (such as 'vivacity' or 'phenomenological presence') or logical characteristics (such as 'confusion' or 'obscurity') criterial for the sensible status of a representation is to lapse into pre-critical conceptions of sensibility.

55.  The opening of the Aesthetic also introduces the idea that intuition depends on affection, which effectively marks out human intuition as a distinctively *sensible* species of cognitive receptivity: 'thought must ultimately relate [ . . . ] *in our case*, to sensibility' (A19/B33, my italics). Sensible receptivity might be contrasted with a non-sensible receptivity for representations through divine implantation or some other 'hyperphysical influx' (cf. B167f.; letter to Herz, 21 February, 1772 (10:131); R5421 and 5424 (c.1776-8) 18:178)). The latter might characterize Aquinas's angels, whose representations derive from those of the divine mind, though they are not acquired through causal affection or sensation, since angels have no bodies (cf. *Summa* Q.55, Art.2). This suggests that the rationale behind Kant's claim may have something to do with the fact that we are embodied, or at least with the fact that our receptive representations are associated with a finite spatiotemporal perspective. At any rate, one significant consequence of the claim that our intuitions depend on causal affection is that it enriches the thought-independence of intuition into a more general form of mental-act-independence. The Introduction already demonstrated that knowable objects are independent of our acts of thinking them. The causal notion of affection implies a further distinctness – not just from our acts of thought, but also from our act of intuiting the object in question. The objects that affect us (i.e. the objects we are given to know), therefore, exist independently of the mind's acts of representing them (in thought *or* intuition). (At this point, acts of reason, acts of productive imagination, etc. are of a piece with acts of the understanding in that they are all classified as spontaneous acts of thought, as opposed to actualizations of receptivity.) This picture is complicated by the fact that the mind (*Gemüt*) can affect itself in inner sense. Yet even self-affection satisfies the relevant type of mental-act-independence, for states represented in inner sense still exist independently of their being thus represented (or subsequently reflected on in second-order thought).

56.  Cf. A23-4/B38-9, A30-2/B46-8, A163/B203f., A526f./B554f.

57.  Cf. A25/B39, A31f./B47, B137n.

# References

Allison, Henry. 1983. *Kant's Transcendental Idealism – An Interpretation and Defense*. New Haven, CT: Yale University Press.

Allison, Henry. 2004. *Kant's Transcendental Idealism – An Interpretation and Defense*. Revised and Enlarged Edition. New Haven, CT: Yale University Press.

Anderson, R. Lanier. 2004. "It Adds Up After All: Kant's Philosophy of Arithmetic in Light of the Traditional Logic." *Philosophy and Phenomenological Research* 69 (3): 501–540.

Anderson, R. Lanier. Forthcoming. *The Poverty of Conceptual Truth: Kant's Analytic/ Synthetic Distinction and the Limits of Metaphysics*. Oxford: Oxford University Press.

Buroker, Jill Vance. 2006. *Kant's 'Critique of Pure Reason'*. Cambridge: Cambridge University Press.

Carson, Emily. 1997. "Kant on Intuition in Geometry." *Canadian Journal of Philosophy* 57 (4): 489–512.

Descartes, René. 1964–1974. *Oeuvres de Descartes*, edited by Charles Adam and Paul Tannery. 12 vols. Revised Edition. Paris: J. Vrin [AT].

Domski, Mary. 2008. "Kant's Argument for the Infinity of Space." In *Recht und Frieden in der Philosophie Kants – Akten des X. internationalen Kant-Kongresses*, edited by Valerio Rohden, Ricardo Rerra, Guido de Almeida, and Margit Ruffing. 2 vols, 149–159. New York: De Gruyter.

Engstrom, Stephen. 2006. "Understanding and Sensibility." *Inquiry* 49 (1): 2–25.

Euclid. 1908. *The Thirteen Books of Euclid's 'Elements'*. Translated and edited by T.L. Heath. 3 vols. Cambridge: Cambridge University Press.

Euler, Leonhard. 1911. "Vollständige Anleitungzur Algebra." In *Opera Omnia*, edited by Rudio Ferdinand, Aldof Krazer, and Paul Stäckel, series 1, 1 vols. Tübingen: Lipsiae & Berolini.

Ewing, A. C. 1950. *A Short Commentary on Kant's 'Critique of Pure Reason'*. 2nd ed. London: Metheun.

Falkenstein, Lorne. 1995. *Kant's Intuitionism*. Toronto: University of Toronto Press.

Friedman, Michael. 1992. *Kant and the Exact Sciences*. Cambridge, MA: Harvard University Press.

Gardner, Sebastian. 1999. *Kant and the 'Critique of Pure Reason'*. New York: Routledge.

Grünbaum, Adolf. 1952. "A Consistent Conception of the Extended Linear Continuum as an Aggregate of Unextended Elements." *Philosophy of Science* 19 (4): 288–306.

Guyer, Paul. 1987. *Kant and the Claims of Knowledge*. Cambridge: Cambridge University Press.

Hollis, Martin. 1967. "Times and Spaces." *Mind* 76 (304): 524–536.

Kant, Immanuel. 1900. *Gesammelte Schriften*, edited by Deutsche Akademie der Wissenschaften. Berlin: de Gruyter.

Kant, Immanuel. 1992. *Theoretical Philosophy 1755–1770*. translated and edited by David Walford, and Ralf Meerbote. Cambridge: Cambridge University Press.

Kästner, Abraham Gotthelf. 1758. *Anfängsgründe der Arithmetik, Geometrie, ebenen und sphärischen Trigonometrie und Perspektiv*. Göttingen: Vandenhoek and Ruprecht.

Kemp Smith, Norman. 1923. *A Commentary to Kant's 'Critique of Pure Reason'*. 2nd ed. London: MacMillan.

Leibniz, Gottfried Wilhelm. 1923. *Sämtliche Schriften und Briefe*. Series 6: *Philosophische Schriften*, edited by the Deutsche Akademie der Wissenschaften. Darmstadt: Akademie-Verlag [A].

Leibniz, Gottfried Wilhelm. 1978. *Die philosophischen Schriften von Gottfried Wilhelm Leibniz*, edited by C. I. Gerhardt. 7 vols. Hildesheim: Georg Olms [G].

Leibniz, GottfriedWilhelm. 1981. *New Essays on Human Understanding*. Translated and edited by Peter Remnant and Jonathan Bennett. Cambridge: Cambridge University Press.

Levey, Samuel. 1998. "Leibniz on Mathematics and the Actually Infinite Division of Matter." *The Philosophical Review* 107 (1): 49–96.

Locke, John. 1959. *Essay Concerning Human Understanding*, edited by Alexander Campbell Fraser. 2 vols. New York: Dover.

Longuenesse, Béatrice. 1998. *Kant and the Capacity to Judge: Sensibility and Discursivity in the Analytic of the Critique of Pure Reason*. Princeton, NJ: Princeton University Press.

Meier, George Friedrich. 1752. *Auszug aus der Vernunftlehre*. Halle: Gebauer.

Melnick, Arthur. 1973. *Kant's Analogies of Experience*. Chicago, IL: University of Chicago Press.

Messina, James. Forthcoming. "Conceptual Analysis and the Essence of Space: Kant's Metaphysical Exposition Revisited." *Archiv für Geschichte der Philosopie*.

Parsons, Charles. 1992. "The Transcendental Aesthetic." In *The Cambridge Companion to Kant*, edited by Paul Guyer, 62–100. Cambridge: Cambridge University Press.

Parsons, Charles. 1998. "Infinity and Kant's Conception of the 'Possibility of Experience'." In *Kant's 'Critique of Pure Reason' – Critical Essays*, edited by Patricia Kitcher, 45–58. Lanham: Rowman & Littlefield.

Parsons, Charles. 2012. "Arithmetic and the Categories." In *From Kant to Husserl: Selected Essays*, 109–121. Cambridge, MA: Harvard University Press.

Paton, H. J. 1936. *Kant's Metaphysic of Experience*. 2 vols. New York: MacMillan.

Pippin, Robert. 1982. *Kant's Theory of Form*. New Haven, CT: Yale University Press.

Quentin, Anthony. 1962. "Times and Spaces." *Philosophy* 37 (140): 130–174.

Rosenberg, Jay F. 2005. *Accessing Kant*. Oxford: Oxford University Press.

Rosenkoetter, Timothy, 2012. "Which Logic is Home to the (So-Called) 'Table of Judgments': An Unconsidered Alternative." Unpublished manuscript.

Schechtman, Anat. 2014. "Descartes' Argument for the Existence of the Idea of an Infinite Being." *Journal of the History of Philosophy* 52 (3): 487–517.

Shabel, Lisa. 2010. "The Transcendental Aesthetic." In *The Cambridge Companion to the 'Critique of Pure Reason'*, edited by Paul Guyer, 93–117. Cambridge: Cambridge University Press.

Smyth, Daniel. n.d. "Infinity and Givenness: Kant's Critical Theory of Sensibility." PhD diss., University of Chicago.

Strawson, P. F. 1966. *The Bounds of Sense*. London: Methuen.

Sutherland, Daniel. 2004. "Kant's Philosophy of Mathematics and the Greek Mathematical Tradition." *The Philosophical Review* 113 (2): 157–201.

Vaihinger, Hans. 1892. *Commentar zu Kants Kritik der reinen Vernunft*. Vol. 2. Stuttgart: Union Deutsche Verlagsgesellschaft.

Wilson, Kirk Dallas. 1975. "Kant on Intuition." *Philosophical Quarterly* 25 (100): 247–265.

# Kant on the Acquisition of Geometrical Concepts

John J. Callanan

*King's College London, London, UK*

It is often maintained that one insight of Kant's Critical philosophy is its recognition of the need to distinguish accounts of knowledge acquisition from knowledge justification. In particular, it is claimed that Kant held that the detailing of a concept's acquisition conditions is insufficient to determine its legitimacy. I argue that this is not the case at least with regard to geometrical concepts. Considered in the light of his pre-Critical writings on the mathematical method, construction in the Critique can be seen to be a form of concept acquisition, one that is related to the modal phenomenology of geometrical judgement.

## 1. Construction and concept acquisition

Kant thought of concepts as falling into three broad classes. As well as empirical concepts, we are in possession of two types of *a priori* concepts.[1] In the first case, there are 'pure sensible' concepts, i.e. *mathematical concepts* (e.g. see A140-1/ B180); second, there are what I'll refer to as *categorial concepts* (i.e. the Categories), concepts that are analogues of traditional metaphysical concepts such as <*cause*>, <*substance*>, etc.[2] That a concept is *a priori* does not entail that it is non-acquired, however. Kant repeatedly states that *all* concepts, *a priori* concepts included, are acquired.[3] For Kant, the acquisition procedure for empirical concepts is conceived along the lines of a Lockean abstraction procedure.[4] Mathematical concepts, I will argue, are acquired through a special process of definition involving a procedure Kant calls 'construction'.[5]

In this paper, I present an account of Kant's account of geometrical construction, understood as a concept acquisition procedure.[6] Construction has not usually been considered as a concept acquisition procedure and one possible reason for this is that Kant's well-known example of construction, the proof procedure of Euclid's I.32 in the Discipline of Pure Reason, seems to begin from a scenario where two inquirers, a philosopher and a mathematician, are *already* in possession of a concept (in this case, the concept <*triangle*>). It therefore seems that Kant is indicating that the construction procedure is a process of validating the already acquired concept by way of Euclidean proof. As such, construction is it seems better characterized as a justification procedure for an already-possessed concept rather than a concept acquisition procedure for that concept.

Another worry regarding treating construction as an acquisition procedure concerns Kant's general Critical approach to epistemic justification. Part of Kant's 'normative turn', it is often supposed, is a general opposition to the justification of a concept's conditions of use by providing a full account of that concept's possession conditions. The *quaestio juris* with regard to a concept's employment in judgement is supposed to have been seen by Kant as not sufficiently secured by the successful answering of a *quaestio facti* with regard to the conditions under which that concept has come to be acquired (A84-5/B116-7). If construction is indeed a process of concept acquisition then – because it seems that Kant certainly does regard construction as a justification procedure – it would entail that a sufficient account of the acquisition conditions for an *a priori* concept can indeed suffice to provide justification for that concept's use. On such an interpretation, Kant's account of construction would offer a significant qualification, if not a reformulation, of his perceived normative ambitions in the first *Critique*.

Despite these concerns, I would claim that a case can be made for interpreting construction as a concept acquisition procedure. The account presupposes a broader methodological perspective on Kant's epistemological project in the first *Critique* and in particular the nature of Kant's 'how possible?' question. In asking how synthetic *a priori* propositions are possible, Kant is asking how it is that we have come to possess the knowledge that we in fact possess.[7] This question should be explored as an expression of a more specific question, namely, how is it that we have come to possess the concepts that figure in those knowledge-generating and knowledge-preserving judgements? Pursuing Kant's epistemological project along these lines can, I claim, provide a fruitful reading of his account of mathematical knowledge.[8]

This is the central feature of Kant's famous discussion at A716-7/B744-5 of the geometer's proof of proposition I.32 of Euclid's *Elements*, that the internal angles of all triangles necessarily equal the sum of two right angles. Specifically, Kant is attempting to show that the full-blooded mathematical concept of triangularity could itself only have been acquired through a proof involving intuition. In this sense, the diagrammatic proof procedure itself constitutes the acquisition conditions for that concept. Only through such reasoning, Kant claims, could the epistemic access to necessarily true propositions be secured. Combining the two claims, I argue that Kant reverses the explanatory order of the Leibnizian account of the relation between concept possession and knowledge. We do not satisfy the possession conditions for a geometrical concept and then use that possessed concept to secure *a priori* knowledge, but rather we meet the possession conditions for that concept by way of securing *a priori* knowledge about that concept's extension.

The standard reading of this example is that it is directed to showing the necessity of intuitions for the acquisition of synthetic *a priori* knowledge. In this case, the requirement is that of the performance of a proof procedure upon a spatial particular in the form of a geometrical diagram. Kitcher's comment captures what can appear to be Kant's methodology here:

> We are supposed to gain a priori knowledge of the elementary properties of triangles by using our grasp on the concept of triangle to construct a mental picture of a triangle and by inspecting this picture with the mind's eye. (1980, 8)

I claim that this reading gets Kant's intentions the wrong way around. The example of I.32 is in fact supposed to show that we must represent an empirical or mental image of a triangle in order to acquire the concept of a triangle. The example is primarily directed towards showing how the geometer acquires the full, developed mathematical concept <triangle>. Central to the account presented here is that it is only through the deployment of concrete representations of particular triangles that the appropriate modal phenomenology involved in proof procedures can be generated. Furthermore, it is only via a procedure that generates the appropriate modal phenomenology that, Kant thinks, the *a priori* content of such concepts and the necessarily true judgements formed with them can be explained.

The aims and context of Kant's discussion of geometrical concept acquisition can only be appreciated against the Pre-Critical development of his views. The questions pursued in the *Critique* concern Kant's submission to the Berlin Academy's Prize Essay competition, the *Inquiry Concerning the Distinctness of the Principles of Natural Theology and Morality*,[9] published in 1764 as the runner-up submission. The question set for the competition was whether metaphysical propositions could be proven and known with certainty comparable with that of the propositions of geometry. The *Inquiry* expressed Kant's growing disillusion at the time for the prospects for metaphysics, though many of the claims from the *Inquiry* survived directly into the 'dogmatic use' section in the *Critique* nearly two decades later, when Kant had regained his optimism. I'll first outline the claims in the *Inquiry* that found their way into the Critical system. Second, I will outline the new model of mathematical cognition presented in the 'dogmatic use' section in the *Critique*. Third, I will present the alternative reading of the example of the Euclidean proof of proposition I.32. Finally, I will consider some of the implications of the reading suggested here.

## 2.  Geometrical concept acquisition in the *Inquiry*

By 1763, when Kant was composing the *Inquiry*, he had already come to hold that all previous metaphysicians had laboured under a misapprehension, namely that they could imitate the methodology of mathematics. That this was in fact impossible would have been clear to them had they paid sufficient attention to how mathematics is actually practised, and specifically to the crucial issue of the conditions under which we come to possess mathematical concepts.[10] In mathematics, we acquire the relevant concepts through a self-conscious, voluntary and creative act of *defining* them, by bringing together already possessed sub-concepts into a synthetic whole:

> There are two ways in which one can arrive at a general concept: either by the *arbitrary combination* of concepts, or by *separating out* that cognition which has been rendered distinct by means of analysis. Mathematics only ever draws up its definitions in the first way. For example, think arbitrarily of four straight lines bounding a plane surface so that the opposite sides are not parallel to each other. Let this figure be called a *trapezium*. The concept which I am defining is not given prior to the definition itself; on the contrary, it only comes into existence as a result of that definition. Whatever the concept of a cone may ordinarily signify, in mathematics the concept is the product of the arbitrary representation of a right-angled triangle which is rotated on one of its sides. In this and in all other cases the definition obviously comes into being as a result of *synthesis* [emphasis in original]. (*Inquiry*, 2: 276)[11]

Kant's account depends on a distinction between concepts that are created and those that are 'given'.[12] That a concept is given for Kant does not entail that it is non-acquired, but rather that it has been acquired on some non-arbitrary grounds. One possibility is that the non-arbitrary grounds are that certain concepts have a pragmatic indispensability to the course of ordinary experience.[13] Thus, there are some concepts that, whatever the particular manner of their acquisition, are routinely acquired for the purposes of the minimal representation of the world by ordinary agents. This is suggested by Kant's use of <*time*> as an example of a given concept – citing Augustine, Kant characterizes it as a concept that we all take ourselves to possess though one of which we also all lack a clear understanding (*Inquiry*, 2: 283–284). The concept is presumably pragmatically indispensable: it and its cognates are required for an enormous range of simple communicative acts, for example; yet this indispensability does not entail that we have a sound understanding of just what time is. It is this circumstance in which we find ourselves – possessing concepts of which we all claim some minimal mastery but of which we lack a full understanding – that prompts inquiry itself. Because metaphysical concepts are given, a definition represents not so much the starting point but rather the end point for inquiry.[14]

By contrast, a voluntarily created concept is marked out by the 'arbitrariness' of this act of creation. An arbitrary combination should not be taken to entail that the content of the concept formed is in any way contingent – it merely marks the fact that the concept's possession is contingent – because it has taken place through a self-conscious decision to form that concept, presumably without being prompted by the pragmatic needs that stimulate the acquisition of given concepts. In metaphysics, we proceed towards a definition through decomposition of a concept into its fundamental sub-concepts, and this presents the most important point of contrast with the method of geometry – 'geometers acquire their concepts by means of *synthesis*, whereas philosophers can only acquire their concepts by means of *analysis* – and that completely changes the method of thought'. (*Inquiry*, 2: 289)[15]

Although it can occasionally appear that the mathematician is putting forward analytic definitions, mathematics always functions in the synthetic definitional manner – to think otherwise 'is always a mistake' (*Inquiry*, 2: 277). If it appears that one is making inferences that follow directly from the meaning of a given concept, it will always turn out that 'in the end nothing is actually inferred from such definitions, or, at any rate, the immediate inferences which he draws ultimately constitute the mathematical definition itself' (*Inquiry*, 2: 277). Thus, in the *Inquiry*, Kant holds that not only are the possession conditions for mathematical concepts the results of an activity of synthesis, but also that those possession conditions can in fact be constituted by the *inferences* that the mathematician performs. If it appears that a mathematician is making an analytic definition by decomposing concepts, it can turn out that this mistaken impression is caused by the fact that what it is to grasp a concept's content is just to be able to perform certain rational rule-governed procedures over particular figures. Kant's own example is instructive here: what it is to grasp the concept <*cone*> is constituted by the subject's grasp of an operation upon a geometrical figure, that of rotating the figure of a triangle along one of its sides. Moreover, there is a kind of reciprocal relation here between two capacities – the capacity to grasp a concept's content on the one hand and the capacity to represent a token of the type expressed by that concept on the other. What is required for a subject to represent a token cone is just that the subject can understand an operational rule regarding the rotation of a triangle, the latter activity constituting the content of the concept <*cone*>. Conversely, what is required for a subject to grasp the concept <*cone*> is just that the subject can perform the operation required for representing a token of that type.

Kant's account of geometrical knowledge hinges upon the epistemic role of 'individual signs' (*Inquiry*, 2: 279). Individual signs are representations whose explicit intentional content is the presentation of a particular (they present the universal '*in concreto*', as Kant puts it). This is in contrast to natural language, which 'represent[s] the universal *in abstracto*' (*Inquiry*, 2: 279), i.e. words are representational vehicles that can purport to capture the generality of their content, e.g. as with 'triangularity'.[16] The signs deployed in metaphysics are invariably signs *in abstracto* in the form of natural language. Ironically though, the fact that natural language consists of such signs actually constitutes an impediment for its task of expressing universal claims. In Kant's view, signs *in abstracto* are clearly inferior to signs *in concreto* with regard to their power to express the universals with which they are concerned. For example, a sign *in concreto* in geometry might be a drawing of a triangle, which provides a concrete instantiation of the universal of triangularity. In the *Inquiry*, Kant seems to be as of yet unperturbed by the generality problem in geometry, i.e. concerning just how the use of a particular example might suffice as an exemplar on the basis of which general claims might be made.[17] Rather, at this point, he seems to take it for granted as a remarkable fact that mathematics, unlike metaphysics, just *is* a practice whereby 'to discover the properties of all circles, one circle is drawn' (*Inquiry*, 2: 278). There is therefore a link made in the pre-Critical period between the possible representation of tokens of a particular type and the possibility of grasping necessary truths about the relevant type.

For this reason, Kant thinks, 'nothing has been more damaging to philosophy than mathematics, and in particular the *imitation* of its method in contexts where it cannot possibly be employed'. (*Inquiry*, 2: 283) Yet this conclusion obscures the crucial insight that Kant brings from the *Inquiry* to the *Critique of Pure Reason*. In the Pre-Critical period, Kant had already held that individual signs are the indispensable means for grasping necessary truths about mathematical universals; according to the Critical model of cognition, *all* grasping of synthetic necessary truth must occur via the presentation of particulars through intuition. The *Inquiry* marks a crucial step in Kant's rejection of the rationalist model of discursive cognition, considered as knowledge gained through discursive representation alone through analysis. It is in the *Inquiry* that he first recognizes the importance of resisting the thought that the appropriate manner of expressing the generality of necessary truths through the exclusive use of signs *in abstracto*, such as the expression of concepts in natural language.[18]

To summarize, Kant was by 1763 already committed to some particular claims regarding the status of geometrical concepts. First, he maintained that the security of propositions in geometry was grounded upon the security of geometrical concepts. Second, he maintained that the security of geometrical concepts was grounded by their having a valid origin, in the literal sense of having secure acquisition conditions. Third, he maintained that the security of those acquisition conditions was achieved by virtue of the acquisition conditions for the specific intensional content essentially involving epistemic acquaintance with token representatives of the relevant extension of the concept acquired. Fourth, he maintained that in some cases, the intension of the relevant concept could itself be constituted by non-syllogistic inferences. This is to say that there are some rational rule-governed representational operations – such as rules for possible triangle rotation – performed just for the purposes of representing tokens of the relevant type. This already presents a radical difference from the nature of empirical concept acquisition, whereby the achievement of representing tokens of a type is explanatorily prior to the achievement of acquiring the concept that is abstracted from those representations. On Kant's account of geometrical concept acquisition, there can be no explanatory priority between the representation of a token of a type and the acquisition of the concept that expresses that type.

## 3.  Intuition and construction

By the time Kant presented his account of mathematical cognition in the first *Critique*, nearly two decades later, his view of knowledge had changed radically. The change could be expressed by contrasting it with the rejected rationalist model of knowledge. The achievement of veridical representation for the rationalist is secured through discursive representation alone, through pure acts of thinking. The transition from the state of lacking knowledge to the state of possessing it is to be understood as the transition from a state of indistinct and obscure thought to clear and distinct thought. The confusion of obscure thought is due to the impurities of sensory representations infecting the capacity of thought itself. Thus, the achievement of knowledge is one that would present no challenge if our discursive capacity were allowed to perform its task unimpeded because the outputs of the human discursive capacity are both necessary and sufficient for knowledge acquisition.

Kant's picture of the mind is, as with most early modern philosophers, considered in terms of certain representational capacities. As well as our discursive capacities, such as the understanding and reason, Kant claims we also possess a range of non-discursive representational capacities. Among those non-discursive representational capacities are the familiar ones of the sensory modalities, memory and imagination. The *Critique* though marks the recognition of the requirement of further, distinct type of non-discursive representational capacity, that of *intuition*.[19] In a broader sense, Kant's project is motivated by the denial of the rationalist picture and by the thought that the necessary co-deployment of both discursive and non-discursive capacities can – on occasion – be the jointly sufficient conditions for knowledge acquisition.

Although our intuitional capacity makes possible the receipt of sensory representational content, intuition also contributes its own representational content.[20] For any given perceptual experience, that experience will contain representational content that is provided solely by the sensory modalities, but it will also contain content that, though capable of being expressed through several sensory modalities,[21] is not itself generated by the senses (either individually or in complex combination). Kant claims that for any given perception of an object, the *spatiotemporal* content of that representation does not come within the sensory content that constitutes the 'matter' of perceptual experience (A20/B34). Yet given that we do in fact represent spatiotemporal features of objects, and assuming that the products of senses exhaust the non-contributed portion of human beings' representational content, it must be the case that the spatiotemporal content within our perceptual representations is *contributed* content (B1-2). For Kant, our intuitional capacity can also be activated either with regard to particular sensory experiences or in the imagination (A713/B741). When this activity of the interaction of our imaginational and intuitional capacities occurs, the representations produced in imaginational space are 'pure' (B3).[22]

In the 'dogmatic use' section of the *Critique* Kant retains many of the previously expressed commitments, most notably his claim that there is an essential difference between the methods of metaphysics and mathematics. By the time of the Critical system, Kant was still of the opinion that the superiority of mathematics was related to its ability to 'go back to the sources of its concepts'. This difference is now expressed in terms of a special procedure that is particular to mathematics, that of *construction*. In explaining why definitions can only be provided in mathematics, he expresses many of the previous claims from the *Inquiry*:

> Thus there remain no other concepts that are fit for being defined than those containing an arbitrary synthesis which can be constructed *a priori*; and thus only mathematics has definitions. For the object that it thinks it also exhibits *a priori* in intuition, and this can surely contain neither more nor less than the concept, since through the explanation of the concept the object is originally given, i.e., without the explanation being derived from anywhere else. (A729-30/B757-8)

Kant's account of construction, I will argue, involves the amalgamation of several of the key claims seen before: construction itself is nothing but the acquisition of concepts through acts of definition, where the latter is understood as manifested by inferential procedures performed upon concrete representations of particulars. Only through this procedure can certain objects even be thought because those concrete representations are constitutive of the proof-procedures that must be performed even to acquire the developed concepts of mathematics. Although before the deployment of individual signs was a mere 'aid to thought', Kant now views the deployment of representations of particulars, in the form of intuitions, as necessary for the acquisition of the relevant conceptual content and the very capacity to think of the essential properties of the objects that fall under such concepts.

It is worth noting that the first *Critique* abounds with unambiguous statements to the effect that mathematical concepts are those that we create.[23] Kant couldn't be more explicit as to the nature of mathematical definition – it is a concept-creation procedure:

> [P]hilosophical definitions come about only as expositions of given concepts, but mathematical ones as constructions of concepts that are originally made, thus the former come about only analytically through analysis (the completeness of which is never apodictically certain) while the latter come about synthetically, and therefore **make** the concept itself, while the former only **explain** it. (A730/B758)

Furthermore, Kant is clear that the task of creating a mathematical concept can be identical to and coterminous with the task of representing a member of that concept's extension. In the conclusion to the section on the Postulates of Empirical Thinking in General, Kant describes a mathematical postulate as follows:

> Now a postulate in mathematics is the practical proposition that contains nothing but the synthesis by which we first give to ourselves an object and generate its concept – e.g., to describe a circle with a given line from a given point on a plane – and such a proposition cannot be proved, because the procedure it requires is precisely that by which we generate the concept of such a figure. (A234/B287)

Similarly, Kant's well-known claim regarding synthesis in §24 of the B-Deduction seems to put a close connection between the very ability to think of something, which requires possession of the concept of that thing, with producing a representation of a token example of the thing in general, if only in the visual imagination. Kant claims both that

> We cannot think of a line without **drawing** it in thought, we cannot think of a circle without **describing** it, we cannot represent the three dimensions of space at all without placing three lines perpendicular to each other at the same point … (B154)

The three examples used here are the concepts <*line*>, <*circle*> and <*three-dimensional space*>. In each case, Kant claims that our capacity to possess this concept is dependent on our capacity to represent a referent of the concept. The capacity to think of a line is dependent on the capacity to draw a token line, and the capacity to wield the concept <*circle*> is dependent on our capacity to present a circular object to consciousness. The third example, that of the capacity to grasp <*three-dimensional space*>, is made possible in a slightly different way. Although our grasp of <*line*> and <*circle*> is secured by the capacity to represent particular lines and circles, our ability to grasp <*three-dimensional space*> is secured by our ability to perform a certain geometrical operation upon a spatial manifold. Nevertheless, there is no doubt about the intended order of explanation here: it is not the case that we can put three lines perpendicular to each other at the same point *because* we have an antecedent conceptual grasp that space is three-dimensional; on the contrary, it is because we can put three particular lines in that particular relation to each other that we are able to secure a conceptual grasp of the propositional content that space is three-dimensional. In all these cases, the *initial* grasp of these conceptual contents is directly tied to specific non-conceptual representational capacities to represent and manipulate spatial particulars in different ways.

In the *Prolegomena*, the distinction between *a priori* and *a posteriori* synthetic judgements is not drawn in terms of what allows for and what does not allow for successful concept formation, but rather in terms of the status of the judgements that issue from the concept that is formed in each case:

> … [J]ust as empirical intuition makes it possible for us, without difficulty, to amplify (synthetically in experience) the concept we form of an object of intuition through new predicates that are presented by intuition itself, so too will pure intuition do the same only with this difference: that in the latter case the synthetic judgment will be *a priori* certain and apodictic, but in the former only *a posteriori* and empirically certain, because the former contains only what is met with in contingent empirical intuition, while the latter contains what necessarily must be met with in pure intuition, since it is, as intuition *a priori*, inseparably bound with the concept *before all experience* or individual perception. (*Prolegomena*, 4: 281)

Here, Kant says that while empirical intuition allows for concept formation from which judgements regarding contingent truths can be made, pure intuition allows for the making of necessarily true judgements. Concepts formed from *a posteriori* individual perceptions afford judgements with *a posteriori* warrant; concepts formed from pure *a priori* intuition afford judgements with *a priori* warrant. The crucial innovation of the Critical model of cognition is to explain how we can access the *necessity* found in the judgements that can be formed by our use of such concepts.

This account of the *a priori* warrant for judgement relates to the account of geometrical construction presented in the Discipline. Here, Kant claims that while neither empirical nor categorial concepts can be defined, mathematical concepts can be defined and this is just because mathematical concepts are created in voluntary acts of the understanding:

> Since therefore neither empirical concepts nor concepts given *a priori* can be defined, there remain none but arbitrarily thought ones for which one can attempt this trick. In such a case I can always define my concept: for I must know what I wanted to think, since I deliberately made it up ... (A729/B758).

Kant thus maintains the same position outlined in the *Inquiry*, namely that the possibility of definition is restricted to elective concepts, ones that have been 'arbitrarily' made up by the subject herself. It is also clear that Kant continues to maintain that the mathematician for just those same reasons secures epistemic certainty without difficulty. However, Kant identifies in the *Critique* a challenge that he did not raise for himself in the *Inquiry*. The challenge is how these two elements, the elective formation conditions of mathematical concepts, and the certainty of mathematical judgement, can be compatible. If, as he maintains, mathematical concepts are simply 'arbitrarily' created concepts, then how can mathematical knowledge be secured at all? Kant is well aware that a perennial source of error in the history of philosophy has been its practitioners' disposition to indulge in the promiscuous formation of concepts without adequate warrant. Such 'invented concepts' can systematically generate false judgements, just because they have been formed arbitrarily and without concern to the very possibility of objects falling under them.[24] One might respond with the claim that elective concepts can have a positive truth value, but that they are true only relative to an arbitrarily summoned-up representation that is the ultimate referent of that concept. This though is to effectively characterize the concept as having a fictional status because the objects that determine the concept's truth conditions are themselves arbitrarily created referents.

The challenge then is to show how mathematical concepts, as given by elective definitions, are not mere fictions as most voluntarily formed concepts are. Kant's strategy for denying that elective definitions are error-theoretic is initially to concede that such acts of definitions are not sufficient for definition of 'a true object'. He contrasts the case of mathematical concept formation with that of empirical concept formation, using the example of <*chronometer*> for the latter, and claiming that

> the object and its possibility are not given through this arbitrary concept; from the concept I do not even know whether it has an object, and my explanation could be better called a declaration (of my project) than a definition of an object. (A729/B757)

With this empirical concept (one that combines the ideas of time measurement and of a mechanical apparatus, say), I can grasp the content of the combined concept without having ever represented its referent. More pressingly, I can grasp the created content without knowing that it does have an actual or even possible object. For an arbitrarily created empirical concept, the conditions for grasping the intensional content and the conditions for representing its extension are distinct. As Kant concludes, the intensional content that is grasped is better thought of as a minimal set of directions for the pragmatic project of discovering the existence of the extension and its properties.

This feature of invented empirical concepts provides the contrast required for Kant's articulation of how an arbitrarily created concept *can* nevertheless be one whose possession conditions ensures epistemic certainty in its use. An arbitrarily created concept can guarantee reference if production of an example of the extension is a necessary condition of the formation of the concept itself. This can occur if and only if the concept's intension itself is understood as a rule for the production of its referents. If a concept is one which can represent its extension in the same task of grasping its intension, then it is a concept that can be defined. This is what Kant claims does in fact occur on occasions of mathematical concept definition. In the passage previously quoted, Kant claims that the only concepts 'that are fit for being defined are those containing an arbitrary synthesis which can be constructed *a priori,* and thus only mathematics has definitions'. (A729/ B757) Kant is explicit here in linking the notions of definition, synthesis and construction. He says that a defined concept is one that 'contains' a rule for synthesis that itself makes possible the construction procedure. Kant's avoidance of the problem of invented concepts requires taking him as holding that the conditions under which the concept is formed are the very conditions under which a possible member of its extension is represented, and that these conditions are the conditions of construction.

## 4. Concept acquisition and content

Kant uses a different geometrical proof to express his point, and the example employed is proposition I.32 of Euclid's *Elements.* Before proceeding to the passage, it's worth first noting that between the composition of the *Inquiry* and the first *Critique,* Leibniz's *Nouveaux Essais* had appeared in print.[25] There he would have seen the definition of 'intuitive knowledge' as knowledge of necessary truths which arises

> when the mind perceived the agreement or disagreement of two ideas immediately by themselves, without the intervention of any other ... In this, the mind is at no pains of proving or examining ... the truth [As the eye sees light, so] the mind perceives, that white is not black, that a circle is not a triangle, that three [is] one and two. [This] knowledge is the clearest and most certain, that human frailty is capable of. (*New Essays*, Book IV, Ch. ii, §1: 361)[26]

When Leibniz comes to the issue of the nature of intuitive and demonstrative knowledge, one of the examples used is proposition I.32:

> Phil .... Now, *demonstrative* knowledge is just a chain of items of intuitive knowledge bearing on 'all the connections of intermediate ideas'. §2 For frequently the mind cannot join, compare or apply its ideas one to another, and it has to avail itself of one or more intermediate ideas to discover the agreement or disagreement which is sought; and this is what we call *reasoning.* For instance, in demonstrating that the three angles of a triangle are equal to the two right angles, one finds other

angles which can be seen to be equal both to the three angles of the triangle and to two right angles. (*New Essays*, Book IV, Ch. ii, §2: 367)[27,28]

Demonstrations lack the infallibility of items of intuitive knowledge just because they consist of chains and not in single units of grasping some certain truth (*New Essays*, Book IV, Ch. ii, §7: 368). The example of the *New Essays* suggests one possible origin of Kant's connection between spatial representation and 'intuitive' inference. For Kant, the deployment of diagrammatic representations is not only a required enabling condition for knowledge of necessary truths but the deployment of those representations is also the explanatory basis of the immediate and certain grasp of that knowledge.[29]

Kant's approach here is similar to the one adopted in the *Inquiry*. He contrasts what a 'philosopher' does, i.e. conceptual analysis, with the practice of the mathematician. The claim is that individual signs are in this case crucial to expressing the relevant necessary truth about triangles:

> Give a philosopher the concept of a triangle, and let him try to find out in his way how the sum of its angles must be related to a right angle. He has nothing but the concept of a figure enclosed by three straight lines, and in it the concept of equally many angles. Now he may reflect on this concept as long as he wants, yet he will never produce anything new. He can analyse and make distinct the concept of a straight line, or of an angle, or of the number three, but he will not come upon any other properties that do not already lie in these concepts. But not let the geometer take up this question. He begins at once to construct a triangle. Since he knows that two right angles together are exactly equal to all the adjacent angles that can be drawn at one point on a straight line, he extend one side of his triangle, and obtains two adjacent angles that together are equal to two right ones. Now he divides the external one of these angles by drawing a line parallel to the opposite side of the triangle, and sees that here there arises an external adjacent angle which is equal to an internal one, etc. In such a way, through a chain of inferences that is always guided by intuition, he arrives at a fully illuminating and at the same time general solution of the question. (A716-7/B745)[30]

This passage is frequently taken by commentators as primarily directed towards showing that the proof procedure for establishing the truth of proposition I.32 is not performed through conceptual analysis alone.[31] In order to perform the task of proving I.32, we must perform the proof outlined in the *Elements*:

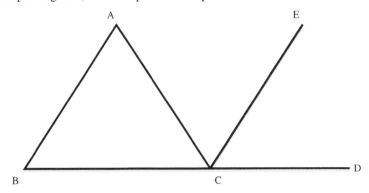

Briefly, the proof proceeds as follows: we construct a triangle ABC, then extend BC to point D, and draw a line CE that is parallel to BA. We see that the angle ∖abc is identical to that at ∖ecd. Because AC is a transversal of two parallel lines, the opposing angles at ∖bac and ∖ace are equal. We see then that the internal angles of ABC are equal to the sum of the three angles ∖ecd, ∖ace and ∖acb (the latter which is held in common). Similarly, we see that those angles are together equal to the sum of two right angles, since we see that those angles together rest upon the straight line BD. Thus, the internal angles of ABC must be equal to the sum of two right angles.

The example is not just directed at demonstrating the limitations inherent in any attempt to prove propositions from the mere analysis of the concept <*triangle*> alone. The example is also directed towards showing the *impossibility of acquiring* the full real concept <*triangle*> that we in fact possess through mere analysis of an initial nominal concept. The claim, I'd suggest, is that while <*sum of interior angles necessarily equal to the sum of two right angles*> is an essential part of the content of the concept <*triangle*>, accessing this content could have only come about through the proof procedure detailed. This alternative reading suggests that Kant's claim is that this content cannot be acquired through mere analysis of the bare nominal concept of a triangle alone. Support for this reading can be found from Kant's metaphysics lectures. There Kant discusses the same example, and characterizes it as one in which we are to think of both the philosopher and geometer beginning with a thin understanding of <*triangle*> and attempting to form a fuller version of that concept:

> Although synthetic judgments can be *a posteriori* judgments, they can also be judgments cognized *a priori*. But even *a priori* reason must add something that did not lie in the concept …

> … in general synthetic judgments are possible only by means of the corresponding intuition and the concept formed and added to this, and that this intuition must happen *a priori* if the judgment is to be posited *a priori*.

> E.g., the three angles of the triangle are equal to two right angles of 180°, is a synthetic judgment: for this proposition cannot be brought out of the analysis of a figure enclosed by three lines, rather it must be made at least in thought experiments for finding it and through that the proof is thought. (*Metaphysik Vigilantius* (K3), 29: 968–969)

Kant evidently takes the metaphysician and geometer to be operating with no more than the *nominal* concept <*triangle*>, captured just as <*figure enclosed by three straight lines*>, i.e. the initial understanding of the concept that Euclid afforded himself in order to begin his proof procedure. Part of what we are supposed to grasp from the discussion of the I.32 example is that there is no available inferential transition from the concept <*figure enclosed by three straight lines*> to <*internal angles necessarily equal to the sum of two right angles*> just through a piece of analysis of the former concept. The latter propositional content cannot be justified just because the latter propositional content cannot be 'brought out of' the former content. Why this is so is because there are steps in the diagrammatic proof that, although properly characterized as achievements of understanding, are not properly characterized as *inferences* in the common Early Modern sense at all, i.e. as steps that might be reconstructed in a piece of syllogistic reasoning with an intermediate middle term. At vital points in the proof of I.32, the steps are not inferred in syllogistic reasoning but are just represented diagrammatically.

The counterfactual claim employed in the thought experiment is one whereby we are to see that, were we (say) the kind of agents Hume characterizes us as being, we would in that case never have acquired the mathematical version of the concept <*triangle*> that we in fact possess. The conceptual content whose origin Kant is seeking to explain is that of <*necessity*>, but the only way we could have acquired *that* conceptual content is if somehow necessity was manifested within the phenomenology of our perceptual or imaginational experience. Just as a matter of fact, both Hume and Kant agree, sensory content doesn't carry this modal inflection within it. But although we can't *sense* that something must be the case, we do in fact *perceive* that something must be the case, as when performing Euclidean proofs. This achievement only occurs, Kant thinks, through the use of individual signs in the performance of inferences that themselves provide the possession conditions of concepts.

This allows for an understanding of the original import of Kant's use of the example of the 'first geometer' in the B-preface to the *Critique* (B*xi–xii*) There Kant lauded the first geometer for recognizing that the nature of the task at hand involved neither 'reading off' properties from a drawn figure, nor reading off from a concept, but instead drawing out from the proof procedure what the geometer had himself put in. The procedure Kant outlines here with regard to proposition I.32 is not that of just reading off properties from pictured *instances* of the initial concept <*triangle*> – all that would do would be to present a series of images of three straight lines laid end-to-end. What Kant is envisaging here is the concept <*triangle*> obtained via a caricatured empiricist concept acquisition process. Were we agents that acquired our mathematical concepts by way of an empiricist model, through a Humean copying or Lockean abstracting from instances of presented sensory particulars, the concepts formed would only contain the content that reflects what we had read off from those presented particulars. But in that case the concept acquired would not include the content <*necessity*> at all because the contents acquired would only have been perceptually presented as contingent (even if uniformly observed) properties of the representations of triangles (A718/ B746). Yet, Kant claims that we come to possess a concept that includes the content that the sum of the internal angles *necessarily* equals 180° degrees as part of the essential intensional content of that concept.

For Kant, the task is to explain the modal phenomenology of geometrical judgement, i.e. how it is that we could have come to make judgements that are 'combined with consciousness of their necessity' (B41). I could of course gain some epistemic confidence that all triangles' internal angles equal 180° inductively, but the judgement that they necessarily do so is a judgement with a distinct intension. Repeated perceptions that I have two hands might make it psychologically compelling for me to assent to that proposition, but no amount of such repetitions would engender in me the distinct judgement that it could not be otherwise. Yet, Kant holds, a single performance of the proof procedure for I.32 can make this distinct component of the judgement cognitively available to me, just because it is a procedure that allows me to perceive that it must be the case. Without such a procedure, we would not grasp the proposition's truth in the way that we in fact grasp it.

The counterfactual reasoning is supposed to hold equally well against Leibnizian rationalism represented by Wolff and Mendelssohn.[32] In his winning submission to the Prize Essay competition, Mendelssohn maintains that we are able to untangle or unpack the concept *<extension>* into all the truths of geometry.[33] Kant's challenge is simply to untangle the conceptual content *<sum of interior angles necessarily equal to the sum of two right angles>* from the initial concept *<figure enclosed by three straight lines>*. Decompose that initial concept all you like, Kant thinks, and the former constituent just won't reveal itself. Analyzing *<figure enclosed by three straight lines>* reveals nothing more than the original constituents: *<figure>*, *<three>*, *<straight>*, *<line>*, etc. The situation for the rationalist is the same it is for the empiricist – were our concept acquisition procedure the one envisaged, we would never have acquired the mathematical concept *<triangle>* that we in fact possess. Moreover, without those possessed concepts, we would never be able to express the propositions that we can in fact express. Neither of these approaches can explain then how these synthetic *a priori* judgements are possible.

## 5. Construction and generality

My claim is not that every occurrence of construction is always an action of concept acquisition, since surely Kant thinks that one may repeat the same proof without thereby acquiring the concept anew. Rather, my claim is that the conditions under which a geometrical concept is 'originally given' are just those construction procedures. In his reply to Kästner, Kant seems to acknowledge this point:

> However, that the possibility of a straight line and a circle can be proved, not mediately through proofs, but only immediately, through the construction of these concepts (which is not, to be sure, empirical), stems from the fact that among all constructions ... some must be the first. [34]

My account might seem vulnerable to the objection that in Kant's own presentation of an example of mathematical construction, that of the I.32 proof procedure, he clearly sets up the case whereby each inquirer, the philosopher and the geometer are *given* the concept triangle before the construction procedure can take place. If this is correct, then of course the procedure itself cannot constitute the conditions for possession of the concept *<triangle>*. However, there are reasons already seen that lead one to think that this is not the case. First, the passage quoted earlier from the *Metaphysik Vigilantius* shows Kant envisaging the case as one whereby the inquirer possesses the content <*figure enclosed by three straight lines*> but does not already possess the content <*internal angles necessarily equal to the sum of two right angles*> and where the goal is to form a new concept including this latter content. Second, it has also been already shown that in the case of empirical concepts, Kant allows that we can have some initial grasp of the content without our thereby possessing the definition of the concept. Kant makes reference to the notion of a 'putative definition' or '**designation**', which he claims is only the elaboration of a minimal set of marks that allow one to identify a subject matter so that one can subsequently conduct experiments on that subject matter, thereby securing some knowledge of the essential marks of the kind under consideration (A728). There is no reason not to think that in the construction example, Kant is similarly imagining two inquirers with the nominal or putative definition of <*triangle*>, one which does not sufficiently count as the definition proper. Third, Kant's analysis of definition is premised on the idea that there is an ordinary or common signification attached to the natural language expression of a conceptual content that is prior to the proper definition. As was seen earlier, in the *Inquiry,* Kant uses the distinction between the 'ordinary signification' and the mathematical definition of the concept <*cone*>:

> The concept which I am defining is not given prior to the definition itself, on the contrary, it only comes into existence as a result of that definition. Whatever the concept of a cone may ordinarily signify, in mathematics the concept is the product of the arbitrary representation of a right-angled triangle which is rotated on one of its sides. (*Inquiry,* 2: 276).

Here, Kant gives an example of a concept of which we might have an ordinary grasp, perhaps based on a simple perceptual individuation capacity, and which gives us an initial intension. However, it does not constitute a grasp of the definition of that concept. The definition of <*cone*> is only grasped conterminously with the production of a cone through a representational operation upon an intuition of a triangle because it is the case both that the definition provides the method for first representing a token of that type and that the representation of a token of that type is required to express the operational rule that forms the definition. Fourth, the example at I.32 is understandable as based on a subject grasping a concept but without a clear grasp of the genuine intension that characterizes the essential marks of its extension. But this is not an unusual cognitive state within Kant's theory of knowledge because this is a cognitive state that familiarly occurs in any grasp of an analytic judgement. In making an analytic judgement, such as that 'all bodies are extended', I am predicating of the subject concept a property that is in fact 'contained' within that subject concept. For such judgements to be possible in a way that presents the illusion of being informative, Kant allows that we can think the predicate concept 'confusedly' in the subject concept (A7/B11). Yet Kant would not deny that such a confused subject does not also have some minimal grasp of the intension of <*body*> in such a case. The notion between there being a graspable but only partially articulated conceptual content is implied by Kant's very notion of analytic judgement. That same state is the one that is attributable to the mathematician prior to the performance of the I.32 proof.

This leads to a second objection to this reading, namely that it threatens to render the relevant propositions *analytic* rather than synthetic *a priori* because on my account, the propositions in question are capable of being cognized just off the basis of gaining a proper understanding of the relevant concept in the subject position of the judgement. However, it doesn't follow that the judgement in such cases is analytic. First, Kant frequently speaks of synthetic knowledge being in one sense knowledge that is made on the basis of some concept possession:

> If one is to judge synthetically about a concept, then one must go beyond this concept, and indeed go to the intuition in which it is given. For if one were to remain with that which is contained in the concept, then the judgment would be merely analytic, an explanation of what is actually contained in the thought. (A721/B749)

Kant here characterizes judging synthetically as judging *about* a concept. He characterizes synthetic *a priori* judgements as ones that identify features that 'belong' to the concept even though they do not 'lie in it' (A718/B746). In the Transcendental Analytic, Kant says that as soon as sensible conditions are involved, 'synthetic judgements that flow *a priori* from pure concepts of the understanding' can be determined (A136/B175). In the Transcendental Aesthetic, Kant similarly claims that there are synthetic *a priori* cognitions that 'actually flow from the given concept' of Space (B40). At A9/B13, Kant characterizes synthetic judgement in relation to causal judgement whereby 'the understanding' seeks 'to discover beyond the concept of *A* a predicate that is foreign to it and which it nevertheless believes to be connected with it'. If we take Kant at his word here then there is nothing contradictory in the claim that in making a synthetic judgement, we are making claims that are in some sense true in virtue of the nature of the concept that features in the subject position of the judgement. The very distinction between analytic and synthetic judgement is not drawn in terms of whether a predicate is connected with a concept in the subject position or not; both types of judgement are based on such a connection. The distinguishing factor is whether the relation of the connection between the concepts is based on containment or some other determining ground.

It cannot be therefore that a judgement is analytic if and only if it is made on the basis of concept possession alone. It is rather that we are making an analytic judgement about that concept on the basis of our possession of that concept by appealing to the intensional features 'contained' in that concept, in Kant's particular notion of containment.[35] This is obscure in itself, and obviously more needs to be said here, not least with regard to the parameters determining Kantian analyticity. Why, for example, would an acquired mathematical concept of <triangle> not become the *new* initial version for our subsequent inquiries? Clearly it does not, since Kant held that what we can analytically infer from the concept <triangle> does not alter as a result of the construction procedure. Yet the thought that it might do so only stems from the thought that somehow the parameters as to what contents are 'contained' in the concept <*triangle*> might alter as a result of the construction procedure, and there is no reason to think that he endorses this latter claim.

Although the construction procedure would be a process enabling possession of a concept and the generation of knowledge in virtue of that possession, it does not follow that the knowledge generated is analytic. Rather construction is a concept-acquisition procedure that involves the acquisition of essential marks of the concept that belong to it and yet are not contained in it. The procedure *can* reveal those non-contained essential intensional marks, but only by virtue of deploying some non-conceptual representational capacities and putting the relevant particular intuitions to use in inferential processes. The resulting judgements will be *a priori* just because the concept acquired expresses necessary truths about the extension and has occurred via the provision of an *a priori* intuition; they will be synthetic because the concept acquired has been acquired through the mediating domain of sensible intuition.

Kant expresses in the *Critique* the confidence he showed in the *Inquiry* that in mathematical cognition we can think 'the universal in the particular'. (A714/B742) Yet it is clear that by this time Kant was not insensitive to the generality problem. In the Schematism section, Kant in fact cites the construction procedure performed on intuitions as the grounds for resolving this problem. There he claims that there are schematic rules for the construction of triangles that allow us to avoid the dependence on particular 'images' of triangles:

> In fact it is not images of objects but schemata that ground our pure sensible concepts. No image of a triangle would ever be adequate to the concept of it. For it would not attain the generality of the concept, which makes this valid for all triangles, right or acute, etc., but would always be limited to one part of this sphere. The schema of the triangle can never exist anywhere except in thought, and signifies a rule of the synthesis of the imagination with regard to pure shapes in space. (A140-1/B180)[36]

Whether we construct a triangle empirically on paper or imaginationally in our mind's eye, there will always be some sensory image employed, either directly or indirectly within our visual imagination. It is the intuitional content imported into any sensory image deployed though that accounts for how the produced individual sign can express general truths about the extension of the concept. It is not the sensory imagistic content of the sign *in concreto* that is being attended to when we reason with triangle images but rather the intuitional content contributed to and entangled within those representations.

This content is only accessed through the construction procedure. One might ask why this should resolve the generality problem though, rather than moving the problematic bump in the rug from the particularity of sensory content to the particularity of intuitional content. Why should the mere appeal to acts of construction grant us the security we need to infer general truths about the class? Kant's answer to these questions must involve his commitment to the claim that

'that which follows from the general conditions of the construction must also hold generally of the object of the constructed concept' (A716/B744). This though just raises the same question anew, plus another. What determines that the conditions of construction are themselves general? Furthermore, on what grounds can one infer from the general nature of the concepts to the general nature of the objects that fall under them?

Addressing these questions is beyond the scope of the present paper.[37] This reading, if correct, does however suggests three points of general relevance to the understanding of Kant's Critical Philosophy, points that I will merely raise here. The first is, as previously mentioned, that it is an oversimplification of Kant's methodology if one attributes to him the view that the acquisition conditions of a concept are irrelevant to the correctness conditions for its use. At least with regard to the example of geometrical concepts, this is not the case. On the contrary, the correctness conditions for such concepts' use are given by the construction conditions for those concepts: a concept applies to the extension that is exemplified on occasions of that concept's construction. It is unclear how we are to understand Kant's general normative turn in epistemology, however, if it does not imply a critique of the investigation into concept acquisition conditions as a general methodology.

Second, there is the related point regarding Kant's inquiry being one into the origin of the concept $<necessity>$. Kant's claim is that this concept serves as a constituent content of both categorial and pure sensible concepts. Yet it is clear, I would claim, that his account of the justification of the use of this concept with regard to geometrical concepts is just to show how the relevant concept is originally formed. This suggests that Kant's general response to Hume should be understood not as limited to the justification of the concept $<cause>$ but rather of the conceptual constituent conceptual content $<necessity>$. Furthermore, it suggests that his strategy for justifying that latter concept is identical to the task of explaining its valid acquisition.

Finally, there is the issue with regard to the 'sense and significance' of *a priori* concepts. Kant appears on occasion to hold that categorial concepts might have an explanatorily prior and purely *logical* content that is capable of being first articulated in isolation from the conditions of sensibility, but that they subsequently receive further 'sense and significance' with their application in the context of possible experience (e.g. A54/B78, A147/B188, A219/B266-7, A239/B298). However, Kant's approach with regard to the *a priori* concepts of geometry involves an identification of the original conditions of their acquisition not with a prior abstract logical formulation of their content in purely discursive terms, but rather with their original generation as rules concretely manifested within spatiotemporal representation. The question raised concerns how we are to understand the explanatory priority relation between abstract and concrete expressions of discursive content for *a priori* concepts in general. My aim here has been to defend that thought that Kant does in fact think that uncovering the general conditions of the acquisition of geometrical concepts can suffice to legitimate those concepts in use. They do so by involving the representations of pure intuition as the fundamental referents of our geometrical knowledge claims. Ultimately then geometrical concepts are the literal products of our reflections upon the most general features of outer sense.[38]

## Notes

1. E.g. A95, A129, B167, A310/B366, A 720/B748. Kant uses 'a priori' to modify a number of terms, such as 'intuition', 'representation', 'principle' 'judgement', 'truth', etc. I don't explore the relationship between this and other uses of the modifier here.

2. I use the angled brackets and italics to indicate mention of the concept. I don't discuss here the concepts <*Space*> and <*Time*>, which also fall under the heading of pure sensible concepts, and which have their own acquisition procedure, as they relate to the pure intuitions of Space and Time.

3. E.g., *On a Discovery*, 8:221. See also *Inaugural Dissertation*, §8, 2:395, §15, 2:406; *Metaphysik Mrongovious*, 29: 760–763.

4. E.g., see the *Blomberg Logic*, §254; *Jäsche Logic*, §3.

5. At A729/B757, having claimed that mathematics proceeds through the particular process of construction, Kant states that only mathematical concepts are apt for this process. As shall be discussed, only mathematics contains *definitions*, Kant thinks, because only mathematical concepts can be acquired through being defined in a construction procedure. Kant does speak of the acquisition procedure for categorial concepts as being of a specific kind, and he refers to this procedure as 'original acquisition' (*On a Discovery*, 8:221). That *a priori* concepts are acquired at all might seem surprising, since whatever else it connotes, 'a priori' surely connotes some sort of 'independence from experience'. One can point out first of all that for Kant, a concept is *a priori* if and only if it issues in an *a priori judgement* when deployed, where the latter is understood as one whose truth conditions are not provided by sensory experience (even if sensory experience nevertheless serves as a necessary enabling condition for the possibility of such judgements – B1-2). This condition does not by itself place any putative restrictions on *a priori* concepts' possession conditions. Nevertheless, the very idea of categorial concepts as acquired might seem to be precluded by the fact that those concepts are for Kant's necessary conditions of the possibility of experience – for discussion see my (2011).

6. I focus on the case of geometry in this paper. Although there are of course notable and important differences between Kant's handling of geometry and his handling of arithmetic and algebra, though there are grounds for thinking that geometrical knowledge was for Kant the paradigmatic case of mathematical knowledge (his mathematical examples are primarily Euclidean examples (e.g. Bxi–xii, A164/B205, A716-7/B745).

7. There are well-known complications here with regard to interpreting the nature of the 'synthetic' and 'analytic' methods supposedly deployed in the *Critique* and the *Prolegomena,* respectively, whereby only the latter is supposed to be pursued upon the assumption of some well-grounded body of knowledge. However, I take it to be well established that Kant frequently argues from the basis of the assumption of some *a priori* knowledge in the form of mathematical knowledge (most notably in the argument for transcendental idealism in the Transcendental Aesthetic) – see B*x*, A4/B8, B4, B20, A38-9/B55-6. For a recent discussion of the meaning of Kant's synthetic method, see Merritt (2006).

8. That Kant is interested in securing epistemological (and not merely psychological) results does not entail that his inquiry cannot be construed as one into the literal origins or sources of our concepts, although it does require re-consideration of the

type of epistemic normativity that might be at stake here, as I argue in my (2011). I take it that the account presented here is in some ways supportive of the picture presented in Longuenesse (1998).

9. Henceforth, the '*Inquiry*'.

10. Kant's insistence here on looking to actual practice in order to determine proper method is repeated in the *Inaugural Dissertation*, where he states that 'in natural science and mathematics, *use gives the method*' (§23, 2:410 – emphasis in original).

11. The essential character of mathematical concepts as 'elective' is noted by Sutherland (2010).

12. See *Jäsche Logic*, §§4–5.

13. However, caution is required here because Kant is not explicit with regard to what in fact determines the parameters for givenness in this sense.

14. As Kant puts it:

In mathematics I begin with the definition of my object, for example, of triangle, or a circle, or whatever. In metaphysics I never begin with a definition. Far from being the first thing I know about the object, it is nearly always the last thing I come to know. In mathematics, namely, I have no concept of my object at all until it is furnished by the definition. In metaphysics I have a concept which is already given to me, although it is a confused one. My task is to search for the distinct, complete and determinate concept. (*Inquiry*, 2:283)

This is a claim that Kant retains in the first *Critique*. There he holds that it is still the case that mathematics begins with definitions, although it is unclear as to whether or not he now holds the achievement of definition in metaphysics to be possible at all. (A729-32/B757-760).

15. It is clear then that in 1763 Kant still viewed the proper method of metaphysics to be that of analysis, a view famously rejected in the *Critique*.

16. This distinction corresponds to types of representational vehicle. Both types of representation can serve to express general content – e.g. <*triangularity*> can be expressed through a picture of a triangle or through the tokening of the word 'triangle' – the difference consists in the manner in which each type of representational vehicle expresses that same representational content. When considered with regard to our performance of mathematical operations, the distinction broadly corresponds to the contemporary one in developmental psychology and neuroscience between *non-symbolic* and *symbolic* numerical cognition, i.e. the respective products of our abilities to represent quantities through dots, strokes, etc. on the one hand and to symbolize those representations in terms of Arabic or Roman numerals or natural language on the other, e.g. see Ansarib, Chee, and Venkatramana (2005), Fias and Verguts (2004), Lipton and Spelke (2005), Spelke (2011). Parsons (1983) gives an illuminating discussion of the possible role of 'concrete tokens' deployed for the purpose of 'verifying general propositions' (136). Though my emphasis on concept acquisition of course differs significantly from the approach pursued there, my account of the epistemic role of signs *in concreto* (which is performed by *intuitions* in the *Critique*) is broadly in accord with Parsons' account of intuition. For differing accounts, see Hintikka (1969), Howell (1973) and Thompson (1972).

17. See Friedman (1985), Shabel (2010), Young (1982).

18. I discuss this theme more in my 'Kant on Signs *in Concreto* in Geometry' (submitted for publication).

19. The notion of intuition itself was made in the *Inaugural Dissertation* – the recognition of its necessary co-deployment with concepts for cognition, i.e. the Discursivity Thesis, was not made until the first *Critique*.

20. This point is stressed in Warren (1998) and Waxman (2005).
21. Either individually or conterminously, as when we can access spatial information through both touch and sight.
22. Although Kant is not clear on this point, I see no reason to take him as claiming that our imaginational access to intuitional content does not involve sensory content, but rather that it involves an indirect reproduction of such content.
23. When Kant is discussing the importance of the distinction of mathematics and philosophy, he notes that mathematicians have rarely philosophized regarding the nature of their own practice. He chastises them for neglecting that task, which he then characterizes as that of accounting '[f]rom whence the concepts of space and time with which they busy themselves … might have been derived' (A725/B753).
24. Typical statements of the dangers of invented concepts can be found at A222-B279.
25. Cassirer (1981) claims that Kant read the *Nouveaux Essais* sometime between its publication in 1765 and the writing of the *Inaugural Dissertation* in 1770 (97–99), though he offers little justification for the claim. Tonelli (1974) adduces evidence for thinking that any familiarity Kant had with the work could not have occurred second-hand through its reception by his contemporaries. I think a case can be made for Kant's first-hand familiarity with the *Nouveaux Essais* (e.g. he refers to the 'Essays of Locke and Leibniz' (4: 257) in the *Prolegomena* 2 years after the publication of the first edition of the first *Critique*) though to do so would be beyond the scope of this paper. In what follows I present one example of the similar themes and modes of expression to be found in both the *Nouveaux Essais* and in Kant's Critical writings.
26. See Leibniz (1996).
27. Ibid.
28. The speaker here is Philalethes, who is Locke's representative, though Leibniz does not have Theophilus – his representative – quarrel on these points. The conception of inference expressed by Locke would have been a common one within Cartesian and Port-Royal Logic. Hume too would have subscribed to it, challenging not the conception of inference at stake, but rather the scope of the knowledge that might be attained through it. For an excellent discussion of these topics, see Owen (1999).
29. The passage is worth considering not least because it gives one likely contender for the source of Kant's focus upon the word 'intuition' (*Anschauung*, but which Kant also refers to with the Latin *intuitus*) that connects it with the epistemic sense of intuitive knowledge found in the rationalist tradition.
30. Whereas Leibniz describes the demonstrative reasoning employed in proving Proposition I.32 as a 'chain of items of intuitive knowledge [*enchaînement des connaissances intuitives*]' Kant's reasoning with regard to the same proposition is held to proceed through a 'chain of inferences [*eine Kette von Schlüssen*] that is always guided by intuition'.
31. E.g. Friedman (1985), Shabel (2003), (2006).
32. I have presented the critique here as if an empiricist account of geometrical concept acquisition is Kant's target. However, I argue in my (2014) that the primary target of the critique is in fact Mendelssohn's rationalist approach. Both positions are, I would claim, effectively criticized in the passage. Shabel (2004) argues that one of the targets here is one who employs empirical methods of proofs with regard to Proposition I.32 and that this target was in fact Wolff, who presented an account whereby the geometer proceeded with particular claims regarding the management of the compass, etc. (209–212). Perhaps Kant does have Wolff's proof in mind here, as it would give a plausible alternative of what it would be to 'read off' properties of a figure. Similarly, Dunlop's (2013) account of Wolff's theory of geometrical concept acquisition might suggest that he is the target here because that account

seems to imply the adequacy of the acquisition of the concept <*triangle*> from occasions of perception of triangle instances (462).

33. See my (2014) for discussion.

34. 'Über Kästner's Abhandlungen,', (20: 411), translated by. D.R. Lachterman, quoted from Lachterman (1989, 53).

35. For discussion of Kant's notion of analyticity and containment see e.g. Anderson (2004) and (2005), de Jong (1995) and Proops (2005).

36. This passage is traditionally thought to be indicative of Kant's familiarity with Berkeley's criticism of Lockean abstract ideas in the *Principles* [e.g. (Guyer 1998, 165)].

37. I have not attempted to give anything like a complete account of the relationship between geometrical concepts, geometrical schemata, and spatiotemporal intuition. Nor have I attempted to adjudicate here with regard to how this account might figure within recent debates in Kant's philosophy of mathematics. However, it is perhaps worth noting some potential relevance in regards to one such recent debates, that between Michael Friedman, and Lisa Shabel concerning the status of diagrammatic reasoning in Kant's philosophy of geometry. One of the points of concern is how generality might be expressed via a particular image contained in a diagram. Friedman's claim is that it is clear that for Kant the generality is contributed by virtue of the conceptual representations the subject possesses prior to the construction procedure involving particular diagrams:

In particular, whereas such diagrammatic accounts of the generality of geometrical propositions, as we have seen, begin with particular concrete diagrams and then endeavor to explain how we can abstract from their irrelevant particular features (specific lengths of sides and angles, say) by relying only on their co-exact features, Kant begins with general concepts as conceived within the Leibnizean (logical) tradition and then shows how to "schematize" them sensibly by means of an intellectual act or function of the pure productive imagination. (Friedman, 2012, 239)

On the interpretation suggested here, Kant's rejection of the Leibnizean logical tradition is more thoroughgoing than Friedman envisages. This is so, I claim, because for Kant the concept is not possessed prior to the schematization process. Rather we acquire the explicit discursive representation in the course of schematizing over intuition. This is the sense in which the concepts are 'originally acquired'. My reading thus supports Manders's (2008) account of 'conceptualization via the diagram construction conditions' (74).

38. For comments on earlier versions of this paper I am grateful to audiences at Humboldt University, University of Amsterdam, University of California at Berkeley and Clare College, Cambridge

## References

Kant's works are cited according to volume and page number of the Akademie edition of *Kants Gesammelte Schriften* (Berlin: Walter de Gruyter, 1902–), except for the *Critique of Pure Reason*, which is cited in accordance with the stand 'A/B' convention. Translations from of the Cambridge Edition of the Collected Works of Kant.

Anderson, R. Lanier. 2004. "It Adds Up After All: Kant's Philosophy of Arithmetic in Light of the Traditional Logic." *Philosophy and Phenomenological Research* LXIX (3): 501–540.

Anderson, R. Lanier. 2005. "The Wolffian Paradigm and its Discontents: Kant's Containment Definition of Analyticity in Historical Context." *Archiv fur Geschichte der Philosophie* 87: 22–74.

Ansarib, D., W. L. Chee, and V. Venkatramana. 2005. "Neural Correlates of Symbolic and Non-Symbolic Arithmetic." *Neuropsychologia* 43: 744–753.

Callanan, John J. 2011. "Normativity and the Acquisition of the Categories." *Bulletin of the Hegel Society of Great Britain* (Special Issue on Kant and Hegel) 63/64: 1–26.

Callanan, John J. 2014. "Mendelssohn and Kant on Mathematics and Metaphysics 2014." *Kant Yearbook* Vol. 6 (1): 1–22.

Callanan, John J. (submitted for publication). "Kant on Signs in Concreto in Geometry."

Cassirer, Ernst. 1981. *Kant's Life and Thought*. Translated by James Haden. New Haven, CT: Yale University Press.

de Jong, Willem R. 1995. "Kant's Analytic Judgments and the Traditional Theory of Concepts." *Journal of the History of Philosophy* 33: 613–641.

Dunlop, Katherine. 2013. "Mathematical Method and Newtonian Science in the Philosophy of Christian Wolff." *Studies in History and Philosophy of Science Part A* 44: 457–469.

Fias, W., and T. Verguts. 2004. "Representation of Number in Animals and Humans: A Neural Model." *Journal of Cognitive Neuroscience* 16 (9): 1493–1504.

Friedman, Michael. 1985. "Kant's Theory of Geometry." *The Philosophical Review* 94: 455–506.

Friedman, Michael. 2012. "Kant on Geometry and Spatial Intuition." *Synthese* 186: 231–255.

Guyer, Paul. 1998. *Kant and the Claims of Knowledge*. Cambridge: Cambridge University Press.

Hintikka, Jaako. 1969. "On Kant's Notion of Intuition (*Anschauung*)." In *Kant's First Critique*, edited by T. Penelhum and J. J. Macintosh, 38–53. Belmont, CA: Wadsworth.

Howell, Robert. 1973. "Intuition, Synthesis and Individuation in the Critique of Pure Reason." *Noûs* 7 (3): 207–232.

Kant, Immanuel. 1900–. *Kants Gesammelte Schriften*, edited by German Academy of Sciences. Berlin: De Gruyter.

Kant, Immanuel. 1992a. *Lectures on Logic*. Translated and edited by J. Michael Young. Cambridge: Cambridge University Press.

Kant, Immanuel. 1992b. *Theoretical Philosophy 1755–1770*. Translated and edited by R. Meerbote and D. Walford. Cambridge: Cambridge University Press.

Kant, Immanuel. 1997a. *Lectures on Metaphysics*. Translated and edited by K. Ameriks and S. Naragon. Cambridge: Cambridge University Press.

Kant, Immanuel. 1997b. *Critique of Pure Reason*. Translated by Paul Guyer and Allen Wood. Cambridge: Cambridge University Press.

Kant, Immanuel. 2002. *Theoretical Philosophy After 1781*. Translated and edited by Henry Allison and Peter Heath. Cambridge: Cambridge University Press.

Kitcher, Philip. 1980. "*A Priori* Knowledge." *The Philosophical Review* 89: 3–23.

Lachterman, D. R. 1989. *The Ethics of Geometry*. London: Routledge.

Leibniz, G. W. 1996. *New Essays on Human Understanding*. Translated by Peter Remnant and Jonathan Bennett. Cambridge: Cambridge University Press.

Lipton, Jennifer S., and Elizabeth S. Spelke. 2005. "Preschool Children's Mapping of Number Words to Nonsymbolic Numerosities." *Child Development* 76 (5): 978–988.

Longuenesse, Beatrice. 1998. *Kant and the Capacity to Judge.* Translated by C.T. Wolfe. Princeton, NJ: Princeton University Press.

Manders, K. 2008. "Diagram-based Geometrical Practice." In *The Philosophy of Mathematical Practice*, edited by P. Mancosu, 65–79. Oxford: Oxford University Press.

Merritt, Melissa McBay. 2006. "Science and the Synthetic Method of the 'Critique of Pure Reason'." *The Review of Metaphysics* 59 (3): 517–539.

Owen, David. 1999. *Hume's Reason.* Oxford: Oxford University Press.

Parsons, Charles. 1983. "Kant's Philosophy of Arithmetic." In *Mathematics and Philosophy: Selected Essays*, 110–149. Ithaca, NY: Cornell University Press.

Proops, Ian. 2005. "Kant's Conception of Analytic Judgment." *Philosophy and Phenomenological Research* LXX: 588–612.

Shabel, Lisa. 2003. *Mathematics in Kant's Critical Philosophy.* London: Routledge.

Shabel, Lisa. 2004. "Kant's 'Argument from Geometry'." *Journal of the History of Philosophy* 42: 195–215.

Shabel, Lisa. 2006. "Kant's Philosophy of Mathematics." In *Cambridge Companion to Kant and Modern Philosophy*, edited by Paul Guyer, 94–128. Cambridge: Cambridge University Press.

Shabel, Lisa. 2010. "The Transcendental Aesthetic." In *Cambridge Companion to Kant's Critique of Pure Reason*, edited by Paul Guyer, 93–117. Cambridge: Cambridge University Press.

Spelke, Elizabeth S. 2011. "Natural Number and Natural Geometry." In *Space, Time and Number in the Brain: Searching for the Foundations of Mathematical Thought*, edited by E. Brannon and S. Dehaene. *Attention & Performance*, XXIV, 287–317. Oxford: Oxford University Press.

Sutherland, Daniel. 2010. "Philosophy, Geometry, and Logic in Leibniz, Wolff, and the Early Kant." In *Discourse on a New Method: Reinvigorating the Marriage of History and Philosophy of Science*, edited by M. Dickson and M. Domski, 155–192. Chicago, IL: Open Court.

Thompson, Manley. 1972. "Singular Terms and Intuitions in Kant's Epistemology." *The Review of Metaphysics* 26 (2): 314–343.

Tonelli, Giorgio. 1974. "Leibniz on Innate Ideas and the Early Reactions to the Publication of the *Nouveaux Essais* (1765)." *Journal of the History of Philosophy* 12 (4): 437–454.

Warren, Daniel. 1998. "Kant and the A Priority of Space." *The Philosophical Review* 107: 179–224.

Waxman, Wayne. 2005. *Kant and the Esmpiricists: Understanding Understanding.* Oxford: Oxford University Press.

Young, J. Michael. 1982. "Kant on the Construction of Arithmetical Concepts." *Kant-Studien* 73: 17–46.

# Kant (vs. Leibniz, Wolff and Lambert) on real definitions in geometry

Jeremy Heis

*Department of Logic and Philosophy of Science, University of California, Irvine, CA, USA*

This paper gives a contextualized reading of Kant's theory of real definitions in geometry. Though Leibniz, Wolff, Lambert and Kant all believe that definitions in geometry must be 'real', they disagree about what a real definition is. These disagreements are made vivid by looking at two of Euclid's definitions. I argue that Kant accepted Euclid's definition of circle and rejected his definition of parallel lines because his conception of mathematics placed uniquely stringent requirements on real definitions in geometry. Leibniz, Wolff and Lambert thus accept definitions that Kant rejects because they assign weaker roles to real definitions.

Two trends have characterized recent work on Kant's philosophy of mathematics. On the one hand, Kant's readers have been providing richer and richer contextual interpretations of his philosophy of mathematics. Of course, interpreters have long emphasized that Kant, in taking mathematical judgements to be synthetic, departed from the Leibnizian and Wolffian views that all mathematical judgements are derivable syllogistically from definitions alone. But starting with works by Friedman (1992) and Shabel (2003), there has been greater attention to how Kant's conception of mathematics was modelled on early modern proof methods, which derived ultimately from Euclid's *Elements*. For example, building on this work, Sutherland (2005) has explored Kant's conception of equality, similarity and congruence in the context of Leibniz's and Wolff's mathematical works, and has recently (Sutherland 2010) examined Kant's reception of Leibniz's geometrical 'analysis of situation' and Wolff's use of similarity in recasting Euclid's geometrical proofs.

On the other hand, Kant interpreters have been expanding their investigations beyond the widely discussed questions of the syntheticity of geometrical axioms and proofs to look at a wider range of topics within Kant's philosophy of mathematics. Recent investigations have taken on Kant's theory of mathematical postulates (Laywine 1998, 2010), Kant's conception of the mathematical method

(Carson 1999, 2006) and Kant's theory of geometrical concepts and definitions (Dunlop 2012). This paper builds on these trends to give a contextualized reading of Kant's theory of real definitions in geometry. In particular, I look at two specific cases, the definitions of *circles* and *parallel lines*, to illustrate how Kant's theory of geometrical definition differed self-consciously in sometimes subtle (but philosophically significant) ways from the theories put forward by Leibniz, Wolff and Lambert. These cases show that Kant's conception of the mathematical method (as resting on the construction of concepts in pure intuition) motivated him to give an extremely demanding theory of geometrical definitions – a theory that ruled out both Leibniz's and Lambert's preferred definitions of *parallel lines*.

Beyond its significance for our understanding of Kant's philosophy of mathematics, the present investigation is historically significant for four additional reasons. First, a striking feature of eighteenth-century German work in the philosophy of mathematics is its explicit engagement with Euclid. Indeed, Wolff, Lambert and Kant each thought of themselves as giving a philosophy of mathematics that was most faithful to the Euclidean model, despite the fact that they disagreed with one another in fundamental ways.[1] (For this reason, it is not enough to say that Kant's philosophy of mathematics is modelled on Euclid's geometry, since this simply raises the question *Whose Euclid?*) An important front in the battle between Euclid's defenders and critics concerned Euclid's definitions. Leibniz, for instance, devoted his longest geometrical work ('In Euclidis πρῶτα') largely to a line-by-line criticism of Euclid's definitions (Leibniz 1858), and one of Salomon Maimon's chief criticisms of Kant centred around what Maimon saw to be Kant's unreflective acceptance of Euclid's definition of *circle* (Freudenthal 2006). Looking at the specific stances Leibniz, Wolff, Lambert and Kant took on two of Euclid's definitions will allow us to see this battle close up.

Second, some commentators (Laywine 2001, 2010; Webb 2006, 219; Hintikka 1969, 43–44) have noted the affinities between Lambert's philosophy of geometry and Kant's own. Looking at the theory of parallel lines, where Kant and Lambert disagree, will put us in a better position to see where the two thinkers (though united in their opposition to Leibniz and Wolff) nevertheless diverge.

Third, a detailed look at Leibniz, Wolff and Lambert's theories of real definitions in geometry will show that these philosophers (like Kant) agree that constructions do in fact have a role in demonstrating the real possibility of defined concepts in geometry. That is, the debate is not whether constructions play such a role, but *how* constructions demonstrate the possibility of defined concepts. This may seem surprising, since one might have thought that it was Kant's unique contribution to the philosophy of geometry to insist on the role of constructions (against Leibniz and Wolff's purely conceptual or 'discursive' conception of mathematics). In fact, I will show, the situation was more complex and interesting: while Kant alone insisted on the role of constructions in *inference*, each philosopher claimed a role for construction in mathematical *definitions* – though each philosopher conceived of this role differently in order to fit their differing conception of mathematical proofs.

Fourth, the debate over the proper real definition of *parallel lines* was central in the most significant eighteenth-century debate in the philosophy of geometry: the status of the theory of parallel lines. Though both the Leibniz and the Kant Nachlass contain writings devoted to the theory of parallels, virtually no scholarly work has been done on these writings.[2] This essay will help rectify this situation. Moreover, the debate over the theory of parallels – and over the status of Euclid's axiom of parallels in particular – has a significance that far transcends the comparatively narrower topic of eighteenth-century German philosophical debates. The axiom of parallels is the fifth postulate in his *Elements*:

> That, if a straight line falling on two straight lines make the interior angles on the same side less than two right angles, the two straight lines, if produced indefinitely, meet on that side on which the angles are less than the two right angles. (Euclid 1925, 202)

From the ancient world through the eighteenth century, many geometers believed that this axiom lacked the self-evidence of the others and thus required a proof. After centuries of debate and failed proofs, it was shown in the late nineteenth century that it cannot be proven from Euclid's other axioms, and most philosophers came to believe that the axiom is either an empirical claim, a conventional stipulation, or an implicit definition. Any of these options are contrary to Kant's view that the axioms of geometry are synthetic a priori truths. For this reason, many philosophers – following logical empiricists such as Reichenbach – concluded from these developments in the theory of parallels not only that Kant's philosophy of mathematics was refuted, but also that the very ideas of pure intuition and synthetic a priori truths were doomed. Given this wider historical context, it is surely significant to understand the eighteenth-century debates over the theory of parallels, especially those that Kant himself participated in.

The history of philosophical debates over the theory of parallels is a vivid illustration of the way that seemingly recondite, and highly specialized debates over mathematical methodology can very quickly open up onto much wider philosophical themes. In particular, Kant's claim that (contrary to Leibniz and Wolff) our cognitive faculty splits into two stems, sensibility and understanding, with the representations of the imagination a result of their joint interaction, is essentially connected to his unique view that mathematical knowledge is rational knowledge from the construction of concepts in pure intuition. A compelling case for this conception of mathematics would *ispo facto* support Kant's conception of our cognitive faculties; conversely, a prior commitment to a certain view of our cognitive faculties would compel a geometer to approach certain concrete mathematical questions in specific ways. A close look at Leibniz, Wolff, Lambert and Kant's take on the very concrete questions of how to define *circle* and *parallel lines* provides a clear illustration of what it looks like when the rubber meets the road.

This paper is organized into six sections. In Section 1, I explain how Kant's theory of real definition in mathematics flows out of his conception of mathematical knowledge as rational knowledge from the construction of concepts. In Section 2, I contrast Kant's and Wolff's take on Euclid's definition of *circle*, turning in the next section to Kant's criticisms of both Euclid's definition of *parallel lines* and the alternative proposed by Leibniz and Wolff. In Section 4, I explain why Kant took to be merely nominal the definition that Leibniz took to be real, and in Section 5 I show why Lambert accepted (whereas Kant rejected) Euclid's definition of *parallel lines*. Section 6, which is an appendix to the main argument of the paper, considers just how strong of a theory of real definitions Kant is committed to.

## 1.  Kant on real definitions in geometry[3]

Kant was committed to a particularly strong thesis about mathematical concepts and definitions. He believed that possessing a concept, having its definition, and being able to construct instances of it were all coeval abilities. On Kant's view, one cannot possess a mathematical concept without knowing its definition, and one cannot know the definition without knowing how to give oneself objects that fall under the concept. Kant's way of putting the first point is to insist that mathematical concepts are 'made', not 'given':

> In mathematics we do not have any concept prior to the definition, as that through which the concept is first given . . . Mathematical definitions can never err. For since the concept is first given through the definition, it contains just that which the definition would think through it. (A731/B759)

His way of putting the second point is to insist that a mathematical definition always contains the construction of the concept. As he put it in a letter to Reinhold:

> the definition, as always in geometry, is *at the same time* the construction of the concept. (19 May 1789, Letter to Reinhold, Ak 11:42)

For Kant, the 'construction of a concept' is 'exhibition of a concept through the (spontaneous) production of a corresponding intuition' (Kant 2002, Ak 8:192). So to construct the concept <circle>[4] is to produce a priori an intuition of a circle. Since <circle> is a geometrical concept, for Kant it follows that one cannot think of a circle without knowing its definition, and one cannot know the definition without being able to construct circles in intuition. Kant in fact says precisely this: 'we cannot think a circle without describing it in thought' (B154), that is, without constructing a particular circle by rotating a line segment around a fixed point in a plane.[5]

Since having the definition of a mathematical concept is always at the same time having the ability to produce intuitions of instances of it, Kant says that all mathematical definitions are 'real', indeed 'genetic' definitions. According to Kant, 'real definitions present the possibility of the object from inner marks'

(*Jäsche Logic* [*JL*], §106).[6] Since mathematical definitions 'exhibit the object in accordance with the concept *in intuition*' (A241-2), they present the *possibility* of an object falling under the defined concept by enabling the mathematician to construct an *actual* instance of the concept. Kant therefore calls them 'genetic' definitions because they 'exhibit the object of the concept *a priori* and *in concreto*' (*JL*, §106).[7] Because anyone who grasps a mathematical concept knows its definition, and because the definition allows one to construct instances of it, it follows that one cannot possess a mathematical concept and still doubt whether or not it has any instances.[8]

Not every concept is made, and not every concept has a real definition. In general, one can possess a concept without knowing what its definition is and whether or not it has instances. Why does Kant endorse this strong thesis for mathematical concepts in particular? The short answer is that these requirements make mathematical knowledge – 'rational cognition from the construction of concepts' – possible. Consider first the thesis that mathematical concepts are 'made', not 'given'. For Kant, that mathematical knowledge is 'rational knowledge from the construction of concepts' means that mathematical proofs proceed by producing a particular object, noting its properties, and then drawing general and a priori conclusions from it (A713/B741). This procedure introduces the danger of illicit generalization. Suppose I want to prove some property of all triangles, say that in any triangle the angle opposite the largest side is the largest angle (Euclid 1925, I.18). To prove this, I draw a particular triangle (let us suppose, an isosceles triangle whose base is the largest side), show that it has the desired property, and generalize from this that all triangles have this property. But how can I take care that in my reasoning I have not illicitly made use of a feature of the triangle (say, its being isosceles) that does not hold of all triangles? What is required is a way of keeping track of precisely those properties that are true of all and only triangles – that is, the definition of *triangle* (A716/B744). Without the definition of *triangle*, I would not be able to reliably make inferences about all triangles from a particular drawn triangle. A mathematical concept without a definition could not then be used for acquiring 'rational cognition from the construction of concepts', which is just to say that it would not be a mathematical concept after all.

Similar reasons explain why Kant thinks that a mathematical concept must contain its own construction. Mathematical concepts are just those concepts that can be deployed in drawing a priori conclusions from the construction of concepts. A concept that did not contain its own construction could not then be used in drawing inferences mathematically: one could only reason about it discursively, as Kant imagines a philosopher would have to do with a properly mathematical concept like <triangle>. What is more, the construction procedure must be *immediately* contained in the definition. For, if it were not, it would have to be *proved* that it is possible to construct instances of the concept. Because Kant believed that mathematical proofs are constructive (and not merely discursive), the proof that a concept can be constructed would itself already require that the concept be constructed. So there could be no (mediate) proof of the reality of a definition. The concept must then have a real definition, and the possibility of constructing instances of it must be (as Kant puts it) a 'practical corollary' (i.e. an immediate consequence: *JL*, §39)[9] of the definition. (This argument that the construction must be *immediately* contained in the definition thus depends essentially on proofs' requiring constructions. A philosopher who conceives of mathematical proofs as non-constructive and composed entirely in words would therefore be free to deny the immediacy condition on geometrical definitions – as we'll see Lambert in fact does.)

One last feature of Kant's theory of mathematical concepts will be especially significant in contrasting his view with Leibniz's and Lambert's. A mathematical proof can be general (despite the fact that it deploys a particular instance) because the particular instance has been constructed according to a procedure that is contained in a definition, and this definition (as I argued above) allows the geometer to keep track of which features of the drawn figure are properly generalizable. This requires that the construction procedure be itself fully general: as Friedman puts it, it must 'yield, with the appropriate inputs, *any and all* instances of these concepts' (2010, 589). Kant calls the 'general procedure' for providing for a concept an individual intuition corresponding to it the 'schema' of the concept. The generality of these schemata secure general proofs:

> In fact it is not images of objects but schemata that ground our pure sensible concepts. No image of a triangle would ever be adequate to the concept of it. For it would not attain the generality of the concept, which makes this valid for all triangles, right or acute, etc., but would always be limited to one part of this sphere. (A140-1/B180)

If a geometer constructed a triangle using a procedure that yields only a subclass of the triangles (say the acute-angled triangles), then she could not reliably infer from her constructed individual to all triangles but only to all the members of the subclass of triangles constructible using her procedure. (In such a case, even though the constructed figure is a triangle, the concept constructed was in fact <acute-angled triangle>, not <triangle>.) Moreover, the procedure contained in the concept not only must be completely general, but the mathematician herself must know it to be so in order for her to be justified in generalizing from the constructed individual.

## 2. Wolff and Kant on the real definition of *circle*

For Kant, mathematics is rational knowledge from construction of concepts. Construction is an activity: the '(spontaneous) production of a corresponding intuition' (Ak 8:192). Mathematics therefore requires the possibility of certain spontaneous acts. The possibility of such an act is guaranteed by a 'postulate': 'a practical, immediately certain proposition or a fundamental proposition which determines a possible action of which it is presupposed that the manner of executing it is immediately certain' (*JL*, §38).[10] A favourite example of Kant's is

Euclid's third postulate: 'To describe a circle with a given line from a given point on a plane' (A234/B287). (An axiom is then a *theoretical*, immediately certain proposition that can be exhibited in intuition (*JL*, §35). A favourite example of Kant's is 'With two straight lines no space can be enclosed.'[11])

Kant has an elegant and satisfying explanation for why postulates are immediately certain. To think the postulate, one must of course possess the concepts contained in it; but the procedure described by the postulate is *itself* the means by which the concepts in question are first generated.

> Now in mathematics a postulate is the practical proposition that contains nothing except the synthesis through which we first give ourselves an object and generate its concept, e.g. to describe a circle with a given line from a given point on a plane; and a proposition of this sort cannot be proved, since the procedure that it demands is precisely that through which we first generate the concept of such a figure. (A234/B287)

To see Kant's point, it will be helpful to lay out Kant's way of expressing Euclid's third postulate and Euclid's definition of circle.

> Euclid's Third Postulate: To describe a circle with a given line from a given point on a plane.

> Circle: A ... line [in a plane] every point of which is the same distance from a single one. (A732/B760)

On Kant's view of mathematical concepts, we cannot possess the concept <circle> without having its definition. But its definition, being genetic, enables me to describe circles in pure intuition *a priori* and *in concreto*. So it is impossible that I should have the concept <circle> and not know that circles can be described with a given line from a given point. The (basic[12]) genetic definitions of mathematics are then virtually interchangeable with postulates.

> The possibility of a circle is ... *given* in the definition of the circle, since the circle is actually constructed by means of the definition, that is, it is exhibited in intuition [ ... ] For I may always draw a circle free hand on the board and put a point in it, and I can demonstrate all the properties of a circle just as well on it, presupposing the (so-called nominal) definition, which is in fact a real definition, even if this circle is not at all like one drawn by rotating a straight line attached to a point. I assume that the points of the circumference are equidistant from the centre point. The proposition 'to inscribe a circle' is *a practical corollary of the definition* (or so-called postulate), which could not be demanded at all if the possibility – yes, the very sort of possibility of the figure – were not already given in the definition. (Letter to Herz, 26 May 1789; Ak 11:53, emphasis added)

In this letter to Hertz, Kant takes as a paradigm real definition Euclid's definition of *circle*, which was Kant's preferred example also in the discussion of mathematical definitions in the *Critique of Pure Reason*. The way that Kant describes this definition to Hertz – as 'the (so-called nominal) definition, which is in fact a real definition' – makes clear that Kant recognizes that his view of Euclid's definition is controversial, and departs from other views that were well known to his audience. Indeed, the parenthetical aside is clearly an allusion to

Christian Wolff, whose mathematical works Kant knew well, having taught them for years to his mathematics students in Königsberg. According to Wolff:

> If a circle is defined through a plane figure returning to itself, the single points of whose perimeter are equally distant from a certain intermediate point; the definition is nominal: for it is not apparent from the definition, whether a plane figure of this kind is possible, consequently whether some notion answers to the definitum, or whether it is actually a sound without meaning [mens]. For truly if the circle is defined through a figure, described by the motion of a straight line around a fixed point in a plane, then from the definition it is patent [patet], that a figure of this kind is possible: this definition is real. (1740, §191)

Wolff here rejects Euclid's definition in favour of one that explicitly describes the procedure for constructing circles:

> A circle is that figure that is described by moving a straight line around a fixed point in a plane.

That all of the points in the circle are equidistant from the centre (which is Euclid's definition) is on Wolff's view not part of the definition, but a consequence of it. In fact, Wolff lists it as an axiom.[13]

This disagreement over Euclid's definition is initially surprising, since Kant and Wolff seem to define real definitions in very similar ways. For Kant, 'real definitions present the possibility of the object from inner marks' (*JL*, §106); for Wolff, they are 'definition[s] through which it is clear that the thing defined is possible' (Wolff 1740, §191, cf. 1741, Introduction §17–§18, 1710, Introduction, §4). But the disagreement is not primarily a difference in the conception of real definition, but a difference in the conception of what it is to grasp a concept. One of the most significant and innovative features of Kant's critical philosophy of mind is the thesis that concepts 'rest on functions', or constitutively include abilities to do certain things (A68/B93). This innovation explains the divergence between Kant and Wolff over Euclid's definition of *circle*. Because Kant identifies the distance between two points as the straight line between them (Reflexion 9, Ak 14: 36) and because one cannot think a line without drawing it in thought (B154), to think of the points equidistant from a given point is *ipso facto* to think of a straight line moved around the given point. The upshot is that 'we cannot think a circle' – that is, understand its Euclidean definition – 'without describing it from a given point' (B154), and Wolff's 'real definition' coincides with Euclid's 'so-called nominal definition'.

## 3. Kant versus Euclid's and Leibniz's definitions of *parallel lines*

As I mentioned in the opening paragraphs of this paper, the most contentious feature of Euclid's geometry was his notorious axiom of parallels. It was virtually the consensus view in eighteenth-century Germany – a consensus shared by Leibniz, Wolff, Lambert and Kant – that Euclid's axiom is not in fact a genuine axiom. Although no one doubted its truth, everyone agreed that it was not self-evident and so not an *axiomatic* truth.[14] The debate over Euclid's axiom in eighteenth-century Germany is rich and interesting, though still largely untold. But it is less well known that there was in the early modern period a different debate, dovetailing with the first, over Euclid's *definition* of *parallel lines*. Euclid defined parallel lines as

> straight lines which, being in the same plane and being produced indefinitely in both directions, do not meet one another in either direction. (Euclid 1925, 190)

Leibniz, Wolff, Lambert and Kant all participated in the debate. As we will see, Leibniz, Wolff and Kant all rejected the definition, whereas Lambert (along with Saccheri 1920, 7, 88–95, 237–41) defended it.

Kant rejects Euclid's definition in Reflexion 6 (Ak 14:31), which dates from the critical period and is the opening reflection in a series of notes (Reflexions 6–11, Ak 14:31–52) on the theory of parallels.[15] Kant begins by restating his now familiar requirement on real definitions in mathematics:

> I think that from a definition that does not at the same time contain in itself the construction of the concept, nothing can be inferred (which would be a synthetic predicate).

Mathematical definitions make immediately patent the procedure for constructing instances of the concept. Any candidate mathematical definition for which the construction procedure is not a 'practical corollary' of the definition, is thus a faulty kind of definition. From this, Kant then draws the surprising conclusion that 'Euclid's definition of parallel lines is of this kind'. Kant's point is that grasping Euclid's definition does not tell you how to construct two non-intersecting straight lines. Euclid's definition of a circle is such that one cannot think it without rotating a line segment around a fixed point in the plane, but Euclid's definition of *parallel lines* provides no guidance whatsoever. A subject could fully grasp the marks <straight line>, <coplanar> and <intersecting>, know full well how to construct straight lines, planes and intersecting lines, understand fully how these marks are put together to form the concept <parallel linesz>, but still not know how to construct them. The definition is thus not genetic. Indeed, since it provides no method for constructing parallels, the definition does not present the possibility of parallels and so is not even a real definition.

Kant's criticism of Euclid here is not unprecedented. He would have known a similar criticism that Leibniz had levelled against Euclid in the *New Essays*:

> [T]he real definition displays the possibility of the definiendum and the nominal does not. For instance, the definition of two parallel straight lines as 'lines in the same plane which do not meet even if extended to infinity' is only nominal, for one could at first question whether that is possible.

Leibniz proposes instead an alternative definition, which he claimed is genuinely real:

> But once we understand that we can draw a straight line in a plane, parallel to a given straight line, by ensuring that the point of the stylus drawing the parallel line remains at the same distance from the given line, we can see at once that the thing is

possible, and why the lines have the property of never meeting, which is their nominal definition. (Leibniz 1996, III.iii.§18, cf. 1858, 201)

Leibniz is suggesting here an alternative definition to Euclid's (which he takes to be merely nominal):

Parallels are 'straight lines in a plane which have everywhere the same distance from one another'.[16]

On Leibniz's view, this definition is real (indeed, genetic) because it tells us how to construct the parallel lines: take a perpendicular of fixed length on a given line, slide the perpendicular along the line, and then the line described by the end point of the moving perpendicular is everywhere equidistant from the given line. Thus, this definition 'carries with it its own possibility' (Leibniz 1858, 202; Definition XXXIV, §6). (Wolff follows Leibniz in defining *parallel lines* in terms of equidistance, and he recasts the theory of parallels using this new definition in his widely read and influential mathematical works.)

Kant, however, thinks that Leibniz's (and Wolff's) definition is no more a real definition than Euclid's. After a series of attempts to reconstruct the theory of parallels using this new definition (Reflexions 7–10), Kant ultimately concludes:

For now the geometric proof [of a proposition in the theory of parallels] rests on [...] a concept of determinate distances and of parallel lines as lines, whose distance is determinate, [a concept] which cannot be constructed, and is therefore not capable of mathematical proof. (Reflexion 10, Ak 45-8)

Leibniz's definition runs afoul at the same point as Euclid's: in constructing it. Granted, it is immediately evident that it is possible to slide a perpendicular line of fixed length along a given straight line and thus trace out a new line everywhere equidistant from the first. But what guarantees that the line traced out is *itself straight*, as Leibniz's definition of *parallel lines* requires? Indeed, without an extra assumption that guarantees that a line everywhere equidistant from a straight line is itself straight, we cannot take ourselves to have constructed parallel lines in Leibniz's sense. As Kant's student Schultz put the point in his 1784 work on parallels, Wolff does not rule out that the concept of everywhere equidistant straight lines is a 'Hirngespinst' – a figment of the brain (7–8, cf. Kant 1998, A157/B196). (As a matter of fact, the assumption that the everywhere equidistant line is straight is just equivalent to the parallel axiom. So the constructibility of Leibniz's definition falls back onto the unsolved problem of finding a proof of the axiom of parallels.) Moreover, even if Leibniz could come up with a proof that shows that the line traced out is in fact straight, this would not make the definition a *real* definition for Kant: a construction procedure which must be *proved* to produce instances of the defined concept does not *itself* show the possibility of the concept. One could easily grasp Leibniz's definition and still doubt whether the line traced out is straight. In Section 1 of this paper, we saw Kant argue that only real mathematical definitions – definitions which themselves show their own possibility by containing their own constructions – can be employed in mathematical proofs, which proceed by constructing concepts. Kant in this note draws the obvious conclusion: Leibniz's definition of *parallel lines* cannot be employed in genuine geometrical proofs.

## 4. Kant versus Leibniz on real definitions in geometry

It will surely be surprising to some – given the 'dogmatism' to which Kant was allegedly prone – to see him rejecting both Euclid's definition of *parallel lines* (which, we will see, Lambert defended) and the alternative definition proposed by Leibniz and Wolff.[17] In the remainder of this paper, I will argue that it is precisely *because* Kant believed (while Leibniz and Lambert did not) that mathematics is knowledge from the construction of concepts that his requirements on the definition of parallel lines are more restrictive than his contemporaries' requirements.

Let us begin with Leibniz. Kant's notion of real definition closely follows Leibniz. Like Kant, he defines a real definition as a definition that 'shows that the thing being defined is possible', and thus also gives the essence, or inner possibility, of a thing (Leibniz 1996, III.iii.18).[18] A special kind of real definition is a causal (or genetic) definition, which 'contains the possible generation of a thing'.[19] Again like Kant, Leibniz claims that the mathematician's ability to form real definitions by putting together geometrical concepts is highly constrained. He emphasizes that 'it isn't within our discretion to put our ideas together as we see fit' (Leibniz 1996, III.iii.15), and calls 'paradoxical properties' those combinations of ideas whose possibility can be doubted (Leibniz 1970, 230). Leibniz gives many examples of merely nominal mathematical definitions: In *Discourse on Metaphysics* (§24), he gives a nominal definition of *endless helix*, and in *New Essays,* two merely nominal definitions of *parabola* (III.x.19). Eerily similar to Kant's Reflexions 5–6, Leibniz (1970, 230) contrasts Euclid's definition of *circle*, which he claims to be properly genetic, from another definition (derived from Euclid III.20) that is merely nominal, even though it provably holds of all and only circles.[20]

Despite their commonalities, the notion of a real definition plays a much different role in Kant's philosophy of geometry than it does in Leibniz's. Kant's restrictions on real definitions in mathematics are motivated by his theory of construction: mathematical concepts are those concepts for which a fully general a priori and certain proof can be given using a representative individual. Leibniz, however, famously believes that the process of reasoning about mathematical truths does not differ in kind from the reasoning about any other necessary truth. He forbids 'admitting into geometry what images tell us', he proposes a new kind of geometry 'even without figures', and believes that all geometrical truths can be reduced to identities through conceptual analysis using the principle of contradiction.[21] Now, reasoning from definitions cannot be done safely 'unless we know first that they are real definitions, that is, they include no contradictions' (Leibniz 1989, 25). On many occasions, Leibniz motivates the necessity of real definitions by criticizing Descartes's argument for the existence of God: we cannot safely infer from the definition of God unless we know that God is possible (Leibniz 1989, 24, 1970, 231). For Leibniz, then, the purpose for introducing mathematical definitions that 'explain the method of production is *merely* to demonstrate the possibility of a thing' (1970, 231, emphasis added). This weaker role for construction in mathematical proof brings with it a weaker requirement on real definitions. A definition that contains a method for constructing some, though not necessarily all, instances of a concept would fulfil Leibniz's requirement. Even one instance of a concept is sufficient to show that the concept is not contradictory. For Kant, though, fully general constructive proofs require that the constructive procedure immediately contained in the concept must be general enough to produce any and all instances of the concept.[22]

Leibniz's real definitions play a different role from Kant's in another respect. Leibniz (1970, 231–232, 1996, IV.vii.1) believed that all necessary truths, including the axioms of geometry, could be reduced to identities once fully adequate definitions have been found. Truths that on Kant's view are expressed by axioms and postulates have to be somehow built into appropriately formulated definitions. For a Kantian, this is putting weight onto the definitions that they cannot possibly bear. Leibniz's and Wolff's own work in the theory of parallels shows this clearly because they more or less smuggled the problem with Euclid's axiom into a definition of parallel lines whose reality is no more evident than the truth of the axiom.

## 5.   Kant versus Lambert on real definitions

In fact, this was precisely Lambert's criticism of Wolff's proofs in the theory of parallels. Lambert begins his justly famous *Theory of Parallel Lines* by criticizing Wolff's definition of *parallel lines*. From his point of view, Wolff 'conceded too much to the definitions; and because he wanted to arrange them suitably for the subject matter [ . . . ] the difficulty is only taken from the axiom and brought into the definition' (Lambert 1895b, §5, §8).[23] Reflection on what goes wrong in Wolff's treatment of parallels (and on what goes right in Euclid's) will motivate a better conception of science, and therefore also a better grounded geometrical theory of parallels.

On Lambert's view, the fallacies in Wolff's treatment of parallels are a symptom of a deeper confusion in Wolff's (and Leibniz's) theory of science.[24] On pain of circularity, not every scientific concept can be defined; some must be simple. For this reason, propositions that contain only undefinable concepts standing in undefinable relations cannot possibly be inferred from definitions. The arbitrary combination of simple concepts into compounds thus needs to be licensed or constrained by certain basic propositions.[25] Lambert (1771, §13; cf. 13 November 1765, Letter to Kant, Ak 10:52) claimed that propositions containing only simples are thus either postulates (which affirm that certain simple concepts stand in certain simple relations) or axioms (which deny other simple relations among simple concepts). The possibility of every compound concept, like Euclid's definition of *parallel lines*, needs to be given a proof that bottoms out in postulates and axioms. On Lambert's view, Leibniz and Wolff therefore inverted the order of dependency among definitions, axioms and postulates. Axioms and postulates are not dependent on definitions as their corollaries, as Leibniz and Wolff claimed, but are preconditions of the possibility of definitions. In any properly produced science, the axioms and postulates should be presented before the definitions, not after (Lambert 1771, §23).

No doubt there was much about this view that accorded with Kant's own. In particular, Lambert's axioms and postulates, which do not follow from definitions and – being composed entirely of simples – cannot be containment truths, would have been for Kant clear examples of *synthetic* a priori truths (3 February 1766, Letter to Kant, Ak 10:65). Nevertheless, Kant could not give his full consent to Lambert's view of science: as we have seen, Kant continued to hold with Wolff (and against Lambert) that *postulates* are practical corollaries of definitions.[26] This disagreement over postulates spills over into a further divergence in their theory of real definitions.

> [W]hat one derives as axioms [Grundsätze] from the definitions according to Wolff are according to Euclid propositions that the definition already presupposes, and from which the definition is formed and proved. In this way the things that appear arbitrary and hypothetical fall away from the definitions, and one is completely assured of the possibility of everything they contain. [ ... ] Definitions produced in this way define the thing itself, and insofar as one brings them forth from basic concepts [Grundbegriffe], one can call them real definitions, which are a priori in the strictest sense. (Lambert 1771, §§23–§§24)

Real definitions are definitions composed of simple concepts whose possibility has been proved from the axioms and postulates. Notice that it is no part of Lambert's view that the possibility of the concept be *immediate from the concept itself*. In fact for Lambert, the proper method (which, again, he thinks is Euclid's) is to first introduce purely nominal definitions and *only later* prove their possibility. After this proof, the 'hypothetical' concept becomes a 'derived concept'.[27]

In fact, for Lambert Euclid's definition of *parallel lines* is a paradigm real definition, since it is first introduced as a nominal definition, and then later becomes a real definition only after Euclid proves its possibility in I.31. Lambert points out that while Wolff assumed without proof that there are everywhere equidistant straight lines, Euclid does not assume without argument that there are non-intersecting lines.[28] He proves in Euclid I.27, without using the parallel axiom, that there are parallel lines (namely, straight lines that make equal alternate angles with some third straight line), and in Euclid I.31, he shows (again without appeal to the parallel axiom) how to construct them. (Euclid's axiom, then, simply shows that this is the *unique* way of constructing parallels: that there are no parallels to a given straight line other than those that make equal alternate angles with a third straight line.) On Kant's view, however, every genuine mathematical concept must itself exhibit its own possibility, and so every mathematical definition must be real from the get-go. Kant would then be unimpressed with the fact that Euclid later proves that there are parallel lines, since a legitimate definition of *parallel lines* would be such that 'the definition, as always in geometry, is *at the same time* the construction of the concept' (19 May 1789, Letter to Reinhold, Ak 11:42). Lambert, on the other hand, is untroubled to admit that Euclid's definition of parallel lines 'does not indeed indicate their possibility' (1895a, #7).

The case of parallel lines illustrates another fundamental difference between Lambert and Kant, Leibniz and Wolff. As we have seen, these other philosophers all insist that every mathematical definition is genetic. Lambert, however, does not. He argues, against Wolff, that one needs to distinguish between the mode of genesis of a concept, and the mode of a genesis of a thing. The mode of genesis of a concept is a proof of the possibility of a concept from axioms and postulates (Lambert 1915, §27, cf. 1771, §24). In this sense, the proof in Euclid I.27 and I.31 gives the mode of genesis of the concept <parallel lines>: it proves, step by step syllogistically from the axioms and postulates, that there are parallel lines. The mode of genesis of the thing, on the other hand, gives the procedure for making an instance. An example of a mode of genesis of a mathematical object is the procedure for producing a conic section (first, produce a cone by rotating a right triangle about one side, then cut the cone with a plane), and an example of a mode of genesis of a natural scientific object is the procedure for producing a salt crystal (first dissolve salt in water, then do such and such to the water...).[29]

Lambert claims that a real definition of a mathematical object requires only the first kind of mode of genesis. Indeed, Lambert claims that a definition that gives the mode of genesis of the thing is appropriate only for a posteriori concepts – concepts whose possibility is shown not by a proof from first principles, but from some other source, such as experience.[30] For example, using the genetic definition of salt crystals, one carries out the procedure described in the definition, observes that something is produced by the procedure, and thereby comes to know a posteriori that the concept <salt crystal> is possible. Lambert is therefore undisturbed by the fact that Euclid's definition of *parallel lines* in no way expresses their mode of genesis.

There is a further difference between Kant's and Lambert's conceptions of geometry that helps explain why Kant rejected the Euclidean definition of *parallel lines* that Lambert accepted. Although every science must on Lambert's view prove the possibility of its compound concepts, this is easier in geometry because the constructed figures themselves demonstrate the possibility of compound concepts.

> It was easy for Euclid to give definitions and to determine the use of words. He could lay the lines, angles and figures before the eyes, and thereby combine words, concepts, and things [Sache] immediately with one another. The word was only the name of the thing, and because one could see it before the eyes, one could not doubt the possibility of the concept. Furthermore, it was the case that Euclid had unstinted freedom to let everything that did not belong to the concept or was not

present in it to fall away from the figure, which is really only a special or singular case of a general proposition, but thereby serves as an example. The figure accordingly represents the concept completely and purely. On the other hand, because the figure does not provide the general possibility of the concept, Euclid had taken care to exposit this [possibility] exactly, and here he uses his postulates, which represent general, unconditioned and simple possibilities that are thinkable for themselves, or things that can be done [Thulichkeiten], and this [possibility] he put forward in the form of problems. (1771, §12)

The possibility of a compound concept is made manifest in the constructed figures licensed by geometrical problems. Lambert's examples include Euclid I.1 (which demonstrates the possibility for <equilateral triangle>), I.22 (for <triangle>) and I.31 (for <parallel lines>).[31] These figures are not merely examples, though, because they are produced by general procedures licensed by postulates. The figures then stand in for the concept completely and purely.

On its surface, this passage is strikingly similar to Kantian passages like A713/B741. Indeed, some commentators have alleged that Lambert's comments on the role of individual constructed figures in geometry were an impetus for Kant's own views on the role of intuitions of individuals in mathematical proofs.[32] However, the mere fact that Lambert – but not Kant! – believed that the construction licensed by Euclid I.31 demonstrated the reality of <parallel lines> should give us pause. On Kant's view, the construction procedure immediately contained in a mathematical concept has to be general in the sense that we can know, with certainty, that each and every instance of that concept can be constructed in this way. Unless we can be certain of the truth of Euclid's axiom, I.31 will not show that Euclid's definition is general *in this sense*. Lambert, however, clearly thinks that Euclid's definition has been proven, and the construction from I.31 is fully general *in the sense that matters to him*, even though Euclid's axiom requires, but does not yet have, a demonstration.

Let me explain these differing notions of generality. Lambert was concerned to justify certain arbitrary combination of concepts in the face of worries that the combined concepts might together contain a hidden contradiction. He imagines a sceptic who denies the possibility of such a combination. There are two senses for Lambert in which Euclidean problems are general – two senses that suffice to refute such a sceptic but which nonetheless fall short of the kind of generality that matters to Kant. For one, Euclidean constructions can be carried out by *any person* in any time or place. In an extended and rich discussion in his 'Essay on the Criterion of Truth', he imagines Euclid responding to a sophist as a metaphysician might respond to a solipsist: 'there is no better way to refute someone who believes something impossible than if one shows him how he can himself bring it about' (§79).[33] Euclid's definition of *parallel lines* is general in this sense, since any geometer can carry out I.31 for herself any time she pleases (Lambert 1918, §20–§21). Second, Lambert highlights that Euclidean problems specify conditions that *always suffice* to provide an instance of the concept. In this sense, Euclid I.22 shows that a triangle can be formed from any three straight lines provided any two are greater than the third (Lambert 1895a, #6, 1915, §79).

Euclid's definition of *parallel lines* fulfils this requirement as well, since a geometer can carry out the construction in I.31 given *any* line segment and *any* point not on that line. Still though, the fact that *any* geometer at *any* time can construct from *any line and point* a new line that does not intersect it in no way shows what Kant wants: that the construction be certainly sufficient to produce *any and all* parallels.[34]

Why in the end do Lambert and Kant demand very different things of Euclid's definition? I have emphasized that Kant's very stringent conditions on mathematical concepts follow from his characteristic view that a mathematical proof works by drawing a fully general, certain and a priori conclusion from a single intuition of an individual. And though Lambert acknowledges the role of individual drawn figures in proving the possibility of compound concepts, his view of mathematical proof is thoroughly Wolffian: the proof of a theorem from axioms and postulates 'depends on the principle of contradiction and in general on the doctrine of syllogisms' (Lambert 1915, §92.14).[35] Because the intuition of the figure plays no role in the proofs themselves, the function of the individual constructed figure is merely to show that the concept is free from contradiction. Since one instance of a concept suffices to show that it is free from contradiction, Lambert does not require that the individual drawn figure be produced from a construction procedure that is general *in Kant's sense*. In Lambert's case, then, just as with Leibniz, we see that Kant's view that mathematics is rational cognition from the construction of concepts made him more, not less, sceptical of his contemporaries' treatment of parallel lines.

Lambert and Kant then disagree fundamentally about the kind of generality that a constructed figure must possess in order for a definition to be real. This disagreement about real definitions should not however obscure their agreement about postulates. Both think of postulates as practical indemonstrable propositions, and both believe (against Leibniz and Wolff) that they are not containment truths (Laywine 1998, 2010). Both further distinguish postulates as practical from axioms as theoretical (Heis, forthcoming, §3). Kant further believes that definitions and practical propositions that assert that a certain construction is possible are virtually interchangeable. In the case of basic concepts such as <circle>, this is the claim (discussed above) that Euclid's third postulate is simply a practical corollary of Euclid's definition of a circle. Although Lambert denies this, this is again a reflection of his different notion of what makes a real definition. A real definition in mathematics for Lambert need not express a construction procedure (it need not be genetic), and so the very tight connection that Kant sees between definitions and practical propositions like postulates is severed.

There is, though, one significant difference between Kant and Lambert's conceptions of postulates, a difference that rests on their differing view of mathematical proof. Lambert argues that the practicality of postulates rests on their fundamental role in establishing that certain combinations of simple concepts are possible, and he criticizes other philosophers who emphasize the practicality of postulates as confusing a superficial property with what is really essential (1915, §48ff., 1771, §18–§20). Euclid, in showing what can be done constructively, was thereby showing what combinations of concepts are possible. And since the possibility of compound, but not simple concepts, needs to be either self-evident or proven, there need to be principles that express the possible combinations of simple concepts. This is the deeper characterization of postulates, which is not explicitly expressed by simply saying that postulates are practical indemonstrable propositions.

Kant, however, does not thematize the difference between simple and compound concepts as Lambert does. And for Kant, it really is essential that the postulates express not just which combinations of concepts are possible, but also what constructions can be executed. After all, since for Kant proofs require the construction of concepts, it would simply not be enough to know that, say, equilateral triangles are possible. To prove anything about equilateral triangles, the geometer needs to produce a representative instance and so needs to know how to actually carry out the construction. Kant's conception of mathematical proof, then, leads him to emphasize the practicality of postulates – an emphasis that Lambert, whose conception of proof is different, thinks confuses the accidental with the essential.

## 6.  The immediacy condition on real definitions

In Section 1 of the paper, I argued that Kant's conception of mathematical methodology required that possessing a concept and knowing its definition are coeval, that every definition exhibits the method for constructing instances of the defined concepts, and that this constructive procedure must be sufficient for producing any and all instances of the concepts. These requirements are sufficient for showing why Kant rejected both the Leibniz/Wolff definition of *parallel lines* in terms of equidistance (without a proof of the axiom of parallels, it is not clear that we can produce any instances of the concept) and the Euclidean definition in terms of non-intersection (this definition does not exhibit any method for constructing parallels, and moreover without a proof of the axiom of parallels, we cannot know that the construction procedure given by Euclid in I.31 is sufficient to produce any and all instances of the concept). If the argument of the paper has made these claims plausible, then the paper has succeeded.

In Section 1, though, I argued that Kant is also committed to a yet stronger claim, that the definition not only must express a construction procedure for the defined concept, but also that the definition must make it *immediately certain* that that construction procedure is possible. Although this stronger claim is not necessary for my argument in the paper, I do believe that Kant was committed to this stronger view, and given the interest of the topic, I would like to end my paper with a defence of the stronger claim.

To begin with, it is incontestable that Kant maintained the immediacy condition for those definitions for which postulates are practical corollaries, since he says so explicitly at Ak 11:53 and A234/B287, in both cases taking as his example the concept <circle>. In note 12, I called such concepts 'basic', and distinguished them from concepts whose construction procedures are not licensed by postulates. A concept of this kind, such as <equilateral triangle> or <parallel lines>, I call 'complex'. The interesting question, then, is whether the immediacy condition should be applied to complex concepts as well.

An interpretation of Kant that maintains the immediacy condition only for basic concepts and not for complex ones would attribute to Kant a principled distinction among concepts akin to Lambert's distinction between *Grundbegriffe* (axiomatic concepts), whose possibility is immediately certain, and *Lehrbegriffe* (derived concepts), whose possibility is proved subsequently to grasping them.[36] On this view, Kant (like Lambert) would allow that one and the same definition would be nominal when it is first grasped, and then would later become real when a proof is provided for it.

I do not believe that Kant can be assimilated to Lambert in this way. For one thing, the plain reading of Kant's texts provides no basis for it. After all, Kant writes:

[T]he object that [mathematics] thinks it also exhibits *a priori* in intuition, and this [object] can surely contain neither more nor less than the concept, since through the definition of the concept the object is originally given. (A713/B741)

A mathematical definition must 'at the same time [zugleich] contain in itself the construction of the concept'. (Reflexion 6)

The definition, as always in geometry, is at the same time the construction of the concept. (19 May 1789, Letter to Reinhold, Ak 11:42)

There is no suggestion in these passages that the immediacy condition is restricted to some concepts and not others. In fact, in Reflexion 6, Kant's claim comes between a discussion of the definition of *circle* and the definition of *parallel lines* – one basic and one complex concept – and Kant in both cases is making the same point. Moreover, I hope to have shown that the case of the definition of *parallel lines* was precisely where the immediacy condition had become an issue in eighteenth-century Germany, and it was the case that Lambert used to illustrate the importance of introducing a difference in kind between those concepts whose possibility was immediately obvious from those concepts whose possibility had to be proven subsequently to their being grasped. In fact, Kant goes on in the very next sentence to endorse the argument (Euclid's definition is not real, because it does not contain its own construction immediately) that Lambert used his distinction to deny. (This is a nice illustration of the advantage at looking at the stances that Kant and his contemporaries took on specific concrete cases. If we just look at frequently discussed passages like Ak 11:53 and A234/B287, where Kant's example is circles, we would never realize that Kant's immediacy condition is quite strong and quite surprising in other cases.) Moreover, I do not think that we can simply attribute to Kant a principled distinction between basic and complex concepts on the grounds that it is obviously the right thing to do and so Kant would have acknowledged it even if he did not do so explicitly. After all, neither Leibniz nor Wolff drew such a distinction – for them, properly formulated mathematical definitions of all concepts satisfy the immediacy condition – and Lambert would not have emphasized it so strongly if he thought it was obvious.

But I think that the argument that Kant held the immediacy condition even for complex concepts goes beyond marshalling passages. Kant's theory of mathematical proof seems to commit him to it. As I argued above, a proof that a concept can be constructed would have to be a mathematical proof, and so – on Kant's view – would already require a construction. A principled distinction between concepts like Lambert's would require a conception of mathematical proof like Lambert's, where we can prove (merely in words) what combinations of concepts are possible.

Still, though, I do believe that Kant's strictures on mathematical definitions are too strong and cannot be made to work in all cases. This is because Kant's position commits him to its being possible to suitably formulate definitions of geometrical concepts so that what were traditionally called 'problems' become obvious. Thus, although Kant clearly does not believe (as Wolff and Leibniz do) that *all* of the axioms and propositions of geometry are derivable from definitions alone, Kant is committed to some seemingly substantive propositions being immediately derivable from definitions. For instance, of the 48 propositions of *Elements* I, there are three propositions that assert that a complex concept is constructible: I.1: to construct an equilateral triangle on a given straight line; I.11: to draw a straight line at right angles to a given straight line from a given point on it; and I.31: to draw a straight line through a given point parallel to a given straight line. In the case of I.31, there was a definition of *parallel lines* well known in the eighteenth century that does satisfy the immediacy condition. Though there is no evidence that Kant was aware of this, Borelli defined parallel lines as:

> any two straights AC, BD, which toward the same parts stand at right angles to a certain straight AB. (Saccheri 1920, 89)

As Saccheri admits, this definition is 'set forth *by a state* (as he says) possible and most evident', since the definition in essence bids the geometer to take the constructive procedure for <perpendicular line>, which Euclid presents in I.11, and repeat it. As Kant would put it: one cannot think <straight line> without drawing straight lines in thought (in accord with Euclid's first two postulates); one cannot think <perpendicular lines> without erecting one straight line from a point of a second straight (in accord with Euclid I.11); and one cannot think Borelli's definition of *parallel lines* without erecting two perpendiculars on a given straight line. Each of the component concepts in Borelli's definition contains its own construction, and the construction procedure for the definition is patent from the way the component concepts are put together. What's more, the case of Borelli demonstrates very clearly that endorsing the immediacy condition even for complex concepts, as I think Kant does, was hardly unprecedented in the early modern period.

What of Euclid I.1?[37] In Euclid I.1, we construct an equilateral triangle on a given base AB by drawing a circle of radius AB centred on A, and then drawing a second circle of radius BA centred on B. An intersection point of the two circles, say C, forms the equilateral triangle ABC. Perhaps surprisingly, on Kant's view it is plausible that the possibility of this construction procedure is evident from thinking the definition of *equilateral triangles* as a trilateral figure whose three sides are all equal. In Section 2, we saw that Kant defended the reality of Euclid's definition of circle (in Kant's terms, a line in a plane every point of which is the same distance from a single one) on the grounds that a distance is just a line segment and one cannot think a line without drawing it. So to think of a line everywhere equidistant from a given point is just to rotate in thought a given line around that point. If Kant's story about circles is granted to him, then it is easily extended to equilateral triangles. To think of an equilateral triangle on the base AB is just to think of that base AB rotated around A and rotated around B until a plane figure is formed. If equidistant points from a given point can only be thought by describing a circle, then to think of three equidistant points is just to think of the intersection of two circles. And so the possibility of I.1 is immediately patent from the definition of *equilateral triangle*.

Even if this story about I.1 is granted, one might still be suspicious that it would work for every case in Euclid's *Elements*. I agree, but I do not think that Kant was committed to that. After all, Kant was perfectly willing to argue that Euclid's definition of parallel lines was illegitimate. He was certainly not alone in that assessment. Indeed, it was the near universal practice of Kant's contemporary geometers to rework Euclid's *Elements* fundamentally in an attempt to get the definitions, axioms, postulates and proofs that would satisfy some ideal of the proper mathematical method. A defence of a certain philosophy of geometry – whether it was given by Kästner, Schultz, Lambert, Leibniz or Wolff – was a normative claim about how geometry should be done, and an expression of a reform campaign for reworking geometry, including its definitions. The point was not to match Euclid exactly – it was to improve the foundations of geometry. And we should expect that many of these reform programmes, despite being philosophically attractive, were ultimately unworkable. (This was certainly the case for Leibniz's and Wolff's philosophies of geometry.)

To confess my own view: I think that the suspicion that a Kantian reconstruction of propositions like *Elements* I.1 is trying to pull a rabbit out of a hat is rooted in a deep problem with Kant's account. The claim that the possibility of a construction procedure is immediately certain from a definition is only as clear as the notion of *immediacy* itself. For Wolff, there is a clear, non-vague notion of immediacy: a proposition is an immediate consequence of a definition if it can be derived from that definition in one syllogism. When Kant rejects the Wolffian account and asserts that mathematical knowledge is rational knowledge from the construction of concepts, he is thereby committed to a new kind of immediacy, *intuitive immediacy*. This is the sense in which a mathematical axiom is supposed to be immediately certain for Kant, and the sense in which a construction procedure is supposed to be immediately certain given the correctly formulated definition. But I worry that this notion of immediacy is vague, and it is this vagueness that makes itself felt when looking at particular cases. Of course, in this respect Kant was no worse than Lambert, who has more or less no positive story of the immediate certainty of axioms and postulates, and Wolff, whose criterion of immediacy is clear but entirely unworkable. But these are large issues and extend far outside the scope of this paper.

## Notes

1.  Of the many passages where Kant aligns himself with Euclid, see for instance, Reflexion 11, where Kant criticizes Wolff for not conducting his proof of Euclid I.29 'in the Euclidean way' (Reflexion 11, 14:52). Wolff aligns himself with Euclid (against his early modern critics, like Ramus) at *Preliminary Discourse* (§131) (in Wolff 1740). Lambert aligns himself with Euclid (against Wolff, whom he thinks distorts Euclid) in Lambert (1895a, 1915, §§78–§§82). Of course, none of these thinkers hold Euclid above criticism: each rejects Euclid's axiom of parallels, for example, and Wolff, Leibniz and Kant all reject Euclid's definition of <parallel lines>.

    Works by Wolff and Lambert will be cited by paragraph number ('§'). Citations of works of Kant besides the *Critique of Pure Reason* are according to the German Academy ('Ak') edition pagination in Kant (1902). I also cite Kant's reflexions by number. For the *Critique of Pure Reason*, I follow the common practice of citing the original page numbers in the first ('A') or second ('B') edition of 1781 and 1787. Passages from Kant's *JL* (edited by Kant's student Jäsche and published under Kant's name in 1800) are also cited by paragraph number ('§'). I use the translation in Kant (1992) both for *JL* and for Kant's other logic lectures. Passages from Kant's correspondence are cited by Ak page number and by date; translations are from Kant (1967).

    On Euclid's influence on Kant, see Shabel (2003) and Friedman (1992); on Lambert's self-conscious Euclideanness, see Laywine (2001, 2010) and Dunlop (2009).
2.  There are notable exceptions: on Kant's notes on parallels, see Adickes (1911) and Webb (2006); on Leibniz, see De Risi (2007); on Adickes's and Webb's readings of Kant, see note 17.
3.  I discuss Kant's theory of real definitions also in Heis (forthcoming, §I). Two paragraphs in this section – and one paragraph from §2 – appear also slightly modified in that paper.
4.  I follow the common practice of referring to concepts in angled brackets and words in quotation marks.
5.  Kant is speaking loosely when he says that one cannot *think* a circle without describing it. Indeed, he is at pains elsewhere to insist that it is possible to think of a mathematical concept without carrying out its construction, as for instance a philosopher would if she were trying to prove Euclid I.32 in a merely discursive way: A716/B744. Kant is being more precise when, in a passage parallel to B154, he says: 'in order to *cognize* something in space, e.g., a line, I must *draw* it' (B138). Kant's point is that one cannot employ the concept <circle> in the uniquely mathematical way – that is, in the way that leads to rational *cognition* – without carrying out the construction and describing a circle.
6.  Kant characterizes 'real definitions' in multiple ways, all of which turn out to be equivalent. In the first *Critique* (A241-2), Kant says real definitions 'make distinct

[the concept's] objective reality', which is equivalent to the 'real possibility' of the concept (see e.g. A220/B268). Elsewhere, he says that real definitions contain the 'essence of the thing' (Vienna Logic, Ak 24:918), which is equivalent to the 'first inner principle of all that belongs to the possibility of a thing' (Kant 2004, Ak 4:467, 2002, Ak 8:229). That mathematical definitions are real, see A242; Blomberg Logic, Ak 24:268; Dohna-Wundlacken Logic, Ak 24:760; *JL* §106, note 2, Ak 9:144; Reflexion 3000, Ak 16:609.

7. On genetic definitions, see also Reflexion 3001, Ak 16:609. That mathematical definitions are genetic: Reflexion 3002, Ak 16:609.

8. Dunlop (2012) makes the same point in her discussion of real definitions in geometry.

9. 26 May 1789, Letter to Herz, Ak 11:53. I quote and discuss this letter in Section 2. In the letter, Kant is referring specifically to the concept <circle>. I take Kant to be further committed to the immediacy condition for all geometrical definitions, not just for geometrical definitions of basic concepts such as <circle>. This further commitment is quite strong, and my reading will require some defense. I return to this issue in the closing section of the paper.

10. On postulates, see also Reflexion 3133, Ak 16:673; Heschel Logic 87 (Kant 1992, 381); 'Über Kästner's Abhandlungen', Ak 20:410-1; Letter to Schultz, 25 November 1788, Ak 10:556.

11. A24, A47/B65, A163/B204, A239-40/B299, A300/B356. Though Kant does not say explicitly in *JL* §35 that axioms are theoretical and never practical, I believe that this was his view. See Heis (forthcoming, §3).

12. By 'basic', I mean those concepts whose definitions have postulates as corollaries. Definitions of complex concepts that are composed from basic concepts are genetic, but their corollaries are not postulates, but problems. The concept <triangle> is complex in this sense, since it is defined as 'a figure enclosed in three straight lines' (Wolff 1716, 1417) and is therefore composed from the basic concept <straight line>.

13. For similar definitions of *circle*, see Wolff (1741, Geometriae, §37, 1710, Introduction §4, 1710, Geometrie, Definition 5). That all radii in a circle are equal is Axiom 3 in Wolff (1710).

14. It is obvious that Leibniz, Wolff and Lambert each rejected Euclid's axiom because each tried to prove it. The argument that Kant also would have rejected Euclid's axiom would require a fuller defense, which I hope to present on another occasion.

15. For a more extended interpretation and discussion of these notes, see (Heis forthcoming).

16. 'Rectae, quae se invicem ubique habent eodem modo' (Leibniz 1858, 201, Definition XXXIV, §3). This definition actually is more abstract than the definition suggested in *New Essays*. Literally, it says that parallel lines are 'straight lines that everywhere have the same situations with respect to one another'. In §9, he notes that this is equivalent to saying that parallels are equidistant straight lines, but declines to use the definition in terms of distance because he does not yet have a definition of the minimal curve from a straight line to a straight line. But we can ignore that subtlety in this paper, as interesting as it may be.

Leibniz almost surely got the idea of defining parallelism in terms of equidistance from Clavius (whose edition of Euclid Leibniz used) and Borelli, whose work he studied closely (see De Risi 2007, 80). On Clavius and Borelli, see Saccheri (1920, 87–91) and Heath's commentary in Euclid (1925, 194).

17. My reading of these notes differs from that proposed by Adickes, who does not see Kant criticizing Euclid's definition (Ak 14:31), and that proposed by Webb (2006,

230–232), who reads Kant as endorsing a 'proof' of Euclid's axiom using Leibniz's definition. For a fuller defence of my reading, see Heis (forthcoming).

18. On real definitions, see Leibniz (1996, III.iii.15, 19); 'Meditations on Knowledge, Truth, and Ideas' (Leibniz 1989, 25–26); 'On Universal Synthesis' (Leibniz 1970, 230–231); Letter to Tschirnhaus (Leibniz 1970, 194); *Discourse on Metaphysics* (Leibniz 1989, §24).

   In this paper, I feel free to cite passages from Leibnizian works that were unknown in the eighteenth century. Although this could be dangerous in contextual histories like this one, in this paper, it will do no damage. All of the doctrines I am ascribing to Leibniz are clearly expressed in *New Essays* and 'Meditations' – both works that Kant and his contemporaries knew very well.

19. *Discourse on Metaphysics* (Leibniz 1989, §24). See also Leibniz (1996, III.iii.18) and *Discourse* (Leibniz 1989, §26, 1970, 230).

20. On the following page, Leibniz provides an additional consideration, not found in Kant: that the same concept can have two real definitions, as – for example – an ellipse can be generated either by sectioning a cone or tracing a curve with a thread whose ends are fixed on the foci. He suggests, however, that there will still be one unique most perfect real definition.

21. See Leibniz (1996, IV.xii.6), supplement to a Letter to Huygens (Leibniz 1970, 250), 'On Contingency' (Leibniz 1989, 28) and 'On Freedom' (Leibniz 1989, 96).

22. I think this point helps explain another fundamental difference between Kant's and Leibniz's notion of real definition. Leibniz (1996, III.iii.18; 1989, 26) allows for *a posteriori* real definitions, since we can know, through experiencing an actual instance, that a concept is possible (*Discourse on Metaphysics* [Leibniz 1989, §24]). But Kant strongly denies that there can be real definitions of empirical concepts, partly because we can never be sure through experience that we have successfully identified marks that will pick out *all* instances of a concept (A727-8/B755-6).

23. A similar complaint against Wolff is levelled by Lambert in 1771 (§11) and in 3 February 1766, Letter to Kant, Ak 10:64.

24. My understanding of Lambert owes much to Laywine (2001, 2010) and Dunlop (2009).

25. On necessity of simple concepts, see Lambert (1764, §653–§654, 1915, §36). Lambert alleges that Wolff did not fully appreciate the role of simple concepts (1771, §11–§18, 1915, §26). For Lambert, Leibniz was more cognizant of the importance of simples than Wolff. Still, though, he lacked a sure criterion for distinguishing simples from compounds, and lacked principles that would license (or preclude) simple combinations of simples (Lambert, 1771, §7–§8). From Lambert's point of view, it is striking that Leibniz (1989, 26) will claim that 'the possibility of a thing is known *a priori* when we resolve it into its requisites, that is, into other notions known to be possible, and we know that there is nothing incompatible among them' (see also *Discourse* [Leibniz 1989, §24]). But how do we know when the resolution has reached 'simple, primitive notions understood in themselves' (Leibniz 1970, 231)? And how would we know that there is nothing incompatible among these simples? Certainly not by their definitions.

26. I do not think, though, that Kant would have agreed with Wolff that axioms are corollaries of definitions. Kant's position is therefore intermediate between Leibniz's and Lambert's. I cannot, however, defend this reading here. See Heis (forthcoming, §3).

27. See Lambert (1764, §650): 'the composition of individual marks is a means of attaining concepts and one can proceed arbitrarily insofar as the possibility of such a concept can be proven *later* (§65ff.). Now as long as the possibility has *not yet* been proved, the concept remains hypothetical' (my emphasis). Lambert says that these

hypothetical concepts (which presumably have only nominal definitions) can later become derived concepts [Lehrbegriffe] (which would then have real definitions).

28. That Euclid's definition can be proved without the axiom of parallels (Lambert 1895b, §8); that Euclid's definition is real (Lambert 1915, §81, 1895a, §7–§8, 1895b, §3, §7, §10) and that Euclid's definition is therefore preferable (Lambert 1915, §79, 1895b, §4–§10, 1771, §12, §23–§24).

29. These are Lambert's examples from (1764, part 1, §63).

30. 'The proof of a derivative concept *a priori* depends on its mode of genesis from axiomatic concepts. An a posteriori proof, however, depends on the mode of genesis of the thing' (Lambert 1915, §92.9–§10). A derivative concept is a concept whose possibility has been proved from the axioms and postulates. This proof is a priori in the sense that it is a logical proof – done purely syllogistically – from what is conceptually prior: axioms and postulates that exhibit the immediately certain (im) possibilities of combining simples.

   Lambert's claim that genetic definitions are appropriate only for a posteriori concepts gives an interesting set of contrasting positions. For Wolff, both empirical and mathematical concepts can have genetic definitions; for Kant, only mathematical concepts do; for Lambert, only empirical concepts do.

31. On <equilateral triangle>, see Lambert (1771, §20, 1915, §79, 1895a, #4); on <triangle>, see Lambert (1915, §79, 1895a, #6); on <parallel lines>, see Lambert (1895b, §3, 1895a, #7).

32. According to Webb (2006, 219), Kant's view 'fits Lambert like a glove'. A similar comment (including speculation about the role of Lambert's 3 February 1766 letter to Kant in shaping Kant's thinking) appears in Hintikka (1969, 43–44).

33. For a penetrating discussion of Lambert's argument, see Dunlop (2009, §5).

34. As an anonymous referee pointed out, it is noteworthy that in the passage quoted from Lambert (1771, §12), Lambert attributes the generality of the figure to the *postulates*, and does not mention the axioms. I have been claiming that for Kant, an instance of the concept <parallel lines> constructed according to Euclid I.31 would be general in Kant's sense, only if we are justified in believing Euclid's *axiom* of parallels. This difference is, I believe, revealing. For Lambert, postulates assert that two or more simple concepts <A> ... <N> can be combined (i.e. that there are instances of <A and ... and N>), while an axiom asserts that two or more simple concepts cannot be combined (i.e. there are no instances of <A and ... and N>). If all that matters for making a definition real is that there provably are instances of the definiendum (as I claim is the case for Lambert), then in general the postulates will be sufficient to show that a definition is real. Axioms, on the other hand, would in general be necessary to show that there are no instances of a concept other than those that meet some condition. For example, we would need axioms to show that though there are instances of the concept <non-intersecting coplanar straights>, the concept <cut by a transversal making interior angles less than two rights> cannot be combined with <non-intersecting coplanar straights>. Postulates would be in general sufficient to show that *all* straights constructed according to the procedure described in I.31 are parallels. Axioms would be necessary to show that *all and only* straights constructed in that way are parallel. Lambert's real definitions need the first condition. Kant's real definitions require the second.

35. See also Lambert (1895b, §11), where he compares giving a proof with solving an algebraic equation, and denies that the 'representation of the thing' can play any role in the proof.

36. That the possibility of only derivative concepts requires proof, whereas axiomatic concepts do not (Lambert 1915, §45, 57, 66, 92.9, 1764, §652). That the possibility

of concepts such as <straight line> does not need to be demonstrated (Lambert 1918, §21).

37. Thanks to Alison Laywine for urging me to make sense of I.1 on my interpretation.

## References

Adickes, Erich, ed. 1911. *Kants Handschriftlicher Nachlass. Mathematik – Physik und Chemie – Physische Geographie*. Volume 14 of Kant, Immanuel. *Gesammelte Schriften*. Berlin: Walter de Gruyter.

Carson, Emily. 1999. "Kant on the method of mathematics." *Journal of the History of Philosophy* 37: 629–652.

Carson, Emily. 2006. "Locke and Kant on mathematical knowledge." In *Intuition and the Axiomatic Method*, edited by Emily Carson and Renate Huber, 3–21. Dordrecht: Springer.

De Risi, Vincenzo. 2007. *Geometry and Monadology*. Basel: Birkhäuser.

Dunlop, Katherine. 2009. "Why Euclid's Geometry Brooked No Doubt: J.H. Lambert on Certainty and the Existence of Models." *Synthese* 167: 33–65.

Dunlop, Katherine. 2012. "Kant and Strawson on the Content of Geometrical Concepts." *Noûs* 46: 86–126.

Euclid. 1925. *The Thirteen Books of Euclid's Elements, Translated from the Text of Heiberg, with Introduction and Commentary*. 2nd ed. Translated and edited by Sir Thomas Heath. Cambridge: Cambridge University Press.

Freudenthal, Gideon. 2006. "Definition and Construction: Maimon's Philosophy of Geometry." . Preprint 317 of the Max Planck Institute for the History of Science, Berlin. http://www.mpiwg-berlin.mpg.de/Preprints/P317.PDF

Friedman, Michael. 1992. *Kant and the Exact Sciences*. Cambridge, MA: Harvard University Press.

Friedman, Michael. 2010. "Synthetic History Reconsidered." In *Discourse on a New Method*, edited by Michael Dickson and Mary Domski, 571–813. Chicago: Open Court.

Heis, Jeremy. Forthcoming. "Kant on Parallel Lines: Definitions, Postulates, and Axioms." In *Kant's Philosophy of Mathematics: Modern Essays. Vol. 1: The Critical Philosophy and Its Background*, edited by Ofra Rechter and Carl Posy. Cambridge: Cambridge University Press.

Hintikka, Hans. 1969. "On Kant's Notion of Intuition (Anschauung)." In *The First Critique*, edited by T. Penelhum and J.J. Macintosh, 38–53. Belmont, CA: Wadsworth.

Kant, Immanuel. 1902. *Gesammelte Schriften*. Edited by the Königlich Preußischen Akademie der Wissenschaft. 29 vols. Berlin: DeGruyter.

Kant, Immanuel. 1967. *Philosophical Correspondence, 1759–99*. Edited and translated by Arnulf Zweig. Chicago, IL: University of Chicago Press.

Kant, Immanuel. 1992. *Lectures on Logic*. Translated and edited by J. Michael Young. Cambridge: Cambridge University Press.

Kant, Immanuel. 1998. *Critique of Pure Reason*. Translated by Paul Guyer and Allen Wood. Cambridge: Cambridge University Press.

Kant, Immanuel. 2002. "On a Discovery Whereby Any New Critique of Pure Reason is to be Made Superfluous by an Earlier One." In *Theoretical Philosophy After 1781*. Translated by Henry Allison. Cambridge: Cambridge University Press.

Kant, Immanuel. 2004. *Metaphysical Foundations of Natural Science*. Translated and edited by Michael Friedman. Cambridge: Cambridge University Press.

Lambert, Johann Heinrich. 1764. *Neues Organon. Band I*. Leipzig: Wendler. Partially translated by Eric Watkins in *Kant's Critique of Pure Reason: Background Source Materials*. Cambridge: CUP, 2009, 257-274.

Lambert, Johann Heinrich. 1771. *Anlage zur Architectonic*. Vol.1. Riga: Hartknock.

Lambert, Johann Heinrich. 1895a. "Letter to Holland." In *Die Theorie der Parallellinien von Euklid bis auf Gauss*, edited by Friedrich Engel and Paul Stäckel, 141–142. Leipzig: Teubner.

Lambert, Johann Heinrich. 1895b. "Theorie der Parallellinien." In *Die Theorie der Parallellinien von Euklid bis auf Gauss*, edited by Friedrich Engel and Paul Stäckel, 152–207. Leipzig: Teubner. Partially translated by William Ewald in *From Kant to Hilbert*. Vol. 1. Oxford: Clarendon Press, 158–167.

Lambert, Johann Heinrich. 1915. "Abhandlung vom Criterium Veritatis." *Kant-Studien. Ergänzungsheft* 36: 7–64. Partially translated by Erick Watkins in *Kant's Critique of Pure Reason: Background Source Materials* (Cambridge: CUP, 2009), 233–257.

Lambert, Johann Heinrich. 1918. "Über die Methode, die Metaphysik, Theologie und Moral richtiger zu beweisen." *Kant-Studien. Ergänzungsheft* 42: 7–36.

Laywine, Alison. 1998. "Problems and Postulates: Kant on Reason and Understanding." *Journal of the History of Philosophy* 36: 279–309.

Laywine, Allison. 2001. "Kant in Reply to Lambert on the Ancestry of Metaphysical Concepts." *Kantian Review* 5: 1–48.

Laywine, Alison. 2010. "Kant and Lambert on Geometrical Postulates in the Reform of Metaphysics." In *Discourse on a New Method*, edited by Michael Dickson and Mary Domski, 113–133. Chicago: Open Court.

Leibniz, Gottfried Wilhelm. 1858. "In Euclidis πρῶτα." In *Leibnizens mathematischen Schriften*, edited by C. I. Gerhardt. Vol. V, 183–211. Halle: Schmidt.

Leibniz, Gottfried Wilhelm. 1970. *Philosophical Papers and Letters*. 2nd ed.. Translated by L.E. Loemker. Dordrecht: Springer.

Leibniz, Gottfried Wilhelm. 1989. *Philosophical Essays*. Translated and edited by Roger Ariew and Daniel Garber. Indianapolis, IN: Hackett.

Leibniz, Gottfried Wilhelm. 1996. *New Essays on Human Understanding*. Edited and translated by Peter Remnant and Jonathan Bennett. Cambridge: Cambridge University Press.

Saccheri, Girolamo. 1920. *Euclides Vindicatus*. Translated by Bruce Halsted. Chicago, IL: Open Court.

Schultz, Johann. 1784. *Entdeckte Theorie der Parallelen*. Königsberg: Kanter.

Shabel, Lisa. 2003. *Mathematics in Kant's Critical Philosophy*. New York: Routledge.

Sutherland, Daniel. 2005. "Kant on Fundamental Geometrical Relations." *Archiv für Geschichte der Philosophie* 87: 117–158.

Sutherland, Daniel. 2010. "Philosophy, Geometry, and Logic in Leibniz, Wolff, and the Early Kant." In *Discourse on a New Method*, edited by Michael Dickson and Mary Domski, 155–192. Chicago: Open Court.

Webb, Judson. 2006. "Hintikka on Aristotelean Cosntructions, Kantian Intuitions, and Peircean Theorems." In *The Philosophy of Jaakko Hintikka*, edited by Auxier and Hahn, 195–265. Chicago, IL: Open Court.

Wolff, Christian. 1710. *Der Anfangs-Gründe aller mathematischen Wissenschaften*. Halle. Reprinted in Wolff, 1962, *Gesammelte Werke*, I.12. Hildesheim: Olms.

Wolff, Christian. 1716. *Mathematisches Lexicon*. Leipzig. Reprinted in Wolff, 1962, Gesammelte Werke, I.11. Hildesheim: Olms.

Wolff, Christian. 1740. *Philosophia Rationalis sive Logica*. Frankfurt and Leipzig. Reprinted in Wolff, 1962, *Gesammelte Werke*, II.1. Hildesheim: Olms. Partially translated by Richard Blackwell as *Preliminary Discourse on Philosophy in General*. New York: Merrill, 1963.

Wolff, Christian. 1741. *Elementa Matheseos Universae*. Halle. Reprinted in Wolff, 1962, *Gesammelte Werke*, II.29. Hildesheim: Olms.

# Definitions of Kant's categories

Tyke Nunez

*Philosophy Department, University of Pittsburgh, Pittsburgh, PA, USA*

The consensus view in the literature is that, according to Kant, definitions in philosophy are impossible. While this is true prior to the advent of transcendental philosophy, I argue that with Kant's Copernican Turn definitions of some philosophical concepts, the categories become possible. Along the way I discuss issues like why Kant introduces the 'Analytic of Concepts' as an analysis of the understanding, how this faculty, as the faculty for judging, provides the principle for the complete exhibition of the categories, how the pure categories relate to the schematized categories, and how the latter can be used on empirical objects.

According to Kant, mathematics offers paradigmatic cases of definitions and starts its inquiries with them. In philosophy, definitions will come at the end of the inquiry, not at the beginning (*KrV*, A730/B757-A731/B758), if they are possible at all. The consensus view in the literature seems to be that Kant thinks definitions in philosophy are impossible.[1] While this is true prior to the advent of Kant's transcendental philosophy, I argue that with his 'Copernican Turn' Kant maintains that we can define some philosophical concepts, the categories, which are the most general concepts of an object.[2] Seeing this will both offer a clue as to why Kant thinks the categories form a complete system that can guide us in the material sciences and clarify how Kant thinks he is proceeding from first principles in the 'Transcendental Analytic', the part of the *First Critique* that does the most work to lay out his positive view.

To begin with §1, I will clarify exactly the nature of the dispute over definitions. Next (§2), I will explicate the requirements on, and kinds of, strict definitions. I will then (§3) look at the main evidence that according to Kant definitions of philosophical concepts are impossible. In §4, I argue that Kant *should* maintain definitions of the categories that are possible because of his deployment of the synthetic method in his analysis of our faculty for judgement. Next (§5), I discuss a few passages where I think Kant is *in fact* sketching definitions of the pure and schematized categories from the end of the 'Transcendental Analytic' of the *First Critique*. In §6, I address an objection to the account that will clarify how the definitions of the schematized categories get their reality.[3]

## 1. The standard view

Definitions in the strict sense, according to Kant, are exhaustive, precise, and original exhibitions of a concept of a thing. I will return to exposit each of these three marks of strict definitions, but before this we should specify what is in dispute. Kant uses the term '*Erklärung*', which gets translated as 'definition' or 'explanation', in a number of different senses and these include a range of possible explanations that may fall short of definitions in the strict sense. This is because 'the German language has for the [Latinate] expressions *exposition, explication, declaration* and *definition* nothing more than the one word "explanation" [*Erklärung*]' (*KrV*, A730/B758).[4] In this essay, I will use the term 'definition' to refer to definitions in Kant's strict sense, whereas I will use 'explanation' to encompass the wider sense of '*Erklärung*'.

What I will be concerned with is definitions of the categories in the strict sense. Although other commentators have allowed that a kind of explanation, an 'exposition', of the categories is possible, they have denied the possibility of defining them strictly. Expositions articulate the other concepts or marks that are thought in the concept. One example is 'space is not an empirical concept' (*KrV*, A23/B38).[5] Expositions differ from strict definitions in that they are cautious about their exhaustiveness (*JL*, §102, §105). And in our example it is clear that only part of what is included in this concept is presented.[6] In contrast with the standard view, I argue that on Kant's account there is a way in which we can achieve strict definitions of the categories, which unlike mere expositions, will be exhaustive.

## 2. Definitions in general

In the *First Critique* Kant characterizes definitions in the following way: 'As the expression itself reveals, to define properly means just to exhibit originally the exhaustive concept of a thing within its boundaries*' (*KrV*, B755). He then expands on this in the footnote:

> * *Exhaustiveness* signifies the clarity and sufficiency of marks; *boundaries*, the precision, that is, that there are no more of these than belong to the exhaustive concept; *original*, however, that this boundary-determination is not derived from anywhere else and thus in need of proof, which would make the supposed definition [*Erklärung*] incapable of standing at the head of all judgments about an object. (*KrV*, B755n)

In this section I will give an exposition of what Kant means by exhaustivity, completeness, and originality, as well as his distinction between nominal and real definitions.

Since Frege, we have largely grown accustomed to extensional adequacy as the mark of sufficiency in definitions. This is only one small component of definitions on Kant's account. In general, however, the first thing to note when considering what Kant means by these terms is that they each concern not so much the extension (*Umfang* or *Sphäre*) of the concept, which includes the concepts and objects *under* it, as the content or intension (*Inhalt*) of the concept, the other concepts contained *in* it.

To exhaustively or completely exhibit the marks in a concept's intension is to exhibit all of its marks. For example, the explanation 'a line whose points are all equidistant from a centre point' gives all the content of the concept *circle* because together the genus *line* and specific difference *whose points are equidistant from a centre point* give all of the marks needed to identify any instance of a circle. Falling short of having all of the marks contained in the concept will mean, however, that the exposition cannot be exchanged for the original concept without a loss in content. For example, *yellow metal* has less content than *gold* – fewer marks in its intension. This means that *yellow metal* is less discriminating, it rules out fewer instances than the richer concept *gold*.

To exhibit the concept within its boundaries is to do so precisely – to not include more marks in the exhibition than belong to the concept's intension. For example, 'a line whose points are all equidistant from a centre point' gives you all of the marks you need to identify circles and no more. Uncontroversially, the addition of *yellow*, as in 'a yellow line whose points are equidistant from a centre point', would make the definition imprecise and damage the concept by excluding instances to which the concept legitimately applies. There is a bit of an ambiguity as to whether precision also excludes attributes of the thing, which are marks that it necessarily has. For example, if instead we were to say 'a curved line ... ', we would add an extra mark, *curved*, which, although it may make applying the concept easier, does not belong to the precise definition of *circle*. Considering the wording of the footnote at A727/B755 on its own, such marks do 'belong to' the exhaustive concept, and so would seem not to be excluded from the precise concept.[7] Of course, a few pages later Kant does explicitly exclude 'curved' from the precise definition of circle (*KrV*, A732/B760), and this 'suggests' that 'precise' in the footnote should be read as excluding them as well.[8]

The final and most difficult condition on an explanation's being a definition is that it being original – that the boundary-determination delimiting the precise concept 'is not derived from anywhere else and thus in need of proof, which would make the supposed definition incapable of standing at the head of all judgments about an object' (*KrV*, B755). The explanation 'a line all of whose points are equidistant from a centre point' is original in that the exhaustive, precise concept of a circle is derived from nowhere else besides it. If it were, then the definition would not be capable of standing at the head of all demonstrations about circles, and so would fail to give us what we want out of such a definition.

Understanding what this amounts to concretely is tricky. If we do not read the footnote at A727/B755 as excluding attributes (like 'curved') from the precise concept, then a clear role for originality remains in determining the boundaries of the concept specified by the definition. For, it would then be originality that ensures only the marks essential to the concept – those which are primitively constitutive of it – are included in its definition, and that other marks, like 'curved', which should be derived from the definition in further theorems, are excluded.

If we read precision as excluding attributes, finding a role for originality is more difficult. For contrast, we can return to our exposition of the concept *gold* as a yellow metal. Gold is a substance that we have discovered. We have not arbitrarily made its concept, but have developed it through the experience of gold, which has included its being a yellow metal. In explaining *gold* as a yellow metal we learn that anything true of all metal or yellow things will be true of it. So we learn something about *gold* that can stand at the head of some judgements about gold. But the collection of judgements that can be made about gold on the basis of it being yellow and metal far from exhaust the judgements that can be made about it, and so we can tell that we do not here have a proper definition. For this we would need to be able to show that the boundary determination of the concept *gold* is derived from exactly where the definition claims and nowhere else. But the explanation *yellow metal* does not make any claims about the boundaries of the concept *gold*, so it could not be original.[9]

According to Kant, however, *circle* is an invented concept that we arbitrarily or electively (*willkürlich*) make. In making this concept we specify the marks something must have in order for it to be a circle – it must be a line; its points must be equidistant from a centre point. The key to how our definition of *circle* can be original is that this explanation of the concept includes a specification of how to make instances of it. For, in describing circles as lines whose points are equidistant from a centre point, this definition gives us a procedure for drawing circles – draw a line around a point, keeping the distance between the line and the point constant. With this procedure we have a rule for applying the concept: anything that we can trace in thought in this way will be circular. This, in turn, ensures the concept will apply to all and only circular things, and thus that the concept's use is exactly and securely what we take it to be.

Accordingly, it is through this procedure for creating circles that the precise, complete boundaries of the concept are secured, and originality for the explanation is achieved.

There is a final objection to our definition of circle that is worth considering. Because the procedure for constructing circles depends on space, an understanding of which is not explicitly articulated in the definition, one might think that our definition is flawed. This dependence is apparent in the conditions on the synthesis of the concept, the action of the imagination in drawing circles. All spatial objects must comply with these conditions on pain of their unreality *qua* spatial object. (For example, the concept *two-sided plane figure* or *round square* are not contradictory concepts, but are also not constructible.) For this reason, this definition may not seem to stand at the head of all judgements about *circles*, without a supplementary articulation of the concept of space and its correlative conditions.[10]

Although we study space in geometry, learning about it through constructing figures, we do not learn about the relation between our concepts of spatial figures and space in geometry, but presuppose it.[11] A thorough investigation of this is beyond what I can offer here. Briefly, however, the general point is that some topics relevant to a science are not themselves treated in the science. For example, the status of mathematics as cognition depends on its relation to appearances (cf. e.g. *KrV*, B147, A157/B196), and this is not studied in mathematics, but philosophy.[12] Furthermore, as regards the specific issue of the relation between geometrical concepts and space, I think it is this kind of relation between space and geometry or its concepts that is at least part of what Kant sets out to establish in his 'Transcendental exposition of the concept of space' (*KrV*, B40–41).[13] This exposition lies outside of geometry and concerns its possibility; it is not a topic for geometry itself.[14] In this way, if we can agree that according to Kant geometry does not treat the nature of the dependence of its concepts on space, then perhaps we can begin to see why no supplementary articulation of the nature of this dependence is necessary for adequate definitions of geometrical concepts.[15]

Kant divides proper complete, precise, and original definitions into merely nominal (or logical)[16] and real definitions, which also secure the application of the concept to objects. A nominal definition of a concept will consist in giving a genus under which the concept stands, along with a specific difference (or *differentia*). The specific difference distinguishes the concept from the other concepts in the community of concepts that both divide the genus and together, taken as a whole, make it up. Nominal definitions are useful for comparing things thought under their definiendum to other things in part because they place the definiendum in a porphyrian tree. One paradigmatic case is 'a human being is a rational animal'.

What is needed for the definition to be real, beyond getting the definition of the concept right as a nominal definition will, is that it also contains a mark, a schema, that makes the concept securely usable in application to objects (cf. *KrV*, A241-2n; *KrV*, B302-3n). This application is secured in one of (at least) three ways, but in each the definition will indicate the nature of the thing, its real essence, which is the ground of its possibility. Paradigmatic cases of real definitions can be found in mathematics: e.g. 'a triangle is a three sided figure'.

In addition to counting as a proper nominal definition, the first grade of real definition will also provide 'a clear *mark* by means of which the *object* (*definitum*) can always be securely cognized, and that makes the concept that is to be explained usable in application' (*KrV*, A241n). In his notes, Kant sometimes calls such definitions which do not generate their objects 'diagnostic' (*diagnostisch/dianoëtisch*).[17] Kant often uses the term 'definition' in a sense that encompasses this grade of real definition (especially in the *First Critique*). This kind of definition grants insight into the real formal possibility of things – their formal real essence – by indicating an element in their discursive form: a discursive mark by means of which their object can always be securely cognized. The possible definitions of the schematized categories will be of this kind.

On the second, stricter grade of real definition, the concept must provide for the generation of its object. 'Mathematical definitions, which exhibit the object in accordance with the concept in *intuition*' are examples (*KrV*, A242n). Usually, Kant uses this generative feature of mathematical concepts to explain why they can be defined so successfully – why they can surely contain neither more nor fewer marks than belong to the concept, and why the explanation of the concept is derived from nowhere else, except the definition (*KrV*, A730/B758) – but he does not intend the meaning of the term 'definition' to include it. This kind of definition grants insight into the formal possibility of things by indicating not only a discursive mark but also a rule for generating intuitive forms, through which their objects can be securely cognized.

Generated mathematical objects, however, are not objects properly speaking – e.g. objects of experience, God – but are merely *a priori* forms of objects (*KrV*, A224/B271). Objects of experience, rather, are the objects that are in fact the ultimate source of meaning for the concepts of mathematics, since without this connection these concepts would be entirely devoid of sense (*KrV*, B299). Accordingly, if there were a cognition that generated its corresponding object (the one to which its sense and significance can ultimately be traced) and not merely a form of this object, then it would be properly understood to attain a further third grade of reality. Such a cognition would comprehend its object, since it would be self-sufficient, generating the object it knows, and our definitions cannot attain this degree of reality, even in mathematics (*JL*, 9:65). If they could, then we could know things as the intuitive intellect does, and we would be able to give not only formal but also material real definitions of objects, through which we would have insight into the real natures of things as they are in themselves.

## 3.  Against definitions of the categories

What Kant says about the possibility of strict definitions in philosophy may seem contradictory, since, at a number of places even in his published writings, he seems to say they are possible, only to say a page or two later that they are not.[18] The most seemingly damning argument against the possibility of properly defining the *a priori* given concepts of philosophy, which all prior commentators take to be decisive, runs:

> Strictly speaking no concept given *a priori* can be defined e.g., substance, cause, right, equity, etc. For I can never be certain that the distinct representation of a (still confused) given concept has been exhaustively developed unless I know that it is adequate to the object. But since the concept of the latter, as it is given, can contain many obscure representations, which we pass by in our analysis though we always use them in application, the exhaustiveness of the analysis of my concept is always doubtful, and by many appropriate examples can only be made *probably* but never *apodictically* certain. (*KrV*, A728/B756-A729/B757)

There are two factors, either of which alone would seem to rule out the possibility of defining any of the *a priori* given concepts of philosophy, even the categories.

First, unlike mathematics whose proper objects are in a sense merely forms of objects produced in *a priori* intuition, the proper objects of the categories are objects of experience and these cannot be defined. Second, the investigation of *a priori* given concepts proceeds by way of analysis, which is a procedure that usually has no guarantee of completeness.

Regarding the first point, I concede that there are two senses in which the categories cannot be defined. First, of course, at least as long as they are considered theoretically, their definitions will not generate objects of experience. Second, the categories 'can be exhibited *in concreto* if one applies them to appearances; for in the latter they have the proper material [*den Stoff*] for a concept of experience [*Erfahrungsbegriff*], which is nothing but a concept of the understanding it *in concreto*' (*KrV*, B595). Considered *in concreto*, as concepts of experience (e.g. as quite general empirical features of bodies), the categories also cannot be defined (*KrV*, A728/B756).[19] Nonetheless, considered as *a priori* given concepts of a natural object in transcendental metaphysics (or the metaphysics of extended matter), I will argue they can be. I will return to a sophisticated form of this objection in §6.

As regards the second point, traditionally when philosophers proceed securely (in a manner ensuring objective validity) they analyse given representations. Accordingly, these analyses are always in danger of having overlooked some feature and in any case are not original, since they do not secure the boundaries of the exposited concept. What will be crucial for responding to this point is seeing exactly how Kant's Copernican use of the synthetic method makes possible a different means of investigating the concepts of philosophy that, at least for some of them, unlike their mere analysis, will secure their boundaries through their original exposition. I will sketch how Kant thinks this is possible in the next section.

In the supposedly decisive passage, I propose that Kant has in mind not so much transcendental, as traditional, philosophy. Traditional philosophy does not have any means for securing the boundaries of its expositions. If, however, there were somehow a way to form legitimate definitions in philosophy that was not analytical, then perhaps the completeness of these expositions could be secured.

## 4. The possibility of definitions of the categories

In this section, I argue that Kant should maintain that an exhaustive exhibition of the categories is possible. To do this I look to Kant's deployment of the synthetic method through his Copernican Turn (§4.1). This will include an analysis of our faculty for judgement, which is the principle from which the system of categories is derived (§4.2). Specifically, this derivation happens by way of the functions of thinking in judgement, which are what make the original exhibition of the categories (§4.3). This original exhibition, in turn, allows the exhaustive and precise exposition of the categories because it shows how they result from a division of the understanding (§4.4). Throughout this section, I will be discussing the possibility of defining the categories in general. It is only in the subsequent sections that I will treat the differences between possible nominal and real definitions of them.

## 4.1   The synthetic method and the Copernican Turn

Traditionally, philosophy had either taken the *a priori* given concepts of metaphysics (like cause) as it found them and analysed the marks contained in them (e.g. Locke, Hume), or it imitated mathematics by inventing its definitions first and then deducing consequences from these (e.g. Spinoza). The first procedure was an analytic one, associated with empiricists, the second, a synthetic one, associated with rationalists.

The analytic method 'begins with the conditioned and grounded and proceeds to principles (*a principiatis ad principia*)', while the synthetic method 'goes from principles to consequences or from the simple to the composite' (*JL*, 9:149). The analytic method thereby generally makes secure, incremental progress in philosophy by beginning with what follows and is evident, the objects, and formulating principles by generalizing from observations of these. The synthetic method, however, which begins with a proposal about first principles, characteristically lacked a way of securing the connection between its concepts and their objects, and as a result of these faulty foundations its orderly systems crumbled.

With his 'Copernican Turn', however, Kant thinks that he has found a way to secure in metaphysics the connection between concept and object by adapting the synthetic method to philosophy's own purposes, instead of merely imitating its use in mathematics. Kant's use of the synthetic method begins with our cognitive faculty as its first principle, not a series of definitions.[20] His strategy is to investigate the structure that objects necessarily have in virtue of their being represented by us. What he recognizes is that mathematics and physics were put on the sure path of a science only when they internalized 'that reason has insight only into what it itself produces according to its own design' (*KrV*, Bxiii). In both, this shift consisted in reason's no longer following nature by merely forming generalizations by induction from experience, but in reason's taking the lead, working from its concepts and principles, investigating according to its plan, and devising experiments to test its hypotheses. The 'Copernican Turn' is Kant's attempt to bring about an analogous transformation in philosophy. With it he tries to 'get farther with the problems of metaphysics by assuming that the objects must conform to our cognition', rather than 'that all our cognition must conform to objects' (*KrV*, Bxvi).

Kant's turn has two parts, corresponding to the passive and active components of our faculty of cognition, our faculties of sensibility and understanding. The first claims that objects must conform to the constitution of our sensibility, not our intuitions to the constitution of the objects themselves. By the end of the 'Transcendental Aesthetic', Kant claims to have made intelligible how this will work, having shown the necessity of the distinction between objects as they appear to us in sensibility and things as they are in themselves apart from this faculty, as well as that we can have apodictic knowledge of the former because of their necessary relation to the form of our sensibility. The second claims that the concepts requisite for intuition to become cognition do not conform to the objects of experience, but that these objects conform to those concepts (*KrV*, Bxvii). In the 'Transcendental Analytic' Kant explicates how this works, attempting to give satisfactory, i.e. apodictic, objectively valid 'proofs of the laws that are the *a priori* ground of nature, as the sum total of objects of experience' (*KrV*, Bxix). It is this second branch of Kant's system that I will focus on, arguing that his way of preceding puts him in possession of the means for defining its *a priori* given concepts.

## 4.2   The faculty for judging as organizing principle

Let us now turn to these concepts of the understanding, which are the fundamental concepts of nature in general, the categories. When philosophy proceeds analytically, it begins with an analysis of experience, collecting seemingly primitive concepts piecemeal, as they are found, used *in concreto*. The particular kind of abstraction that it employs picks 'out from ordinary [*gemeinen*] cognition the concepts that are not based on any particular experience and yet are present in all cognition from experience' (*Prol.*, 4:322–323). Accordingly, these concepts constitute the mere form of connections in experience. They are rules for experience in general, analogous to 'rules for the actual use of words in general' in a universal grammar applicable to all languages, culled from a language in use (*Prol.*, 4:323).

When assembling his list of categories, according to Kant, Aristotle proceeded through such an analysis of experience. Because he had no common principle to guide him in his search, he haphazardly 'rounded them up as he stumbled on them' (*KrV*, A81/B107; cf. A67/B92). The completeness of his rhapsodic list – that there were just these fundamental concepts and no others – could supposedly be inferred only through induction, through his not having found more that also belong to the list.

With his Copernican hypothesis, which dictates one look first to the nature of our faculties for *a priori* insight into objects, Kant takes himself to be able to introduce order among the Aristotelian categories,[21] and to be able to explain why those categories that he includes in his list, but no others, belong there. For, first he distinguishes 'the pure elementary concepts of sensibility (space and time) from those of the understanding' (*Prol.*, 4:323), which allows him to separate out the concepts in Aristotle's list of categories that are modes of pure sensibility (*KrV*, A81/B107).[22] Kant then 'cast about for an act of the understanding that contains all the rest and that differentiates itself only through various modifications or moments' (*Prol.*, 4:323).[23] This act of the understanding consists in judgement. The functions in judgement are all of the possible modifications or moments of this single act, which together make up the faculty for judging. It is to this faculty for judging that Kant traces all actions of the understanding which, in the first section of the 'Analytic of Concepts', he identifies with it: 'the *understanding* in general can be represented as a *faculty for judging*' (*KrV*, A69/B94). And immediately after presenting the table of the pure concepts of the understanding Kant claims 'this division is systematically generated from a communal principle [*gemeinschaftlichen princip*], namely the faculty for judging (which is the same as the faculty for thinking)' (*KrV*, A80/B106).

It is the understanding, as a faculty for judgement, that Kant announces he will be attempting an analysis or anatomy (*Zergliederung*) of at the outset of that chapter (*KrV*, A65/B90).[24] Unlike traditional philosophic analyses of *a priori* given concepts, which merely bring their content to distinctness, Kant's analysis of our faculty undertakes to discover how its *a priori* given concepts are possible by investigating their origin and birthplace (*KrV*, A65/B90). In the introduction to the First Division of the 'Transcendental Logic', Kant stresses that the table of elementary concepts of the pure understanding, the categories, must be complete and exhaust the entire field of this faculty, as well as be a precise exhibition of its elementary concepts that separates these cleanly from those concepts derived from them (*KrV*, A64/B89). For it is through the precise and exhaustive exhibition of these fundamental concepts that metaphysics, as 'the science of the first principles of human cognition' (*KrV*, A843/B871), acquires that systematic unity 'which first makes ordinary cognition into science' (*KrV*, A832/B860). Accordingly, Kant says:

> Now this completeness of a science cannot soundly be supposed from a rough calculation of an aggregate put together by experiments [*durch Versuche*]; hence it is possible only by means of an *idea of a whole* of the *a priori* cognition of the understanding, and through the determinate division [*bestimmte Abtheilung*] of concepts that such an idea makes out, consequently only through their *connection in a system*. (*KrV*, A64/B89)

Kant does not attempt to secure the completeness requisite for the successful exhibition of his categories, or for the system of principles that follows from it, from an analysis of these concepts, in accord with the method of traditional philosophy. Rather, he does this through the division of the concepts that are determinately situated within, and together constitute, an '*idea of a whole* of the *a priori* cognition of the understanding'. In this way, it is through articulating the original connection of these concepts in the faculty of understanding, the faculty for judgement, that he will secure their complete and precise exhibition. And so it would seem, already at the outset of the Transcendental Analytic, he is announcing that something very much like strict definitions of his elementary concepts of the understanding will be given, if his analysis is successful.

### 4.3  The analysis of the faculty for judging

Pausing to take stock, we have seen in §2 that since the originality of a definition ensures that the boundary conditions of a concept can be derived from nowhere else, the originality will also secure the precision and exhaustiveness of the exhibition. Furthermore, we have found a strategy by which Kant might be able to work from the first principle of our faculty for knowledge, in particular its active stem, the understanding, in order to secure such an original exhibition of its fundamental concepts. This exhibition will proceed by an anatomy of the faculty for judging that divides it into its pure concepts, the categories, which will stand in connection in a system.

When examining our example *circle* we found that the possibility of its original exhibition had its seat in its schema. At the outset of the 'Transcendental Analytic', however, Kant declares that besides completeness in the exhibition of its fundamental concepts, an adequate analysis of the understanding will also respect the self-sufficient nature of it, treating it as much as possible in isolation from sensibility. Accordingly, we should not look to the schemata of the categories for the source of the originality of their definitions, grounded as they are in our *a priori* intuition of time.

Rather, Kant looks to general logic, which is the science of the laws of the understanding considered in isolation from sensibility, for a clue as to how we might give an original exhibition of the categories, the pure concepts of the understanding.[25] This science abstracts away from the content of thoughts and their connections to objects and attends only to the act of thinking in them. From it Kant is able to derive a table of the functions of thinking in judgement, which will consist in the acts that together compose the generic act of thinking in any judgement (*KrV*, A70/B95).[26] What Kant then goes on to maintain is that 'the same function that gives unity to the different representations *in a judgment*, also gives unity to the mere synthesis of different representations *in an intuition*, which, expressed generally, is called the pure concept of the understanding' (*KrV*, A79/B104–105). For this reason he takes the table of the functions of thinking in judgement to be a clue to the structure of the table of categories.[27]

Returning to our question of how the faculty for judging can be the principle of the table of categories, we are now in a position to see why the functions deployed merely in thinking are not enough to explain the possibility of judgement. In merely thinking we may combine our concepts in whatever way we please. But when we judge, we make a claim about how things are. Specifically, in experience we make a judgement that any knower presented with the same manifold of intuition should be in a position to make. On Kant's account it is the functions of synthesis of the categories uniting this manifold for every knower that ensures they too could make this judgement. And it is because these are at work in the constitution of objects of experience that it is possible for our judgements to be about how things are.[28] In this way, the categories, no less than the logical functions of thinking, will be elements in the faculty for judgement, since it is their combination of the manifold of intuition that makes judgement, over and above mere thought, possible.

In particular, Kant explains that the categories 'are concepts of an object in general, by means of which its intuition is regarded as *determined* with regard to one of the *logical functions* for judgments' (*KrV*, B128). In merely thinking, there is a certain freedom in how we order the concepts involved. Both 'all bodies are divisible' and 'something divisible is a body' are perfectly acceptable according to the laws of logic, and justified by our intuitions of bodies. If I bring the concept of body under the concept of substance, however, it is determined that its empirical intuition in experience, the immediate representation of a body, 'must always be considered as subject, never as predicate' (*KrV*, B129). For, a substance is 'something that can occur solely as subject (without being predicate of anything)' (*KrV*, A242/B300), and by determining the intuition of a body through it, I judge that all finite knowers should also judge this intuition to be of a substance, a persisting substrata of predication across changes in its accidents.

### 4.4 Division and the community of the categories

I will be developing this account of the categories below by examining what the definitions of the categories look like. Before turning to this, there is a final element in his analysis of the faculty for judgement relevant to why he should think these definitions are possible. Although the marks uncovered in the common analysis of a concept do not stand in a special relation to one another that allows their boundaries to be determined, things are otherwise with the fundamental concepts that are originally and constitutively connected in an analysed faculty.

The analysis of the understanding consists in its division. This is the key to understanding how the boundaries of its elementary concepts are determined. Normally, what one divides is a concept. According to Kant, concepts stand in a porphyrian tree: there are higher genus (e.g. life form), and lower species (animal, plant, etc.) concepts. The sphere or extension of a genus concept is completely divided into the species concepts that are under it. Such species concepts, which are under a common genus and together exhaust its extension, stand in a community of concepts under their genus. The members of such a community are in different senses both independent of one another and interdependent on one another. They are independent in that none of them is superordinated to the others – none of them includes the content or intension (*Inhalt*) of any of the others in its content. They are, however, interdependent insofar as some individual or representation falling under one of them cannot fall under any of its siblings. The spheres of such concepts 'reciprocally determine' each other in that the sphere of one of the siblings is the complement of the sum of the spheres of the other siblings under the common genus.[29]

When we divide the faculty of the understanding into its functions or actions, there is an analogous reciprocal determination among the fundamental concepts, except that this determination is not among their extensions, but their content or intension. This disanalogy, of course, radically transforms the nature of the reciprocal determination. Rather than their extensions together dividing the sphere of a higher concept, these categories will together constitute the *a priori* cognition of the faculty for judgement, are determined by the idea of this cognition, and are thereby connected together into a system of such cognition.

To get some idea of how this works, consider first the logical functions of thinking in judgement. Although I will not dwell on it, we can see why Kant might have thought that these form a systematic, exhaustive, exhibition of the moments in any act of thinking in judgement. For example, categorical judgements like 'Every S is P' have a four-part structure insofar as they are connected together with our other judgements: a subject concept, a predicate concept, a copulative relation between these, and a connection of this judgement to the rest of our knowledge. These four parts or moments in judging correspond to the four headings in the table of the functions of thinking in judgement. And in general, although these moments will be instantiated in different ways in different judgements, there is some plausibility to the thought that all judgements will involve these and only these acts of thinking in some way or other.

Now the categories, as 'the rules of the pure thinking of an object' (*KrV*, A55/ B80), are determinations of intuition with regard to the logical functions of thinking in judgement, and so also stand in a system. These form a system because without each of these ways of determinately thinking about intuition linking together as they do, it would not be possible to determinately think or judge about objects. Their boundaries reciprocally determine one another insofar as they each have their place within this system that makes judgements about an object possible, and these moments in judging about an object cannot be exchanged with one another.

Here is a very brief sketch. According to Kant, every knowable object is extensive, is real, can be considered as a substance, causally interacts with other substances, and contemporaneously stands in a community with other substances that together make up the whole of nature. Without objects being extensive magnitudes, without their having a duration or filling space, we would not have intuitions of them, and would not be able to situate them in space and time, which are themselves extensive magnitudes (cf. *KrV*, A162/B202). Without their being real, objects would not affect us and so could not be perceived (cf. *KrV*, A166/ B207). Without a persisting object underlying these perceptions, or relating to one another successively and simultaneously, these perceptions would not stand in necessary connections, and their object would not be knowable as an object that interacts with and is situated in nature through them (cf. *KrV*, A176/B218). Finally, without relating this object to our system of knowledge as possible, actual, or necessary, we cannot determinately know it at all (cf. *KrV*, A219/ B266). Although this was quick, in these headings we find a system of ways that objects might be, which correspond to possible judgements about them. On Kant's account, it is through the categories that we recognize these kinds of features and make these kinds of judgements.

Returning to the way Kant maintains these are situated within the faculty for judgement, the understanding, both the categories and the mere functions of thinking in judgement are *a priori* elements in this faculty that are necessarily at work in every judgement about an object and, accordingly, both will have their place in an analysis of this faculty. Unlike the mere functions, however, the categories will count as cognition because of their essential connection to objects. Nonetheless, both will compose systems, and they will be parallel because of how the categories depend on the functions of thinking.

In each of these systems the way in which the elements reciprocally determine one another, as distinct, yet necessary activities in making judgements, is not as straightforward as when we divide a concept into its subordinate concepts. This should be unsurprising, however, because dividing a faculty is more difficult than partitioning a genus into species. It requires long and careful attention to the elements in an exercise of the faculty, distinguishing which of these are necessary or contingent, and articulating the structure of the necessary elements. Specifically, it requires seeing how the necessary actions in thinking differ from the necessary actions in thinking of objects and getting clear on how the elements within each of these overall actions fit together. The result of this examination is an understanding of these elements, as the actions or functions that together constitute the faculty, where the nature of each is revealed through its place in the overall generic act of the faculty. Accordingly, it is in this way that we should understand what Kant means by an analysis of the faculty of judgement, and it is in this way that such an analysis makes the precise, original boundary determinations of the categories possible.

## 5.   Definitions of the categories

In the 'Phenomena and Noumena' chapter, at the end of the 'Transcendental Analytic', after having surveyed the understanding and determined the place of each of its parts, Kant casts back 'yet another glance at the map of the land that we would now leave', in part by taking up the subject of the definitions of the categories (*KrV*, A235/B294-A236/B295). This is his fullest treatment of the topic, and here I take him to be sketching both merely logical definitions of the pure categories and real definitions of the schematized categories.

First, a textual note. The passages in this chapter on the definitions of the categories are among the most worked over across the two editions, and the main passage that seems to give the logical definitions of the pure categories is omitted in the B edition. It is tempting, then, to think that Kant removed these passages because he wanted to avoid taking on the commitments they entail. It is more plausible, however, that he thought they were particularly obscure, and that some of the work done in the deleted sections of A241–242 and A244–246 was unnecessary for the argument of the chapter. This is especially true with the paragraph at A244–246, which seems like a digression. Furthermore, he could not have been too shy about these commitments, since he covers some of the same ground as A245–246 in the B edition's 'explanation of the categories' at B128– 129. More strongly, however, in other unaltered passages Kant seems to indicate what real definitions of the categories would look like, and to be in a position to give these, one must also be able to give corresponding nominal definitions.

Kant treats the topic of merely logical definitions of the pure categories in the removed paragraph at A244–A246. There he gives an explanation similar to the one we examined at the end of §4.3, from B128. He says, the pure categories 'are nothing other than the representations of things in general insofar as the manifold of their intuition must be thought through one or other of these logical functions [i.e. the logical functions of judgment in general]' (*KrV*, A245).[30] In this specification, Kant is, however, focusing in on the logical component of the 'categories' and abstracting away from their sensible condition, since the pure 'categories' omit this, and contain 'nothing but the logical function for bringing the manifold under a concept' (*KrV*, A245). Returning to the case of substance, this means that we are considering only the way of determining the order in a manifold of intuition in general – that the intuition is of a substance that can only be thought of as subject, never as predicate, in a judgement – without specifying that this manifold is spatio-temporal. Accordingly, it means that we are abstracting away from the way in which the manifold of intuition will determinately represent the object as a permanent substrata across changes in its state. For, permanence is a temporal quality, and changes only happen in time. So, considered merely as a pure category, substance is only 'that which, in relation to intuition, must be the ultimate subject of all other determinations' (*KrV*, A246). And to consider the pure category is to consider merely the way the functions of thinking in judgement will determine any manifold of intuition in general that is thought through the category, be it spatio-temporal or not, abstracting away from the specific nature of the intuition in question.

A little later, Kant fills out his description of a pure category by saying that in it 'no object is determined, rather only the thought of an object in general is expressed in accordance with different *modi*' (*KrV*, A247/B304). I take this to suggest that we might think of 'thought of an object in general', or 'the determinate thinking of the manifold of intuition of things in general' (now abstracting away from the sensible condition) to be the genus of the nominal definitions of the pure categories, and their differentia to be the logical functions of thinking in judgement, which are the different *modi*.

This suggestion is supported by what I take to be examples:

> Magnitude [*Größe*] is the determination that must be thought only through a judgment that has quantity [*Quantität*] (*judicium commune* [i.e. is plural or universal]); reality, that which, can be thought only through an affirmative judgment; substance, that which, in relation to the intuition, must be the ultimate subject of all other determinations.[31] (*KrV*, A245–246)

In each of these cases we have a bare logical specification of a kind of feature of an object, through the kind of judgement that can be used to think of this kind of feature. When we make these judgements we determine the object, insofar as we think of it as being one way rather than another. We do this by representing the manifold of its intuition as determinately united together and as related to the object. Insofar as the category indicates this unity among the manifold of intuition that must be thought in accord with one or the other of the functions of thinking in judgement, the category is what makes the representation of this unity possible. And the way in which the category represents the object, through this representation of the unity of the manifold of intuition, is that it represents the manifold of intuition not merely as united this way for me, the one who is judging, but for everyone. That is, through the category, I represent the manifold not only as combined that way for me in my mind but as though it would be combined in that way for everyone to whom it was available. And because through the category I represent that this same unity of the manifold would be present for every possible knower who comes into contact with this manifold, I attribute this unifying element among the manifold to the object.

Returning to our examples, in determining the manifold of intuition through substance, I represent that in it which must be the ultimate subject of other predications or determinations. This will be the object in which these predications inhere for everyone, and so will be the ultimate subject of predication not just for me, but for everyone. In determining the manifold through the category of reality, we represent a feature of the object that must be represented through an affirmative judgement. This will concern a way the object is, a positive determination, and will be a judgement that anyone similarly affected by the object can make. In determining the manifold through the category of quantity, we represent a feature of the object that must be represented through a general judgement (cf. *KrV*, A71/B96), and so will concern the magnitude of the intuition unified.

Now, admittedly, Kant does not announce the explanations of the pure categories I quoted above as nominal definitions. Nonetheless, I think these explanations fit the bill. For, each specifies a general mode (or way) in which we can think of an object, determining its manifold of intuition. These modes correspond to the logical functions of thinking in judgement, since they are different ways of thinking of an object in general. Still, they are merely nominal because they leave entirely open how the manifold of intuition is given, and so tell us nothing about how to apply the concepts they define.

The main text that speaks in favour of maintaining that Kant thought real definitions of the schematized categories were possible is an unaltered passage where he gives examples of quantity and reality:

> No one can define (i.e. give a real definition of) the concept of magnitude in general, except by something like this: that it is the determination of a thing through which it can be thought how many units are posited in it. Only this how-many-times is grounded on successive repetition, thus on time and the synthesis (of the homogeneous) in it. Reality, in contrast to negation, can be defined only if one thinks of a time (as the sum total of all being) that is either filled by it or empty.[32] (*KrV*, A242/B300)

The difference between these and the nominal definitions of the pure categories is that they secure their application to empirical, i.e. spatio-temporal, objects of experience, through their reliance on their temporal schemata. Accordingly, these are the same concepts of an object in general, but now with a rule for how they can be applied to possible objects of experience, which is a rule for how an object can actually be determined through them. In this way, I think the genus of the real definitions of the categories will be the same as that of the nominal definitions I sketched, but the differentia will have a further specification. This specification is that they now include *how* the manifold in intuition in general is to be thought determinately through a concept.

If we bring this together with the preceding, we might, for example, think of the proper real definition of magnitude as, 'magnitude is the determination that must be thought only through a judgement that has quantity, where this states how many units are posited in that which is determined'.[33] Since this how-many-times is grounded on successive repetition, it will be grounded on time, and include the schema for quantity, which specifies how we can make these judgements. In this case the genus is still the determinate combination of the manifold through the concept of an object in general, but the differentia specifies not only that the determination must be thought through a judgement that has quantity but also how that determinate combination takes place.

We can perhaps see how this definition would work concretely by examining the way Kant thinks quantity, in particular plurality or magnitude (*Prol.*, 4:303), determines a manifold of intuition in a given case. In the *Prolegomena* Kant says,

> The principle: a straight line is the shortest between two points, presupposes that the line has been subsumed under the concept of magnitude [*Größe* ], which is surely no mere intuition, but has its seat solely in the understanding and serves to determine the intuition (of the line) with respect to such judgements as may be passed on it as regards the quantity of these judgements, namely plurality [*Vielheit* ] (as *judicia plurativa*), since through such judgements it is understood that in a given intuition a homogeneous plurality is contained. (*Prol.*, 4:301–302)

Asserting 'a straight line is the shortest between two points' requires having considered the possible lines that could connect any two points and having recognized that in all cases the shortest must be straight. Here, although we need not have explicitly formulated what the units are, by claiming that a straight line is the shortest, we think of the possible pairs of points, compare the possible lines connecting each pair, and claim that no matter the unit, there will be fewer in the straight line connecting the points, than in any other possible line. Of course, this is a sophisticated process with many components. For our purposes, what is significant is that in thinking of the straight lines as *shorter*, we think of them as having a magnitude (a *Größe*), which Kant takes to involve a plurality (*Vielheit*), and whatever the unit measure, there will be fewer of these in this shortest line, than in the others. So, for the principle 'a straight line is the shortest between two points', when we consider the intuitions on which the judgement is grounded, the relevant determination of these is through the category of magnitude. This is because it is through this category that we can think about how many units would be contained in each line, determine the difference between these relative lengths, and so figure out which line is the shortest. In this way, in seeing how the category of magnitude determines the manifold of intuition on which this principle rests, we can see that it determines this manifold in accord with the logical function of quantity, where the quantity in question is determined by how many units are posited in the intuition.

Finally, before leaving this section, because the real definitions of the categories specify a way of applying them through their schemata, I think these definitions also express the real essence of an object of possible experience in general, insofar as it is thought through the understanding. To spell out why this is in detail would be a large task. For, part of Kant's Copernican Turn is that the categories are representations that make the objects cognized through them possible (cf. *KrV*, Bxvii). Specifically, how this works comes out through the way Kant understands the nature of objects, especially objects of experience. An object 'is that in the concept of which the manifold of a given intuition is *united*' (*KrV*, B137). And the categories are the concepts of an object in general that make such unification possible. In this way, they express the real essence of an object of possible experience in general, which is a partial explanation of the nature of appearances. To really understand how this works, however, we would need to explicate the inner workings of the 'Transcendental Deduction of the Categories' and delve deeper into Kant's Transcendental Idealism than we can here.

## 6. An objection

There are many open questions about the nature of these definitions, but I take the preceding to have sufficiently shown that Kant thinks definitions of the categories are possible. In this final section, I would like to address an objection that will clarify the exact nature of the reality secured for the definitions of the schematized categories.

The objection is that merely schematizing the categories does not guarantee that their definitions are real. In sensation things are represented immediately as real (cf. *KrV*, B147). But the schematization of the categories does not guarantee that there will be any sensation, any given real appearance, in which they are operative. So their schematization alone is insufficient for securing their reality; for this, sensation is also needed.

This line of objection seems to find support in the 'General Note on the System of Principles'. There Kant says that

> we can not have insight into the possibility of any thing in accordance with the mere categories, but we must always have available an intuition in order for it to display the objective reality of the pure concept of the understanding. (*KrV*, B288)

Furthermore, to establish the objective reality of the categories 'we do not need merely intuitions, but always outer intuitions' (*KrV*, B291), and not only these, but even the motion of a point in space is required (*KrV*, B292). Motion and matter as the movable in space, however, are empirical concepts (*MAN*, 4:470; 4:472; 4:480; 4:482), derived from moving spatial objects. Thus, at least some experience of such objects is required for securing the reality of the schematized categories and experience always involves sensation.

There are a few distinct issues wrapped up in this objection that need to be separated. The first is whether the temporally schematized categories require spatial exhibition for their reality. The second is whether this exhibition must be through motion. The third is whether motion in this sense is empirical. The fourth is whether sensation is required for this exhibition, or whether the intuition in question can be *a priori*.

On the first, Kant is unambiguous that the categories, and thus their definitions, require spatial exhibition for their reality (*KrV*, B293). In 'The General Note', Kant pays particular attention to the categories under the heading of relation, arguing that outer intuition is required to show the objective reality of the concept of substance, as well as the concepts of cause and community which depend on it. This is because 'space alone persistently determines, while time, however, and thus everything that is in inner sense, constantly flows' (*KrV*, B291), and so for the exhibition of these concepts we need an intuition *in space* (of matter). At the end of the 'Note', Kant briefly addresses the mathematical categories, claiming that

> it can just as easily be established that the possibility of things as *magnitudes*, and thus the objective reality of the category of magnitude, can also be exhibited only in outer intuition, and that by means of that alone can it subsequently be applied to inner sense. (*KrV*, B293)

Presumably, this is because extensive magnitudes must at least be drawn *a priori* order to be cognized, and in this way presuppose space in their exhibition. I take this remark to encompass *reality* as well because the real in appearances, sensation, has an intensive magnitude, which is cognized objectively through *reality* by uniformly reducing the intensity of the sensation to zero and then bringing it back up, thereby exhibiting it extensively and allowing the measurement of its degree (cf. *KrV*, A166/B207*ff.*; A143/B182).

Regarding motion, I think there is also a case to be made that it must be involved in securing the reality of the categories. With respect to it, however, Kant only explicitly discusses the concept of causality. To exhibit alteration, as the intuition corresponding to causality, he thinks we must take motion as our example. Specifically, he thinks that alteration, as 'the combination of contradictorily opposed determinations in the existence of one and the same thing' (*KrV*, B291), is incomprehensible without an example, an intuition. This intuition, perhaps surprisingly, 'is the motion of a point in space' (*KrV*, B292), and cannot be merely an alteration in inner sense. It is Kant's reasoning for this that makes me think there is a case to be made that motion must be involved in exhibiting the reality of the other categories as well:

> in order to subsequently make even inner alterations thinkable, we must be able to grasp time, as the form of inner sense, figuratively through a line, and grasp the inner alteration through the drawing of this line (motion), and thus grasp the successive existence of ourself in different states through outer intuition. (*KrV*, B292)

The nascent case here is that because all of the schemata of the categories are temporal, this reasoning generalizes: because we must represent time figuratively, the way that we bring any of the categories to an example will itself involve drawing a line, and thus motion.

Even if it turns out that motion is required for the exhibition of any of the categories, regarding the third issue, I do not think the motion in question has to be empirical, but can be *a priori*, at least as long as the nature of its object is left undetermined as to whether it is corporeal or thinking matter, as it is in the *First Critique*. The key passage for assessing this aspect of the objection is the footnote at B155:

> Motion of an *object* in space does not belong in a pure science, thus also not in geometry; for that something is movable cannot be cognized *a priori* but only through experience. But motion, as *description* of a space, is a pure act of the successive synthesis of the manifold in outer intuition in general through productive imagination, and belongs not only to geometry but even transcendental philosophy.

The question is whether the motion that may be required for the exhibition of the schematized categories is that of an object or of the description of a space. The kind of motion required is that of a point through space (cf. *KrV*, B154, B292). This motion describes a line. A point is the boundary of a line, and a line is a space (as well as the boundary of a space; *Prol.*, 4:354). This does not, however, make the point an object. Insofar as a point is an 'object', it is a mathematical one, a locus in space, but not an object of experience. Thus, this motion is not of this or that body, but describes the space of a line *a priori*, and is foundational to geometry, and even transcendental philosophy, insofar as it is required for the figurative exhibition of time itself, but need not be empirical.

Although the motion in space required for the potential exhibition of the categories merely describes space, and so is *a priori*, it is not entirely clear that this is sufficient for showing the reality of the categories. For, while this exhibits the function of the power of judgement, whereby an object is subsumed under a category, the category acquires its 'object, thus its objective validity, only through an empirical intuition', which is the '*data* for possible experience' (*KrV*, A239/B298). Nonetheless, I think that in whatever sense in which we can have real definitions of mathematical concepts without needing experience, in an analogous sense we can have real definitions of the categories without it.

This is because although it can seem as though mathematics is independent of experience in a stronger sense than the categories, it is not. The two depend on experience in similar ways. Although the concepts of mathematics, its principles, and its objects 'are generated in the mind completely *a priori*, they would still not signify anything at all if we could not always exhibit their significance in appearances (empirical objects)' (*KrV*, A240/B299). Kant does go on to claim that mathematics makes its abstract concepts sensible, i.e. display their objects in intuition, 'by means of the construction of the figure, which is an appearance present to the senses (even though brought about *a priori*)' (*KrV*, A240/B299), which is something that philosophy cannot do.

Nonetheless, for the concepts of both, the '*use* and relation to supposed objects can in the end be sought nowhere but in experience, the possibility of which (as far as its form is concerned) is contained in them *a priori*' (*KrV*, A240/B299). In this way, although the concepts of mathematics can be made sensible *a priori* while the categories cannot, as far as their dependence on experience for their potential reality, the concepts of mathematics and the categories are on a par.

Specifically, as concepts of the form of appearances, both kinds of concepts depend on experience for their significance, since for their reality both must contain 'a clear *mark* by means of which the *object* (*definitum*) can always be securely cognized' (*KrV*, A241n). This mark, however, need not guarantee the existence of the object; for all it contains, the object may be merely possible. Does the reality of a definition then depend on a potentially contingent fact, external to the explanation? I doubt this. The definitions of mathematics and the categories would indicate the real essence of appearances, not as regards their matter, but their form, even if there were no existing empirical things. This is not to deny that both would not be possible without the fact of experience. On the contrary, if there were no sensation to enliven our faculties, and if the resulting activity were not ordered into objective thought, then we would not have cognition or experience at all, let alone be able to formulate definitions of concepts that are forms of possible experience.

## 7. Conclusion

With his Copernican Turn, Kant is able to trace the pure forms of empirical objects back to their original sources in our faculties for knowledge. Specifically, with his analysis of the understanding, he traces the pure concepts of this faculty, the categories, back to their origin through the functions of thinking in judgement and thereby makes possible an original, thus complete and precise, but merely logical exhibition of these concepts. These merely logical definitions acquire their reality, and thus securely cognize their objects, only through their application to objects of experience through their schemata and the conditions of our sensibility, since this is the only kind of intuition by which we can be given objects for knowledge. In this way real definitions of the categories were possible on Kant's account.

**Acknowledgements**

In preparing the final version of this essay I received helpful comments from Lisa Shabel and Emily Carson. I have presented earlier versions of this material to audiences at a workshop on the table of categories at Merton College, Oxford, and the 2011 Princeton-Penn-Columbia Conference in the History of Philosophy. In addition, this essay has also benefited from the feedback of Ralf Bader, John McDowell, and Karl Schafer. I also owe special thanks to Steve Engstrom and Anja Jauernig, both for comments on the paper and insights gained through their seminars on Kant.

## Notes

1. There are, to my knowledge, nine substantive treatments of the subject of Kant's view of definitions. In all of them, the question concerning the possibility of philosophical definitions is answered negatively. (Beck 1956) is the essay which most directly addresses the subject of definition and most of the subsequent treatments follow his to a greater or lesser degree. The others are: Stuhlmann-Laeisz (1976, §6), Capozzi (1981, 424), von Wolff-Metternich (1995, §4: 1.1), Carson (1999, §4), Maddy (1999, §2), Dunlop (2005, ch. 5), Dunlop (2011, §2), and Rosenkoetter (2009, 200–201). Of these, Rosenkoetter's comes closest to admitting the possibility of defining the categories, since, although he maintains Kant rejects this, he nonetheless holds that the categories together give a real definition of 'the object as such [*Gegenstand überhaupt*]'. For an illuminating treatment of Kant's views on definition in the pre-critical period in relation to those of Leibniz and Wolff, see Sutherland (2010).

2. I do not think that this exhausts the philosophical definitions that become possible according to Kant. In particular, he also thinks that many concepts in morals and in the metaphysics of extended matter can be defined, and this is so in a more robust way than the categories, because these definitions also make possible the construction of their objects (albeit in quite different senses). At one point it even sounds as if Kant thinks he can define the concept of time as 'the order of things, in so far as they follow one after the other' (*Busolt Logic*, 24:659), but it is not clear to me how he would secure this definition. I will not discuss these further kinds of philosophical definitions here.

3. All of my references to Kant's works will be to the Akademie Ausgabe (vol:page number), except in the case of references to the *Critique of Pure Reason*, which will be cited using the pagination of the first (A) and second (B) editions. I will also usually abbreviate the work in question as follows: *KrV (Critique of Pure Reason)*, *Prol. (The Prolegomena to Any Future Metaphysics)*, *MAN (Metaphysical Foundations of Natural Science)*, *JL* (Jäsche Logic), *Refl. (Reflexionen)*. Translations are my own, but were done in consultation with the Cambridge editions. When interpreting Kant's logical views, we are faced with the problem that Kant did not himself author a treatise on logic. Rather, what we have are fragmentary notes contained in the *Reexionen zur Logik* and various transcripts of his lectures on logic taken by students. Of these, the Logic prepared by Jäsche, which was authorized by Kant, and prepared consulting his handwritten notes, stands out as the privileged one. Here I am agreeing with Young, among others (Kant and Young 2004, xix). For more on the respectability of Kant's various logical remarks see Kant and Young (2004, xvii–xix) and Boswell (1988).

4. In Kant's German, '*Exposition, Explication, Declaration und Definition*', compared with '*Erklärung*'.

5. Another example is the exposition he gives of the faculty of desire. This is the faculty of a being to be the cause of the object of one of its representations, through that representation (*KpV*, 5:9n). In the end, we might discover that the concept of this faculty includes that it is always determined through pleasure, but this is left open in Kant's exposition.

6. We need not dwell on the differences between the other various kinds of explanations, but briefly: descriptions are expositions that are not precise (*JL*, §105); declarations are arbitrarily (*willkürlich*) invented concepts for which it is not certain whether the object can be made (*KrV*, A729/B757). '*Explication*', which seems to be the specification or spelling out of the content of an expression (*Refl.*, XVI:577, 2922, 2923; XVI:579, 2931), is a less used term. It is very close to '*exposition*' as the making distinct of a concept, although expositions can be given either of concepts or

of appearances, while explications cannot be given of appearances. '*Explication*' is often contrasted with '*declaration*' (*Refl.*, XVI:585, 2950). Finally, *Erörterung* is a term Kant will gloss as *Exposition* (*JL*, §105), and I do not take him to distinguish these.

7. The possibility of reading the footnote in this way was suggested to me by Stephen Engstrom.

8. Beyond the *First Critique*, most of the logical works, even the pre-critical ones, characterize precision through an analogy with putting a fraction into minimal terms (e.g. *Bloomberg Logik*, 24:263–264; *Refl.*, 2979). The examples Kant gives of marks that would be excluded from such a definition in minimal terms include *curved*, or are often similar, excluding divisible from 'the body is extended' (*Logik Pölitz*, 24:575). All of which suggests that if he did intend precision in the weaker sense in the footnote, he was breaking with prior usage.

9. The chemical formula for the element, however, comes closer to giving a boundary determination that allows the concept to stand at the head of all judgements about gold things. Nonetheless, the sub-atomic structure of certain samples of the element may lead to those samples exhibiting novel behaviour that goes beyond that which is fixed by its atomic structure. Accordingly, although an explanation of gold in terms of its atomic structure, which presumably fixes most of its macroscopic and mesoscopic properties, comes very close to fixing the precise boundaries of the concept, this too will not suffice, strictly speaking, for a definition. This would be no surprise to Kant, who thought empirical concepts were not definable (*KrV*, A727/B755).

10. We find something similar in the case of invented concepts of objects of experience like *Schiffsuhr*, a clock precise enough for the computation of longitude. Such a thing had not been invented in Kant's day. Until we have built one, until we have proved that it can be produced in accord with the conditions of objects of experience in general, we do not know that this arbitrarily made concept has a really possible object (cf. *KrV*, A729/B757). It may, like a perpetual motion machine, not be physically constructible.

   Kant will sometimes call specifications of how to make empirical objects, like instructions for building such a clock, definitions. One example is 'the definition of cinnabar: mercury and sulfur sublimated produces cinnabar' (*Busolt Logic*, 24:660). Such a definition, made out of empirical concepts, however, is not a counterexample to Kant's claim that empirical concepts cannot be defined. This is because these are technical concepts for making things, not the kind of empirical concept we might mistakenly want to define in an empirical science.

   I think one way at this distinction is to note that although the matter for this definition is empirical, the definiendum is made *a priori*, since the concepts are put together through an act of will, not through an exposition of given appearances. Both Beck (1956, 184) and Dunlop (2011, 96), however, count this kind of invented concept as *a posteriori* made. I take this latter class, however, to contain only those concepts that we arrive at through hypothesis in empirical natural science, and which we test against appearances through observation (*JL*, 9:141). My reason is that I take the nature of the synthesis involved in the creation of the concept to be more important than the kind of matter combined, in determining whether a concept is *a priori or a posteriori* for Kant. Of course, I, nonetheless, allow that there is an important difference between *Schiffsuhr* and *circle* insofar as the matter of the concepts and the conditions on construction are empirical or *a priori*, respectively.

11. One place where the kind of relation I have in mind comes out is in a passage of Kant's on biangles: 'In the concept of a figure that is enclosed between two straight lines there is no contradiction, for the concepts of two straight lines and their

intersection contain no negation of a figure; rather the impossibility rests not on the concept itself, but on its construction in space, i.e. on the conditions of space and its determinations; but these in turn have their objective reality, i.e. they pertain to possible things, because they contain in themselves *a priori* the form of experience in general' (*KrV*, A220–221/B268). Circles, like biangles, are connected to space. But instead of this connection accounting for their impossibility, circles are possible. And in both cases, the possibility or impossibility of the figure in question depends 'on the conditions of space and its determinations' – whether the concept describes a possible limitation of space or not (cf. *KrV*, A619/B647). It is the relation between the concept circle or biangle, and space that I suspect is taken for granted in geometry, according to Kant. For one discussion of the topic in the secondary literature, which also situates it within a wider discussion of the transcendental exposition of space, see Shabel (2010, esp. 102–108).

12. The relation of mathematics to appearances is studied in philosophy because the principles of mathematics are made possible through the principles of the pure understanding (*KrV*, A162/B202), the study of which belongs to philosophy. Specifically, Kant says that the application to experience of the principles of mathematics (which are derived from intuition, not the understanding) still always rests on the pure understanding (*KrV*, A159/B199). And the principle of the pure understanding in question is 'all appearances are, as regards their intuition, *extensive magnitudes*' (*KrV*, B202). He goes on to call this the 'transcendental principle of the mathematics of appearances' (*KrV*, A165/B206), and explain that this is the principle which ensures that mathematics governs appearances, objects of experience. So it is this principle, treated in philosophy, that ensures the objective validity of mathematics.

13. Introducing his topic Kant says: 'I understand by a *transcendental exposition* the explanation of a concept as a principle from which insight into the possibility of other synthetic *a priori* cognitions can be gained. For this it is required 1) that such cognitions actually flow from the given concept, and 2) that these cognitions are only possible under the presupposition of a given way of explaining this concept' (*KrV*, B40). In the surrounding exposition, the concept in question is space, and the other synthetic *a priori* cognitions are those of geometry. I take the first paragraph of the exposition (the one spanning B40–B41) to be concerned with establishing this connection between our representation of space and our cognition in geometry.

14. Another place where the dependence of geometry on philosophy for securing the relation between its constructions and space comes to the fore is in Kant's discussion with Eberhard. For example, in one suggestive remark Kant says, 'the question, however, as to how this single infinite space is given, or how we have it, does not occur to the geometrician, but concerns merely the metaphysician' (20:420–421; Kant and Allison 1973, English trans., 176). I take the surrounding context to fill out this remark in the direction I am suggesting.

15. Before leaving the discussion of completeness, precision, and originality, I should note that my focus has been on the account of definitions given in the *First Critique*. In the works on logic, Kant approaches definitions from a slightly different angle. For example, in the *Jäsche Logic* he says, 'a definition is a sufficiently distinct, and precise concept [*zureichend deutlicher und abgemessener Begriff*]' (*JL*, §99; cf. *KrV*, B759). I take the differences between the terms used in the logical works and the *First Critique* to be largely insignificant, but in the *Logic* there is a shift in focus and perhaps a loosening. '*Abgemessen*' I take to be the Germanic equivalent of the Latinate '*Präcision*'. '*Zureichend deutlichkeit*' will have two sides, sufficient extensive and intensive distinctness (cf. *JL*, Intro §VIII, esp. XI:62–63). The former will roughly correspond to exhaustiveness, while the latter will be closely linked to

originality. In shifting to talking about sufficiency of distinctness, Kant is de-emphasizing the explanatory elements of the definition – its originality and exhaustivity – and focusing on the logical form granted through these elements. Furthermore, in the context of general logic, which will govern all sciences, because sufficiency is relative to a use, we can perhaps see the point coming to the fore that what exactly is required for strict definitions is particular to a science, insofar as what will count as sufficient may be different in different sciences.

Another reoccurring theme in the logical works is the requirements or perfections of definitions which sometimes track the four headings of the functions of thinking (cf. *JL*, §107, 9:144; *Busolt Logik*, 24:658–660; *Refl.*, 16:588–600; *Wiener Logik*, 24:921–922; *Pölitz Logik*, 24:574–575; *DW-Logik*, 24:759–760; *Philippi Logik*, 24:458; *Blomberg Logik*, 24:263*ff*). Some requirements these give, which do not get explicitly touched on in the Critique, are that definitions should not be tautologies, they should not be circular, and they should not explain the obscure by the equally obscure.

16. Beck distinguishes 'nominal' and 'logical', using 'logical' as the name for specifically analytic nominal definitions. I do not see evidence for thinking this follows Kant's usage. Although it is true that Kant will often say 'logical nominal definition', what we have here are two adjectives describing the same kind of definition, not a specific kind of nominal definition. Kant tends to use 'logical' when he wants to emphasize that the definition is of a thing's concept, or specifies a logical essence, and tends to use 'nominal' when he wants to emphasize that the definition is of a name or a word, and although he will use both to contrast with '*Real-*' or '*Sach-Erklärungen*', '*Nominal-*' or '*Namen-Erklärungen*' seems to be his preference. Nonetheless, I do not think these preferences in use constitute a distinction in kind as Beck does.

17. *Refl.*, XVI:609–610, 3001, 3002, 3003.

18. For example, on the one hand, at B757, he seems to rule them out entirely – 'thus there remain no other concepts that are fit for being defined than those containing an arbitrary synthesis which can be constructed *a priori*, and thus only mathematics has definitions' – on the other, a page later, at B758, he suggests that they are possible and that 'in philosophy the definition, as distinctness made precise, must conclude rather than begin the work'.

19. Although in the pre-critical 1772 *Philippi Logik*, Kant seems to think that *body*, unlike other empirical concepts, is general enough to be defined (24:457).

20. There are two terms that get translated into English as 'principle': *Grundsatz* and *Princip*. Kant usually uses *Grundsatz* narrowly to speak of fundamental principles that can be formulated into judgements, and which can be laws of nature. *Princip*, however, often means something closer to Aristotle's *arche* or 'starting points', and can include faculties or concepts, as well as *Grundsätze*.

21. The original ten: substance, quality, quantity, relation, action, affection, time, place, position, and state; as well as the post-predicaments: opposition, priority, simultaneity, motion, and possession.

22. Time, place, position, priority, and simultaneity.

23. It should be noted that Kant's notion of 'action' is not our modern notion of intentional action (e.g. fixing a water heater). This is especially clear when he talks of the actions of our cognitive faculties. What he means by 'act' is, rather, the more traditional philosophical, scholastic sense of the term, as in, 'When one substance modifies another, the first *acts* on the second'. The 'acts' of the understanding will be modifications of our intellect – of the order possessed by the whole of our cognition.

24. While Kant tends to use *Analyse* mainly to speak of the analysis of concepts, he will use *Zergliederung* to mean both the analysis of a faculty and of a concept. In this way, I think the situation is similar to that of *Grundsatz* and *Princip* (see footnote 20).

25. After presenting his table of categories, Kant distinguishes them as *ursprünglich stammbegriffe* – original root (or stem) concepts – from derived (*abgeleitete*) pure concepts of the understanding, which he calls predicables. In this way, the categories are a class of pure concepts of the understanding – those that are elementary. In the first instance, however, Kant has the categories in mind when speaking of these pure concepts, and I will follow this usage here.

26. Kant explains what he means by functions as 'the unity of the act of ordering different representations under a communal [*gemeinschaftlichen*] one' (*KrV*, A68/B93). We need not get bogged down in the intricacies of the relation between the unity of an act and an act.

27. I will not dwell on the metaphysical deduction of the categories from the functions of thinking in judgement, although to really see how the original, precise, and exhaustive exhibition of the categories is working, this would have to be done. The issues here, however, are vast and many interpreters have dealt with these topics in greater detail than I could in this short essay. In what follows, I will only be briefly raising those points relevant to my case, and it goes without saying that even on these, much more could be, and has been, said.

28. In other work I hope to develop a fuller account of these two aspects of the categories and their roles in making both cognitions and their objects possible.

29. The predicate concepts of a disjunctive judgement stand in a conceptual community with one another under the subject concept (cf. *KrV*, A74/B99). Paradigmatic examples will be 'Every triangle is either right, acute, or obtuse', 'Every cat is either a calico, or a non-calico', or 'Every animal is either a mammal, a reptile, a fish, … '. In these cases, all of the predicates, when taken together, will exhaust the sphere of the divided subject concept and if, e.g. some triangle is right then it is not acute.

30. Here and at B128, Kant is giving an explanation of the categories, which are concepts. Accordingly, these can seem like 'higher order' explanations, explanations about concepts rather than objects. I do not think Kant thinks about distinctions between 'orders' as we do, and he tends not to discuss concepts of concepts or judgements about concepts as such. Regardless, I do not think the general 'explanation' of the categories at B128 (or the one here) gives a strict (i.e. complete, precise, and original) definition of the categories, but only points the way towards them, by indicating how these concepts work. For, as we saw in the supposedly decisive passage, strict definitions must be adequate to their object (*KrV*, A728/B756), and in defining the categories what is at issue with their strict nominal or real definitions is the way in which an object is thinkable or cognizable through them. In this way, these definitions will indicate what is specific to each of them, and explanations of how they generally function will be inadequate.

31. More examples are buried in his discussion of the real definitions of the categories on the preceding pages: leaving persistence out of substance we have 'the logical representation of the subject, which I try to realize by representing to myself something that can occur solely as subject (without being a predicate of anything)', in the pure category of cause we will only find 'that it is something that allows an inference to the existence of something else', and the pure concept of community will only contain the thought of 'reciprocal causality in the relation of substances to each other' (*KrV*, A242–243/B300–301).

32. For the rest – substance, cause, community, possibility, existence, and modality – Kant does not sketch a real definition of the schematized versions in making his argument, although it seems clear that he thinks he could. Instead, he contents himself with making the case that (at least for substance and cause) without their schemata, not only would we lack all knowledge of the conditions under which the pure category can be attributed to any sort of thing but also that no consequences can

be inferred from it, since we cannot know whether we can determine any object through it. At another place, in his elucidation of the postulates, he does claim that these postulates offer definitions or explanations (*Erklärungen*) of possibility, actuality, and necessity (*KrV*, A219/B266). Although I think these plausibly are proper real definitions, I do not take this passage to be decisive.

33. For other examples, I think we would need to stray farther from the text in bringing together what I take to be Kant's logical definitions at A246 and the real specifications that came before at A242/B300. For this reason, I will leave considering what these might look like to the reader.

## References

Beck, L. W. 1956. "Kant's Theory of Definition." *Philosophical Review* 65 (2): 179–191.

Boswell, T. 1988. "On the Textual Authenticity of Kant's Logic." *History and Philosophy of Logic* 9 (2): 193–203.

Capozzi, M. 1981. "Kant on Mathematical Definitions." In *Italian Studies in the Philosophy of Science*. Vol. 47, edited by M. D. Chiara, 423–452. Dordrecht: Boston Studies in the Philosophy of Science.

Carson, E. 1999. "Kant on the Method of Mathematics." *Journal of the History of Philosophy* 37 (4).

Dunlop, K. L. 2005. "Kant on the Reality of Mathematical Definitions." PhD diss., University of California, Los Angeles, California, United States.

Dunlop, K. 2011. "Kant and Strawson on the Content of Geometrical Concepts." *Noûs*.

Kant, I. 1979. *Werke (24 Bd.) Akademie Textausgabe*. Berlin: Walter de Gruyter GmbH.

Kant, I., and H. Allison. 1973. *The Kant-Eberhard Controversy: An English Translation*. Baltimore, MD: Johns Hopkins University Press.

Kant, I., H. Allison, P. Heath, G. Hatfield, and M. Friedman. 2010. *Theoretical Philosophy after 1781*, (The Cambridge Edition of the Works of Immanuel Kant in Translation). Cambridge: Cambridge University Press.

Kant, I., P. Guyer, and A. Wood. 1999. *Critique of Pure Reason* (The Cambridge Edition of the Works of Immanuel Kant in Translation). Cambridge: Cambridge University Press.

Kant, I., and J. Young. 2004. *Lectures on Logic* (The Cambridge Edition of the Works of Immanuel Kant). Cambridge: Cambridge University Press.

Maddy, P. 1999. "Logic and the Discursive Intellect." *Notre Dame Journal of Formal Logic* 40 (1): 94–115.

Rosenkoetter, T. 2009. "Truth Criteria and the Very Project of a Transcendental Logic." *Archiv für Geschichte der Philosophie* 91 (2): 193–236.

Shabel, L. 2010. "The Transcendental Aesthetic." In *The Cambridge Companion to Kant's Critique of Pure Reason*, edited by P. Guyer. Cambridge: Cambridge University Press.

Stuhlmann-Laeisz, R. 1976. *Kants Logik (Quellen und Studien zur Philosophie)*. Vol. 9. Berlin: Walter de Gruyter.

Sutherland, D. 2010. "Philosophy, Geometry, and Logic in Leibniz, Wolff, and the Early Kant." In *Discourse on a New Method: Reinvigorating the Marriage of History and Philosophy of Science*, edited by M. D. M. Dickson, 155–192. Chicago, IL: Open Court.

von Wolff-Metternich, B. -S. 1995. *Die Überwindung des Mathematischen Erkenntnisideals (Quellen und Studien zur Philosophie)*. Vol. 39. Berlin: Walter de Gruyter.

# Arbitrary combination and the use of signs in mathematics: Kant's 1763 Prize Essay and its Wolffian background

Katherine Dunlop

*Department of Philosophy, University of Texas at Austin, Austin, TX, USA*

In his 1763 Prize Essay, Kant is thought to endorse a version of formalism on which mathematical concepts need not apply to extramental objects. Against this reading, I argue that the Prize Essay has sufficient resources to explain how the objective reference of mathematical concepts is secured. This account of mathematical concepts' objective reference employs material from Wolffian philosophy. On my reading, Kant's 1763 view still falls short of his Critical view in that it does not explain the universal, unconditional applicability of mathematical concepts.

For scholars of Kant's philosophy of mathematics, Kant's "Inquiry Concerning the Distinctness of the Principles of Natural Theology and Morality" is of special interest for the partial way in which it anticipates his Critical view. Kant wrote the essay for a contest sponsored by the Berlin Academy of Sciences. The assigned topic was whether metaphysical truths "admit of distinct proofs to the same degree as geometrical truths", and if not, what is the "genuine nature" and degree of their certainty. In his essay, Kant draws a distinction between the methods of metaphysics and mathematics that is based in part on an account of mathematics' use of definitions. Both the distinction and the theory of definition are incorporated, as elements whose importance has been recognized,[1] into Kant's Critical view. Yet, the Prize Essay (as I will call it) is missing the most distinctive feature of the Critical view, namely that objects of mathematical reasoning are "given" to us in "pure intuition".

This raises a puzzle concerning the relationship of these doctrines and the importance of each to the first *Critique*. In particular, we must ask what Kant achieves by introducing the notion of intuition, whose "immediacy" and "particularity" is supposed to distinguish it from concepts (which, for Kant, always represent in a general manner). Since Kant makes explicit that all our intuition comes from sensibility, giving intuition a role in mathematics makes it necessary for him to explain how sensible representation can be sufficiently "pure" (independent of sense experience) to support reasoning to a-priori conclusions. On Jaakko Hintikka's well-known treatment of the issue, the notion of sensible intuition makes no essential contribution to Kant's view of mathematical method. The Prize Essay already specifies what is distinctive to the mathematical method, which is to consider particular representatives of general concepts (Hintikka 1967, 164, cited as reprinted in Hintikka 1974), and Kant conceives these representations of individuals as sensible in the *Critique* only because he acquiesces in the Aristotelian view that "we can come to know individual objects" only through the senses (Hintikka 1965, 132, cited as reprinted in Hintikka 1974). But other interpreters (such as Emily Carson and Paul Guyer) hold that the introduction of intuition is crucial, for in the earlier writings, Kant fails to account for mathematical concepts' applicability to objects of experience – what he will later call the concepts' "objective reality".

Part II of this paper argues that the Prize Essay contains the materials for an account of how mathematical concepts apply to extramental objects (more precisely, objects which do not depend for their existence on intellectual activity). This account of the concepts' objective reference is based on Kant's view of the use of signs in mathematical reasoning, specifically as representing "the universal *in concreto*" (2:278), through sensible tokens. Like Engfer (1982, 64–65), I take such representation to fulfill the role played by intuition in the *Critique*. Yet there remains a crucial difference between the views of the Prize Essay and the *Critique*, which is that in the latter Kant distinguishes between pure and empirical intuition. Because Kant takes pure intuition to express necessary conditions on empirical intuition (and thereby all perceptual experience), by showing that objects corresponding to mathematical concepts can be exhibited in pure intuition, he is able to guarantee the concepts' necessary applicability to *all* objects of experience. Whether such a guarantee is required to establish a concept's "objective reality" depends on how this technical term is understood. As Kant uses it in the *Critique* (such as at A155/B194 and A221/B268), "objective reality" seems to mean merely that an object corresponding to the concept can be given in intuition, but in the *Prolegomena*, mathematics' objective reality seems to be tantamount to the "necessary validity" of its propositions "for everything that may be found in space" (4:287). The important point, in any case, is that in the Prize Essay, Kant can show the former but not the latter.

However, some of the best scholarship on the Prize Essay (in particular, Rechter 2006) questions whether it gives a role to the perceptibility of mathematical symbols. So it is important to see how perception of signs figures in the view that Kant sought to displace. The Academy's question served to focus attention on Christian Wolff's programme to institute mathematical method in philosophy. To establish his conclusion – that philosophy should follow a different method than mathematics – Kant must consider how the use of signs features in Wolff's account of mathematical method and its application (to philosophy). Part I of this paper argues that Wolff sees an important role for concrete representation of sign-tokens (in the senses and imagination). Wolff holds, specifically, that the procedures of symbolic representation can establish the possibility (meaning at least the logical consistency) of concepts, and his view supplies Kant with the resources required for an account of concepts' objective reference.

I think this aspect of Wolff's view is not more widely appreciated[2] in part because it is assumed that because he closely follows Leibniz, Wolff must deny non-rational faculties a role in mathematical cognition. So I begin (in Section 1.1) by explaining how an important role for the perception of sign-tokens can be seen as consistent with writings by Leibniz that were known in the eighteenth century.

## 1. Wolff and Leibniz on the importance of signs

### 1.1 Leibniz's view that symbols "Represent Abstract Reasoning to the Imagination"

For the Academy's members, the modelling of metaphysics on geometry represented a striving for certainty, opposed to fallibilism and naïve reliance on the senses. They located its sources in seventeenth-century rationalism,[3] prominent among them Leibniz's notion of symbolic or "blind" cognition. I will now argue that the conception associated with Leibniz, of how the use of symbols furthers cognition, leaves room for a contribution by non-rational faculties, even if Leibniz does not stress its importance. To be clear, I am not offering my own interpretation of Leibniz's thought, but merely trying to explain how it might have been understood on the basis of texts available in the eighteenth century.

As Leibniz explains in *Meditations on Knowledge, Truth, and Ideas*, we may "make use of signs in place of the things themselves", as for instance "when I think about a chiliagon, that is, a polygon with a thousand ... sides, I don't always consider the nature of a side ... or of thousand-foldedness, but in my mind I use these words in place of the ideas I have of these things". He follows Descartes in taking the chiliagon as an example of an object whose complexity outruns sensibility's scope and precision, and must instead be represented with the aid of the intellect.[4] Leibniz claims that arithmetic and algebra exemplify this mode of thought (1989, 25). He elaborates on its utility for metaphysics, specifically its potential to eliminate error, in a letter to Henry Oldenburg of 1675 (published in 1699).[5] Leibniz speaks of a criterion that "renders truth stable, visible, and irresistible, so to speak, as on a mechanical basis". Algebra, which has "accomplished this much ... that truths can be grasped as if pictured on paper with the aid of a machine", is said to owe its usefulness to "a higher science, which I now usually call a *combinatorial characteristic*". Leibniz claims that "if this basis for philosophizing is accepted, ... we shall have as certain knowledge of God and the mind as we now have of figures and numbers" (1969, 166). From these remarks, it appears that sense perception can have a role in certain knowledge (even knowledge of things inaccessible to the senses), when it is directed to symbols that are generated and interpreted according to rules which are strict enough that a machine could implement them. (This feature of symbols could be seen to explain their epistemological significance, in the context of Leibniz's view of sense-perception as "confused" representation: it is precisely because we can give a complete account of the generation of symbols that they are not confused in the way perceptions are,[6] and therefore can be a basis for scientific knowledge.)

There is evidence that Leibniz holds not merely that certain knowledge can be attained with the help of non-rational faculties (the senses and imagination), but that it requires their use. In "Replique de M. Leibniz aux Réflexions continuës dans la seconde Edition du *Dictionnaire Critique*, de Mr. Bayle, Article Rorarius, sur le Systême de l'Harmonie préétablie", Leibniz insists that the "most abstract reasonings" are always accompanied by bodily movements corresponding to them, "by means of the symbols that represent them to the imagination" (Desmaizeaux, ed. 1720, 399; similar views are expressed on p. 409). The fact that this parallelism is accomplished through the same symbols that facilitate thought for the reasoner shows that their use is in a way necessary.[7] It also suggests that bodily involvement (specifically of the imagination) may be epistemologically significant in the same way as the use of symbols; that is, not merely as a necessary condition on abstract thought, but as a resource that reasoners can exploit.

In the letter to Oldenburg, Leibniz appears to grant that the access to "truth" yielded by the senses (as in algebra) can amount to certainty. Yet, it seems to be part of our understanding of Leibniz's rationalism that neither perception nor imagination can provide us with certainty (whatever else they may contribute to knowledge). Indeed, in the *Nouveaux Essais* Leibniz explicitly denies that "general certainty is provided in mathematics", specifically in geometry, by perceptual experience of figures.[8] (He is objecting to Locke's view that "particular demonstrations", carried out using drawn figures, can show propositions to be generally true, namely when we perceive the immutability of the "agreement of ideas"[9] that constitutes their truth (IV.i.9).) Leibniz maintains that

> Geometers do not derive their proofs from diagrams, although the expository approach makes it seem so. The cogency of the demonstration is independent of the diagram, whose only role is to make it easier to understand what is meant and to fix one's attention. It is universal propositions, i.e. definitions and axioms and theorems which have already been demonstrated, that make up the reasoning, and they would sustain it even if there were no diagram. (IV.i.9, 360–361)[10]

Leibniz reinforces the separation between what "makes up" and "sustains" reasoning, and what merely facilitates our comprehension, by distinguishing "ideas" from "images". The chiliagon is an example of an object of which "I cannot have [an] image", because "one's senses and imagination would have to be sharper and more practiced if they were to distinguish such a figure from one which had one side less". Leibniz claims, however, that I have an idea of the chiliagon, deriving from my understanding of the number 1000, through which "I know [its] nature and properties very well". He concludes that "knowledge of figures does not depend upon the imagination, any more than knowledge of numbers does, though imagination may be a help" (II.xxix.13, 261; *cp.* IV.ii.15, 375).

It is reasonable to suppose that the need to articulate his differences with Locke in the *Nouveaux Essais* prompts Leibniz's most careful statement of his view (of experience's role in knowledge). Then to the extent that the other eighteenth-century texts conflict with the *Nouveaux Essais*, they must be said to give a distorted view of Leibniz's thought.[11] Even in that case, however, it can be allowed that they would have appeared definitive until the publication of the *Nouveaux Essais* in 1765.[12]

The *Nouveaux Essais* itself, however, supplies additional evidence that perception of symbols is both necessary and conducive to abstract thought. Leibniz begins his answer to I.i.5 of Locke's *Essay* by granting to Locke that the rationale for counting a certain truth as innately known will extend to many other truths, so that "on this view the whole of arithmetic and of geometry should be regarded as innate", and "one could construct these sciences in one's study and even with one's eyes closed". But Leibniz then concedes that "if one had never seen or touched anything, one would not bring to mind the relevant ideas."

> For it is an admirable arrangement on the part of nature that we cannot have abstract thoughts which have no need of something sensible, even if it be merely symbols such as the shapes of letters, or sounds; ... If sensible traces were not required, the pre-established harmony between body and soul ... would not obtain. (I.i.5 77)

It thus appears that symbols may be essential to reasoning, even if images in general are not. Later, Leibniz explains how symbolism can further reasoning by providing "an example which can be grasped by the senses and which can even serve as a check". He proposes that in the final stages of solving an algebra problem (to find two numbers whose sum is 10 and whose difference is 6), "instead of putting $2a = 16$ and $2b = 4$, I could have written $2a = 10 + 6$ and $2b = 10 - 6$; this would have given me $a = \frac{1}{2}(10 + 6)$ and $b = \frac{1}{2}(10-6)$". Then the very same symbols 10 and 6 would be taken both "for the numbers they ordinarily signify", thus furnishing an example for the senses to grasp, and "for general numbers like the letters $x$ and $y$, so as to get a more general truth or method". Leibniz elaborates that it is "very helpful to use numbers in place of letters in extended calculations, for avoiding mistakes and even for carrying out checks (e.g. by casting out nines) in mid-calculation without waiting for the final result". What makes this possible is that "the numbers [are selected] shrewdly, so that the assumptions turn out true in the particular case" (IV.vii.6, 410–411). Here, it is crucial that the symbols concretely exemplify the general notions they express. (It is perhaps unclear how the inscriptions "10" and "6" could be perceived as "the numbers they signify", but the numeral strings of the formal systems discussed below obviously could fulfill this function.)

Leibniz concludes that numerical symbols "are more suitable than letters, even in algebra", because they are "useful in displaying connections and patterns which the mind would not be able to sort out so well by letters alone" (IV.vii.6, 410–411). Insofar as this suggests that some results can be found *only* by means of symbols that exemplify what they express, it seems to go beyond his concession that imagination "may be a help" to the intellect (II.xxix.13 261).

To be sure, these passages hardly do more than hint at a role for symbols in mathematical thought. But for my purpose, it is enough to show that Wolff could have granted them an important role without taking himself to disagree with Leibniz. The additional evidence from the *Nouveaux Essais* (which was of course not available to Wolff) suggests that such a reading might not involve much strain.

## 1.2   Wolff on the "symbolic art"

Leibniz's conception of a universal "characteristic" is the backdrop to the remarks about symbolism quoted earlier. In the letter to Oldenburg, Leibniz says of the "combinatorial characteristic" which subsumes algebra that he "cannot encompass the method in a few words" but hopes "sometime, given health and leisure, to explain its remarkable force and power by rules and examples" (1969, 166). Similarly, the "Replique de M. Leibniz ..." ends with an allusion to "a calculus more important than arithmetic and geometry, and which depends on the analysis of ideas. This would be a universal characteristic, whose construction seems to me one of the most important things that anyone might attempt" (Desmaizeaux ed. 1720, 455–456). The collection in which it appeared also contains a letter (to Nicolas Remond) in which Leibniz speaks of the universal characteristic as an unfulfilled ambition:

> [I]f I had been less distracted, or if I were younger or had talented young men to help me, I should still hope to create a universal symbolistic [*spécieuse générale*] in which all truths of reason would be reduced to a kind of calculus. At the same time this could be a kind of universal language or writing, though infinitely different from all such languages which have so far been proposed, for the characters and the words themselves would give directions to reason ... (Desmaizeux ed. 1720, 131–132; translation Loemker, Leibniz 1969, 654)

Wolff writes, with reference to the letter to Oldenburg, that Leibniz had "a notion" [*Begriff*] of a "symbolic art" that he called "*ars characteristica combinatoria*" ([1751] 1983, §324, 180). As I will now explain, Wolff seems to regard Leibniz's failure to develop the universal characteristic as a missed opportunity to remedy the haphazard presentation of his views.

In his introduction to an edition of the Leibniz-Clarke correspondence, Wolff complains that Leibniz "never put metaphysical truths in a correct order, but rather only spoke of them every now and again, as he had occasion, in the common manner" ([1737] 1981, 279). He recommends working through his own (German-language) textbook of metaphysics in order to better understand Leibniz's views (280–281).

Wolff suggests that a better grasp of metaphysical doctrines, Leibniz's in particular, would be achieved by applying algebra to philosophy. Both in his essay introducing the Leibniz-Clarke correspondence, and in the preface to (the second edition of) his German metaphysics textbook, Wolff first claims to give better explanations of certain Leibnizian doctrines (such as pre-established harmony), then proceeds to explain how "one can elucidate [*erlaütern*] metaphysical concepts more than a little through examples from mathematics, especially from algebra" ([1737] 1981, 289). Wolff uses algebra to show how a contradiction can arise from a combination of concepts, and thus to illustrate (by contrast) the notion of possibility. In his introduction to the Leibniz-Clarke correspondence, he claims to show by this means how mathematics can elucidate metaphysical concepts, and to improve on proofs that Leibniz draws "from very general grounds", which "pertain to things foreign to the imagination" (283).

Wolff shares Leibniz's conception of the *ars characteristica combinatoria* as a sort of generalized algebra that "visibly" exhibits connections which make judgments true. In his German textbook of metaphysics, Wolff claims that algebra exemplifies the "symbolic art", which permits one to "hold up to one another composite signs that are indifferent to the concepts", see from them [*daraus ersehen*] how things (thought by means of the concepts) are related to one another, and thus "put before the eyes what is to be met with in a thing" and how things are to be distinguished from one another ([1751] 1983, §324, 179).[13]

These views of Wolff's together suggest that Leibniz's failure to develop the *ars characteristica combinatoria* explains why he did not give his views an orderly, specifically an algebraic, presentation.

Wolff does not explain how algebraic reasoning can apply within philosophy more precisely than in terms of "equivalent substitutions" ([1738] 1968, 214–215). Specifically, the containment relations (of predicates in subjects) that make judgments true are exhibited by replacing given concepts with "equivalent" combinations of simpler constituents.[14] But Wolff holds that concepts can be ordered according to generality in "Porphyrian" trees (so-called after Porphryry's Introduction to Aristotle's *Categories*), so that every (decomposable) concept has for its constituents (a) the genus that contains it as a species and (b) a "specific difference" characterizing the objects that fall under it.[15] (It is worth noting that the concept can be defined in terms of its genus and specific difference, so these can substitute for the concept as its *definiens*.) By laying out this structure, the procedure of equivalent substitutions makes it possible to generalize knowledge about particular species of things, both "upwards" to higher genera and "laterally" to other species contained in the genus. Wolff regards this sort of expansion of knowledge as integral to philosophy (1963, 25–26).[16]

In spite of its vagueness, Wolff's explanation of how artful symbolization leads to new discoveries is illuminating. To show how "examples from mathematics, especially from algebra", can elucidate metaphysical concepts, Wolff explains the genus–species relationship in terms of the variable places in a formula for generating "polygonal numbers",[17] that is, for finding partial sums of certain arithmetical sequences ([1751] 1983, "*Vorrede zu den andern Auflage*", *n.p.*). Corresponding to the genus and species are the variables that specify, respectively, a sequence and the number of terms to be added. As an expression for the genus, this formula (with the variables set to appropriate values) can replace terms for particular polygonal numbers. But the formula also expresses in a general way every combination of primitives (i.e. natural numbers) that yields a polygonal number. So it can be taken to express the conditions under which polygonal numbers are possible. Since Wolff understands possibility as the absence of contradiction ([1751] 1983, §12, 7), the condition for possibility in general is that no predicate is combined with its negation. Without spelling out precisely how natural numbers are to play the role of predicates and their negations,[18] we can take the formula to illustrate, in a general way, a means for showing which combinations of primitives issue in objects that are possible (i.e. whose descriptions are not contradictory).

The relationships among things in virtue of which they can be combined to yield a possible object are surely among the most important of those "put before the eyes" by means of symbols. This role for symbolization is (I will suggest) emphasized and more sharply articulated by Kant in "Distinctness".

Of course, combinations are guaranteed to be consistent only so long as the combined elements have no components that contradict one another. As long as the elements retain some complexity, contradiction is still a threat. Wolff claims that the *ars characteristica combinatoria*, which, if realized, would verify possibility in the way just indicated, is "most difficult" to discover because few if any concepts have been resolved into their ultimate constituents (elements with no further compositional structure) ([1738] 1968, §301, 215). One way to put the point is that we do not have ideas to which to assign the "characters", or primitive signs, of the *ars*.[19] But Wolff is generally sceptical of both the possibility of and the need for complete resolution. In one of his most prominent works, he claims "it is no ways necessary, and very rarely possible, for us to bring ... analysis to a conclusion, that is, to carry it on until we come to such notions as cannot be farther analyzed; but we may be satisfied to carry it on so far as is necessary to our present purpose".[20] (Thus, an objection often attributed to Kant,[21] that the absence of complete analyses makes it impossible for philosophy to emulate mathematics' use of signs, is somewhat off-target.) Accordingly, while Wolff conceives the *ars characteristica combinatoria* as an intermediary through which mathematics ("especially algebra") can be applied to philosophy ([1751] 1983, "*Vorrede zu den andern Auflage*", *n.p.*), he does not think this application requires its full implementation. For, he holds algebra's usefulness for philosophy is already clear from actual examples, even though the *ars* exists only as an idea.

It is reasonably clear how the substitutional procedure typical of algebra could serve the purposes Wolff has in mind, even without thoroughgoing analysis. By factoring an equation into several of lower degree, we may be able to tell whether its roots are complex, negative etc. without yet solving to find the roots. Such information has implications for the constructibility of the roots, and hence, as I will now explain, for their possibility.

Lisa Shabel has shown that in the early modern period in general, and in Wolff's *Elementa Matheseos Universae* ([1742] 1968) in particular, algebra is not considered an autonomous discipline but rather "an art or method which aids in the solution of certain arithmetic or geometric problems" (Shabel 1998, 600). The confinement of algebra to a subsidiary role seems to arise from the view that mathematics' subject matter consists in magnitudes, as classically conceived, for these are already the objects of geometry and arithmetic (and there are no other properly mathematical objects for algebra to study). On the classical understanding, both geometrical and arithmetical magnitudes are "spatially and temporally extended, particular, and relatively concrete", as Daniel Sutherland puts it (2004, 158). In ([1742] 1968), Wolff conceives numbers in terms of their relation (commensurability or incommensurability) with unity, which he prefers to represent as a line segment, so that his understanding of "the objects of arithmetic ... ultimately relies on traditionally geometric concepts", as Shabel explains (1998, 600). Since Wolff thus requires arithmetical quantities (as well as geometrical ones) to be geometrically represented, a symbolic expression of an unknown quantity (*e.g.* an equation "in" $x$ or $y$) is not yet a solution to the arithmetical or geometrical problem to which algebra is applied. Given such a formula, it still remains to construct the quantity geometrically (*cf.* Shabel 1998, 605).

If the quantity cannot be geometrically represented, the problem has no solution. For Wolff, negative quantities are examples of quantities that can be designated in symbols but cannot be geometrically constructed.[22] Since we are now used to representing such quantities graphically (in the lower and left-hand quadrants of a coordinate plane), a better example might be the square root of a negative number.[23] Conversely, a quantity that is designated as positive or as the root of a positive number is thereby shown to be possible, or at least to pass an important test for possibility. This is another way in which symbolization might be taken to reveal the conditions under which a combination results in a possible object.

## 1.3   Wolff on analysis and synthesis

To the extent that Wolff's interest in symbolization has been overlooked,[24] there are several likely reasons. First, as important as the *ars characteristica combinatoria* may be for metaphysics, it is peripheral to Wolff's overall theory of cognition.[25] Second, Wolff inherits from Leibniz a multivalent distinction between "analysis" and "synthesis" that complicates his understanding of the mathematical method (and its application to philosophy).

As Leroy Loemker helpfully summarizes, Leibniz has both a "combinatorial" understanding of analysis and synthesis, on which they "operate with logical entities treated as terms", and a "Euclidean" one, "which deals with propositions in the form of theorems" and the "components" of a deductive structure (namely axioms, definitions and already-proved theorems) (1966, 146; *cf.* Engfer 1982,

195). The combinatorial notion of analysis is the early modern interpretation of the method described by the fourth century A.D. commentator Pappus, which takes what is sought as "assumed" and designates both this unknown and the known quantities by signs, so that they can be treated indifferently (Engfer *loc. cit.*). It is illustrated by algebra (in which quantities are designated by combinations of signs for the known and unknown quantities and for operations performed on them). Loemker finds that Leibniz "never gave a careful account of the relation between these two models" (*loc. cit.*), but in his employment of them, the former seems to dominate.[26] For on the Euclidean model, axioms, postulates, and definitions all have the same standing, as primitives. But Leibniz sometimes[27] calls for axioms and postulates to be "demonstrated", by which he means, reduced to identities. Such reduction is achieved by successively substituting definitions for *definienda*. Insofar as concepts are defined as combinations of simpler concepts, a prior application of analysis in the algebraic or combinatorial sense is required to ground the starting points for synthesis in the Euclidean sense.

In contrast, Wolff's most prominent remarks on mathematical method characterize it in terms of deductive organization: specifically, as proceeding from definitions to axioms [*Grundsätze*] and thence to theorems ([1750] 1999, §1, 5), so that all terms used are "accurately defined", and only what is "sufficiently demonstrated" is "admitted as true" (1963, 76). This conception, modelled on Euclid's *Elements* rather than algebra, does not involve substituting signs for things. On this basis, one could doubt whether symbolic representation of relationships is part of the method that Wolff seeks to export from mathematics into philosophy.

But just as Leibniz seems to hold that analysis and synthesis apply, in the same general manner, both to terms and to propositions, so Wolff fails to sharply distinguish between *ars characteristica combinatoria* and the logical structure of theories. For he claims that the conclusions of syllogisms are formed by combining (terms in) their premises ([1740] 1983, §§333–335, 290), in what seems very similar to the manner in which, in the *ars*, composite signs (and presumably the concepts corresponding to them) are formed from primitive ones (expressing "primitive or irresoluble concepts" [*notiones*], [1738] 1968, §300, 214). More generally, on his view theories are ordered deductively so as to proceed from what is simplest (and also best known or most certain), at the level of concepts as well as of propositions. Accordingly, Wolff's logic textbooks (as was typical for the eighteenth century) first treat the formation of concepts, then the formation of judgments and inferences; and a principal way of forming concepts is by combining simpler concepts ([1754] 1965, I.30–33, 139–140). However, Wolff does distinguish deductive structure from systematic symbolic representation to the extent that he claims to have organized philosophy deductively at the propositional level, whereas he regards the art of using and combining signs as a not-yet-actual discipline. Deductive structure and symbolic representation are also distinguished insofar as on Wolff's view, deductive organization at the conceptual level is made possible by systematic symbolization. Since the symbolic system is not yet in place, the compositional ordering of concepts, from simpler constituents to more complex wholes, lacks a fully rational basis (as Wolff seems to admit when he says analysis is carried "only so far as necessary for our present purpose").

Wolff's account of analytic and synthetic method in mathematics also supports the view that mathematics' method of proof – the source of the certainty that he aims to export – can be understood (at a sufficiently deep level) to encompass the symbolic techniques characteristic of algebra. Wolff conceives the methods as ways of ordering knowledge, that is as static patterns, rather than as directives or routines.[28] He identifies the "analytic method" with the order by which truths are or can be discovered ([1740] 1983, §885, 633) and claims it is followed in algebra ([1716] 1965, 51–52), which implies that artful use of signs is part of the method. Wolff claims that "synthetic method" orders truths in the sequence in which they can be most easily understood and demonstrated, and was used by the ancient geometers ([1740] 1983, §885, 633). As Wolff describes the "mathematical" method, in the passages quoted earlier, it appears to coincide with the synthetic. Wolff claims that it does – but only if what is "mixed" is prescinded from [si a mixta discesseris] (§887, 634). This clearly leaves room for an analytic component within mathematical method. More generally, Wolff takes the method to deliver both an expansion of scientific knowledge and certainty (1963, 76), so that the ordering it imposes is both analytic and synthetic at once.

## 2. Definition and the use of signs in Kant's Prize Essay

### 2.1 Weak and strong arbitariness in the Prize Essay

Kant's essay is organized into four "Reflections", only the last two of which directly address the Academy's question. The First Reflection is a "general comparison" of the methods of mathematics and philosophy. It shows, according to the summary with which the Second Reflection begins, that the differences between philosophical and mathematical cognition are "substantial and essential".

The three main points of the summary roughly correspond to the section headings of the First Reflection:

(1) In mathematics, concepts are made available through definitions. In philosophy, they are "given" in a confused manner (2:283).

In the Leibnizian tradition in which Kant is working, "confusion" is opposed to "distinctness", which can be understood as articulation into discrete (but not necessarily irresoluble) components and is tantamount to definability. So, in philosophy, definitions are only to be sought. But they are not required as starting points, because

(2) In philosophy, "immediately certain characteristic marks" of the object take the place of definitions, as a basis for indemonstrable propositions.

Finally,

(3) In mathematics, signs are employed with a certain [*sicher*] significance, which is just what "one wished to attribute" to them. But in philosophy they acquire their meaning through "linguistic usage" (2:284), and often the same word is used to express concepts among which "considerable" differences are "concealed" (2:285).

The feature of mathematical symbolism highlighted here – that signs acquire their meanings by stipulation – reflects a certain arbitrariness in mathematical thought. This arbitrariness also manifests in its definitions, and is supposed to explain their accessibility. The first substantive claims of the Prize Essay are that concepts can be attained "either by *arbitrarily combining* concepts, or by *separating out* cognition which has been rendered distinct" by analysis, and that mathematical concepts are attained the first way. (For example, the trapezium is defined by "thinking arbitrarily of four straight lines bounding a plane surface so that the opposite sides are not parallel to one another" [2:276].) The results of combination and separation are expressed by the concepts' definitions. Kant indicates that in mathematics, arbitrary combination is allowable because there are no antecedently given concepts with which the definitions must agree.

There are clear suggestions in Kant's pre-Critical work that mathematical concepts are arbitrary not only in the way they are made distinct, but also in not having to correspond to anything not produced by the mind's combinatory activity. Notes of his logic lectures indicate that mathematical concepts can be defined through combination precisely because they are arbitrary in this stronger sense. When a concept is up to me to "make up", it "has no other reality than merely what my fabrication wants[;] consequently I can always put all the parts that I name into a thing[,] and these must constitute the complete ... concept of the thing, *for the whole thing is actual only by means of my will*" (24:268, emphasis added). Another set of notes asserts that a mathematician "thinks everything that suffices to distinguish the thing from all others, for [it] is not a thing outside him ... but rather a thing in pure reason, which he thinks of arbitrarily and in conformity with which he attaches certain determinations" that distinguish it (24:125). While these accounts of Kant's view are not authoritative, the same line of thought can be discerned in the Prize Essay itself. Kant claims that mathematics does not define a given concept but rather "defines an object by means of arbitrary combination" (2:280).[29]

The differences between mathematics and philosophy in regard to first principles and use of signs are neatly explained by supposing mathematics to be arbitrary in the stronger sense.

In comparing mathematics' and philosophy's "indemonstrable propositions", Kant "seems to speak of their claims to truth in different ways", as Emily Carson puts it (1999, 639). He describes the latter as "fundamental truths", but says that the former are "regarded as immediately certain" and "immediately presupposed as true" (2:281), roles which they may fulfill without being true (or certain). The simplest way to understand this prospect is, in accordance with the usual definition of truth,[30] as a lack of correspondence between the propositions and objects (not produced by the mind's own combinatory activity).

While in his summary Kant focuses on the absence of equivocation, it seems the real usefulness of mathematics' notational conventions is that signs can replace "universal concepts of the things themselves" in reasoning (2:279). (Wolff also stresses the advantages of signs for general reasoning; see [[1751] 1983] §319, 177.) Kant claims that the words in which philosophical reasoning is carried out "can neither show in their composition the constituent concepts of which the whole idea, indicated by the word, consists" nor "indicate in their combinations the relations of the philosophical thoughts to each other" (2:279). So to discover these relationships, the things themselves must be considered – but always "in their abstract representation", to ensure generality (2:291). Whatever "abstract representation" is supposed to mean, it is clearly supposed to contrast with representation through signs. For Kant claims that in arithmetic and algebra, by contrast, one first "posits not things themselves but their signs, together with the special designations of their increase or decrease, their relations *etc*." and thereafter follows "easy and certain rules" for combining and transforming the signs (2:278). That the rules have been followed can be known "with the degree of assurance characteristic of seeing something with one's own eyes", merely by thinking signs "in their particular cognition which, in this case, is sensible in character" (2:291). In philosophical reasoning, on the other hand, constant vigilance is required to make sure that no element of an abstract concept is overlooked.

The important point for present purposes is that for Kant, certainty is attained by verifying that the signs are produced according to the rules.[31] But at no stage of this process are the signs compared with "things themselves". So following the rules will not guarantee correspondence with objects (other than those produced by combination). Hence, it seems, knowing them to have been correctly applied suffices for certainty only because no such correspondence is required.[32]

## 2.2   *Problems with the strong understanding of arbitrariness*

As was noted earlier, the supposition that mathematical concepts are arbitrary in the stronger sense (of not having to correspond to anything besides what the mind's combinatory activity produces) explains some of Kant's contrasts between mathematics and philosophy, and appears to enjoy strong textual support. But it also faces several problems. First, it makes one of Kant's main argumentative moves appear bizarre. The fourth section of the First Reflection argues that "the object of mathematics is easy and simple, whereas that of philosophy is difficult and involved". To compare the "objects" of mathematics and philosophy clearly requires that each has an object. Now, Kant claims that with respect to its object (namely quantity), mathematics only considers "how many times something is posited". According to transcripts of his logic lectures, in mathematics positing does not "actually establish anything at all that lies in the object, but only what I arbitrarily wish and want to have in it" (24:230). So it could be argued that mathematics' so-called object is really an arbitrary product of stipulation. But if that were mathematics' distinguishing feature, then the comparison between objects, in terms of their simplicity, would be an unnecessary detour on the way to Kant's conclusion. Rather, the thesis for which Kant is arguing (namely that mathematics achieves certainly more readily) would seem to follow immediately from the view that mathematical concepts differ from philosophical ones in not having to correspond to an extramental reality.

The supposition that Kant understands "arbitrariness" in the strong sense also makes Kant's position unstable. First, it fails to explain the overall shape of his opposition to Wolffianism. Carson shows that in other work of this period, Kant rejects metaphysicians' efforts "to turn mathematical concepts into subtle fictions, which have little truth to them outside the field of mathematics" (2:167). His example is that the view that space consists of simple parts, which is based on "ambiguous consciousness" of a concept of space "thought in an entirely abstract fashion", is defended against objections "by raising a specious objection against [mathematics], and claiming that its fundamental concepts have not been derived from the true nature of space at all, but arbitrarily invented" (2:168). But on the interpretation we are now considering, this charge is hardly "specious", and Kant's denial that the concepts can be treated as "fictions" seems not to rest on any argument. Second, Kant does not supply a reason why arbitrary invention is legitimate in mathematics but not metaphysics.

Finally, this interpretation fails to explain how Kant's view of definitions in mathematics survives the Critical turn. In the Prize Essay, Kant does not yet hold that processes of mathematical construction and formal conditions of sense perception are represented in pure intuition. In the *Critique*, he offers this as an account of how mathematics "is in its complete precision applicable to objects of experience" (A165/B206). But while affirming mathematics' applicability, he continues to hold that its concepts can be defined because they are "arbitrarily thought" (A729/B757). Since on the strong interpretation "arbitrariness" is incompatible with applicability, it must be supposed to have a different meaning here. But the remarkably close correspondence between the Prize Essay and these paragraphs of the *Critique* tells against such a shift.[33]

### 2.3 The Formalism of the Prize Essay

Having shown how the absence of an account of mathematics' truth (and especially its concepts' applicability) seems to undermine Kant's objections to Wolffianism, Carson concludes that in the early 1760s, he lacks the resources to distinguish his position from "the formalism [he] so obviously opposes". His position thus "threatens to collapse into" the formalism he rejects (1999, 643). The understanding of formalism as an analogy to a game – on which mathematics consists of moves made according to arbitrary rules and yields no knowledge of objects (not introduced by those rules) – is standard.[34] But "formalism" is also used with a wider meaning, to refer to any of a family of views that advocate a nonrepresentational role for language in mathematical reasoning.[35] Allowing signs to function nonrepresentationally in some context is consistent with requiring results thus found to be interpretable in a domain, which avoids the "collapse" into the view Kant opposes. That arithmetic and algebra are ultimately subject to a condition of interpretability is suggested by Kant's claim that in operating with their signs, "the things signified are completely forgotten ... until eventually, when the conclusion is drawn, the meaning of the symbolic conclusion is deciphered" (2:278).

It can be held that the reference of arithmetical discourse is secured precisely by operations on numeral tokens that do not treat them as referring. Thus, Charles Parsons observes in his classic (1969) that the generation of sequences of numeral tokens satisfies the existential presuppositions of arithmetical identities. Parsons argues that on Kant's mature view, intuition gives content to arithmetic by representing such procedures (concretely, in space and time). Ofra Rechter's detailed investigation of the Prize Essay shows it to contain some elements of such a view. This, I suggest, is the account of mathematics' objective reference that Kant needs to distinguish his view from the Wolffians'.

Rechter shows that in lectures given while he worked on the Prize Essay, Kant treats the Arabic numerals as a system in which the symbols "1" through "9" are defined by means of the successor operation, and further rules associate the concatenation of numerals with addition and multiplication (so that the numeral at place $n$ represents a multiple of $10^{n-1}$ and the string represents their sum). Rechter argues that addition and multiplication according to the usual algorithms (on which addends and multiplicands are vertically aligned in columns) match Kant's description of a procedure in which one "operates with signs according to easy and certain rules". During calculation, the "things symbolized" are "forgotten" in the sense that the numeral occupying a certain place is not regarded as a multiple of that power of 10. But the numeral is treated as an argument for addition or multiplication, which are defined on numbers, therefore as referring (rather than as a "meaningless" symbol or "mere syntactic entity"; [2006] 36). The system's rules guarantee reference for the symbols, specifically, a unique significance for every well-formed numeral string.

Here I will not argue on behalf of Parsons and Rechter, but will assume that their accounts are basically correct. It follows that in both the Prize Essay and the *Critique*, consideration of the role of signs in mathematics reveals why some measure of arbitrary invention is appropriate, without assuming that mathematical cognition pertains only to objects created by the mind's combinatory activity.

But in the Prize Essay, the view that quantity is constructed "symbolically" in pure intuition (A717/B745) is not yet in place. Indeed, Kant does not seem to clearly articulate a role for sensible representation in mathematical cognition. This is why Hintikka attributes to him the view that mathematical method involves merely "the use of individual instances", which is only a "starting point" for the theory that mathematics requires sensible intuition (1967, 164).

The role Rechter finds for sensibility is sharply delimited. Rechter emphasizes that Kant attributes the "degree of assurance" characteristic of perception to knowledge *that* the rules have been applied. On her interpretation, only the soundness of the system itself can rule out error. Hence, the basis of mathematics' certainty is the "metatheoretical" claim that false statements cannot be derived by (correct) application of the rules. But this could not itself be perceptually evident; Rechter claims that Kant only "likens it to" the deliverances of perception (2006, 25).

Now in the *Critique*, algebra is introduced to illustrate how mathematics proceeds "through a chain of inferences that is always guided by intuition" (A717/B745). Kant claims that algebraic treatment of equations "displays by signs in intuition the concepts [and] secures all *inferences* against mistakes by placing each of them before one's eyes" (A734/B762, emphasis added). In the Prize Essay, algebra (called the "general arithmetic of indeterminate magnitudes") and ordinary arithmetic are introduced under the heading of how universals are examined (through signs) in "analyses, proofs, and inferences". But the operations which Kant says are symbolized are just "increase or decrease" (along with unspecified "relations"), not the full complement of arithmetical operations as in the *Critique*.[36] According to the Prize Essay, one operates "with these signs ... by means of substitution, combination, subtraction" (2:278) and unspecified "transformations" (presumably those governed by laws of associativity and commutativity), rather than "exhibit[ing] all the procedure by which magnitude is generated and altered" (as in the *Critique*, A717/B745). The differences in phrasing suggest that in the Prize Essay, Kant may understand these operations as concatenations and deletions by which *terms* are formed, rather than as transformations that take us from one *equation* to another. In that case, perception of sign tokens might function to secure the reference, rather than the truth, of arithmetical formulae.

Rechter gives reasons to doubt it does even this.[37] She notes that Kant does not "moot" the question of the numerals' reference "by identifying the 'object of the concept' with the posited sign, regarding its reference reflexively, as it were" (2006, 37). Instead, he seems to assume that numerals "behave like names" in that they are fixedly assigned to particular denotata (which there is no reason to regard as perceptually given).[38]

I think Rechter is right that for Kant, the abstraction from reference that arithmetical thought involves is not so radical as to permit the assignment of new referents to signs. But it may abstract from the numerals' reference in the weaker sense of treating the numerals merely as sign tokens, without identifying the tokens as the signs' referents. This would be "formalism" in the wider sense introduced above. In this case, perception of the tokens can validate the existential assumptions of arithmetical formulae (and thus play a greater role than Rechter allows). This can be the case if, specifically, the operations by which tokens are formed (which Kant calls "combination" and "subtraction") are isomorphic to those by which numbers themselves are generated (which Kant designates as "increase" and "decrease").[39] If the latter are understood as addition and subtraction of *unity*, then the Arabic numerals do not meet this condition. Still, it is satisfied by the "strokes and points" to which Kant adverts at A240/B299, as by the "strings" of modern formal arithmetic. The generation of a token would verify the existence of the signified number in the sense that the collection of concrete objects thus produced would be a bearer of the number; perception of such a collection is the most obvious candidate to qualify as perceptual verification of the number's existence. This is a sharpening of Wolff's view that mathematical signs can literally show, in their relations, that a combination is such as to produce a possible object.

Even if the error ruled out by the use of signs involves reference failure rather than falsity, the claim that it is ruled out would again be metatheoretic (or more precisely metalinguistic). And we granted that in general, such claims are not perceptually evident. But in this case, perception needs to only show that for every step of combination or subtraction, an increase or decrease (by 1) of number is possible. I suggest that it verifies this possibility just by showing that a stroke can be added to or removed from a collection. The claim is, however, supposed to hold no matter how large the collection (of strokes) to which combination or subtraction is applied, and we must set aside the delicate issue of the extent to which claims involving indefinite iteration can be perceptually evident.

On this reading of the Prize Essay, the use of signs in arithmetic corresponds closely to their use in geometry. Specifically, in both cases, signs serve as "sensible means to cognition" (2:291) by making it possible to perceptually verify that finite configurations can be extended. Kant claims that the infinite divisibility of space can be "recognized with the greatest certainty" by means of a symbol [*Symbolo*], if one "takes a straight line standing vertically between two parallel lines" and draws lines from a point on one of the parallels "to intersect the other two lines" (2:279). Here, the segments into which the vertical line is divided (at any given stage) are themselves divided (in the next stage) by the construction of new lines, which form ever more acute angles with the first parallel and intersect the second at ever greater distances. Thus, the possibility of extending the second parallel establishes the possibility of dividing the vertical line. The only clear way in which the use of a diagram could be held to make this reasoning maximally certain is by making evident (to perception) that the second parallel can be indefinitely extended. In the same way, I have suggested, it is perceptually evident through the use of signs that any given number can be increased (through successive addition of units).[40]

## 2.4 Why metaphysics cannot follow mathematics' method

We are primarily concerned with the Prize Essay's account of mathematical method, not its argument that metaphysics cannot follow the method. But it will be worthwhile to consider this argument's main moves. Kant first links the use of signs in mathematics to the thoroughness with which we grasp mathematical concepts. This is relevant in considering how Kant moves beyond the view of the Prize Essay and will be touched on below. Kant then links the latter feature of mathematical thought to the simplicity of mathematics' object.

We have now seen how the use of signs can legitimate the procedure of arbitrary combination by which arithmetical concepts are defined: by allowing for perceptual verification of the existence of objects (or properties, such as indefinite extensibility) corresponding to the concepts. At a superficial level, it is clear why this cannot occur in metaphysics. Philosophical reasoning is carried out in words, which (Kant claims) cannot "show in their composition the constituent concepts of which the whole idea, indicated by the word, consists" (2:279). But this just describes philosophy's current state of development. The real issue is whether words could be made dispensable by a system of the kind envisioned by Leibniz and Wolff, whose signs *would* "show in their composition" the constituent structure of concepts.

Kant does not pose the question in these terms. But he evidently holds that if a sign system were to mirror the constituent structure of (and relations between) philosophical concepts, we would not recognize that it did.[41] His view is that concepts are "given" in a "confused" manner to the philosopher, whose "task" or "business" is to make them distinct (2:278). Specifically, the philosopher is to abstract out "characteristic marks", which must be "collated with each other" to eliminate redundancy, "combined together", and "compared with" the given concept "in all kinds of contexts" (2:276–7), to see whether they capture its application conditions. So a "given" concept's constituent structure can be discerned only through strenuous effort. Moreover, a full account of its structure (i.e. a definition) is the culmination of philosophical research; as Kant puts it, "far from being the first thing I know about the object, the definition is nearly always the last thing I come to know" in philosophy (2:283). So, if we were in position to adopt signs exhibiting the structure of philosophical concepts, there would be no philosophical work left for them to facilitate.

Kant's assertion that mathematical and philosophical concepts are apprehended in different ways appears subject to the same objection as his original claim that the disciplines use different kinds of linguistic representations. Namely, the confusion that impairs grasp of philosophical concepts might only reflect philosophy's incomplete development. It seems that Kant must do more to show that the confusion will persist as long as philosophy has work to do.

Kant offers considerations to "put beyond doubt" that in philosophy, what is "initially and immediately perceived" in a thing must "serve as an indemonstrable fundamental judgment" on which to base its definition (2:282). These seem designed to show specifically that the confusion that attends philosophical concepts is ineluctable. Kant blames the confusion on the multifariousness and obscurity of philosophy's subject matter. There are "infinitely many qualities which constitute the real object of philosophy", and while mathematics' sole object, quantity, is "easy and simple", it is "an extremely strenuous business" to distinguish between the qualities (that is, to specify the marks that comprise their distinct concepts, or definitions) (2:282).[42]

This section of the Prize Essay contains a striking anticipation of Kant's Critical view. Having asserted that mathematical concepts do not require analysis prior to being defined, Kant concludes that mathematics does not define a concept through analysis, but rather "defines an object by means of arbitrary combination" (2:280). This has been understood to mean that mathematical objects themselves result from the mind's combinatory activity. But that interpretation does not accord with Kant's explanation of why philosophy does not share mathematics' power to define concepts. His explanation contrasts the objects of mathematics and metaphysics in terms of their simplicity and ease of identification, not the causes of their existence. (And Kant does not assert that these properties of mathematical objects follow from their creation by the mind.) Now the *Critique* maintains that definitions of ("arbitrarily thought") mathematical concepts also define "true objects" (A729/B757), because a mathematical definition both renders a concept distinct and "exhibits the object in accordance with the concept *in intuition*" (A242 *n.*). Hence, it proves the concept's objective reality (A242*n.*). The earlier claim that mathematics "defines an object" can also be taken to mean that mathematical definitions prove the concepts' objective reality. To be sure, it cannot mean that the object is presented in pure intuition; but an alternative theory of how the reference of mathematical terms is guaranteed is (I have argued) available to Kant. In particular, if the natural numbers are defined in terms of the successor relation, then each number's possibility can be verified by constructing the formula that expresses its definition.

On this reading, mathematics' special ability to "define an object" is explained by its object's simplicity and identifiability. For what makes it possible to show (by means of a definition) that there can be objects corresponding to a concept is our thorough and exact grasp of what it is to be such an object. Precisely this is lacking in metaphysics.

### 3. Conclusion: From the Prize Essay to the *Critique*

Once it is acknowledged that the view of Kant's Prize Essay is not formalist in the narrow sense, it becomes pressing to explain why concepts that we arbitrarily define should be applicable to empirical objects. Commentators such as Carson and Guyer find that Kant fails to reconcile the applicability of mathematical concepts with their definability (although they differ in that Guyer takes the definability thesis to preclude an account of the concepts' application, while Carson only notes that the question of applicability is not addressed).[43] For both, the Critical doctrine of construction in pure intuition is supposed to resolve the tension. Kant claims that we can define only concepts that "contain an arbitrary synthesis that can be constructed *a priori*" (A729/B757), and specifically that manifolds "belonging to" mathematical concepts are "put together in pure intuition". To drive home the point, he claims that a mathematical concept "already contains a pure intuition in itself" (A719/B747).

But the notion of intuition is introduced by way of a sharp distinction with concepts, according to which it does not make sense for an intuition or a manifold constructible in it to belong to a concept. Kant's view, more carefully stated, is that the constructible manifold belongs in the first instance to the concept's *schema*.[44] The schema is defined as a "general procedure for providing a concept with" a corresponding intuition (A140/B179). Kant gives as example the schema of the concept *triangle*, which he describes as "a rule of the synthesis of imagination with regard to pure shapes in space" (A141/B180). The schema is supposed to "stand in homogeneity", as a "third thing", between intuition and concept. Schemata are homogeneous with concepts in that both have the generality characteristic of rules. They are homogeneous with intuitions insofar as, as instructions for the imagination, they issue in images (concrete representation of particular objects); in fact, they are introduced to provide "pure concepts of the understanding" with the sensible content they lack.[45]

Kant's considered view seems to be that the provision of schemata also explains the applicability of arbitrarily synthesized concepts.[46] In her survey "Kant's Philosophy of Mathematics", Lisa Shabel observes that in the *Critique* Kant takes mathematical concepts to be "arbitrary" in the sense that "one knows precisely what [their] content is" since they are "deliberately made up", rather than "given through the nature of the understanding or through experience".[47] So far, this seems to be an understanding of "arbitrariness" that does not require correspondence with extramental reality. But Shabel goes on to explain our knowledge of mathematical concepts' content in terms of the "arbitrary syntheses" that they contain, as for instance "in the case of a triangle, one considers the concept *figure* together with the concepts *straight line* and *three*, and then proceeds to effect the synthesis of these concepts by exhibiting an object corresponding to this new [composite] concept", which (object) can be either drawn or merely imagined (2006, 99). Hence, our access to a mathematical concept's content is explained in terms of our ability to construct an object falling under the concept. Shabel makes the further observation, however, that the "processes" of "concept construction" (i.e. of the exhibition of objects corresponding to the concepts) are supposed to have a generality and universality that grounds the connection between predicate and subject concepts in mathematical propositions, hence the generality and universality of mathematical propositions (108). Kant "points us to the Schematism for an explanation of the universality of the act of construction" (109), an explanation which lies in our "awareness that the production of... a concrete representation of the general concept *triangle* depends on a general 'rule of synthesis' for the production of any such figure, that is, its schema" (111–112). So the syntheses which themselves explain the applicability of arbitrarily defined concepts are ultimately to be explained in terms of schemata.

But Kant's account of the schema is so brief and obscure that even with the help of Shabel's insightful exposition, it remains unclear how this single item can both share a concept's generality and ensure that there can be intuitions falling under the concept. (Kant himself makes it harder to see how the schema can ensure objective reference, to intuitively given particulars, when he describes the schema as "really only [*eigentlich nur*] ... the sensible concept of an object" (A146/B186).) So as long as "arbitrariness" is thought to be incompatible with applicability, as on the strong reading, it will remain unclear how the *Critique* advances beyond the view of the Prize Essay.

I have argued that in the Prize Essay, Kant has a way to explain the applicability of arithmetical concepts, specifically to sign tokens. The problem facing this view is to explain the *universal* applicability of mathematical concepts to (all) objects of experience.[48] Throughout his *oeuvre*, Kant stresses that the "marks" or characteristics of which mathematical concepts are composed are easily discerned, and on his view, these criteria apply with definiteness in all cases.[49] (I argue in (2012) that this is explained by our prerogative to include in the concepts only criteria whose satisfaction can be "assumed", leaving out those whose satisfaction would have to be empirically investigated.) In this respect, mathematical concepts differ not only from metaphysical, but also from empirical ones (as Kant makes clear in the *Critique*, A727-8/B755-6). Both critics and defenders of the kind of formalism outlined in Section 2.3 acknowledge that the difference makes it difficult to explain mathematics' application to ordinary empirical objects. In particular, the concepts of such objects lack sharp application conditions and do not permit us always to distinguish and re-identify objects falling under them. Stephan Körner explains that in virtue of these limitations of the concepts, we may "say sometimes with equal correctness, of a fruit growing on an apple-tree, that it is one apple and also that it is two apples grown together" (1968, 107). These features of the concepts thus entail that statements such as "one apple and one apple make two apples" are not "self-evidently true". But then a formalist cannot treat such a statement as "the result of replacing in [the juxtaposition of strokes] described by '1 + 1 = 2' each stroke by an apple and the juxtaposition of strokes by the juxtaposition of apples", for in that situation "either both statements will be self-evidently true or neither will be" (106). Since statements about empirical objects cannot be regarded as "replacement instances" of statements of formal arithmetic, it becomes a challenge to ground the former sort of statement on the latter.

In my view, what solves the problem of universal applicability in the *Critique* is not the constructibility of mathematical objects in pure intuition, but the view that pure intuition is the form of empirical intuition. Its "constant form" is thus "a necessary condition of all the relations within which objects can be intuited" (A27/B43), which ensures that what mathematics says about pure intuition is "undeniably valid of" empirical intuition (A165/B206). I suggest that only here, in his account of mathematics' application, does Kant go beyond the resources already available in the Wolffian tradition. For in the Prize Essay, on the reading offered here, we already see how these resources suffice to explain the objective reference of mathematical concepts – at least, to ensure that the concepts apply to the signs used to express them.

## Acknowledgements

For very helpful comments on drafts, I am grateful to Emily Carson, Lisa Shabel and participants in the workshop "Algebra, Magnitude, and Continuity" at the Max-Planck-Institut für Wissenschaftsgeschichte, Berlin, July 2014. I owe special thanks to Vincenzo De Risi for convening the workshop and for his comments. I am also indebted to Al Martinich for his generous assistance in translating Wolff's Latin texts.

## Notes

1. On the importance of Kant's conception of mathematical method, see Shabel (2006) and Hintikka (1965, 1967 cited as reprinted in Hintikka 1974). On his theory of definition, see Capozzi (1981) and Dunlop (2012). Carson (1999) is a thorough account of the continuities between the Prize Essay and the first *Critique*.
2. Despite the excellent studies cited in this paper, in particular the work of H. W. Arndt. See also Sepper (2012).
3. See Basso (2011) and Zac (1974).
4. Descartes writes in the Sixth Meditation that "if I want to think of a chiliagon, although I *understand* that it is a figure consisting of a thousand sides just as well as I understand [a] triangle to be a three-sided figure, I do not *imagine* the thousand sides or *see them as if they were present before me*" as I do the three sides of the triangle (AT VII 72/CSM II 50, emphasis added). Descartes proceeds to argue that whatever "confused representation of some figure" I might happen to "construct in my mind" would not be a chiliagon, because it would "in no way [differ] from the representation I should form if I were thinking of a myriagon, or any figure with very many sides", and would be "useless for recognizing the properties that distinguish a chiliagon from other polygons".
5. In the third volume of John Wallis's *Opera Mathematica* (Oxford: Sheldonian Theater), 620–622. Cited by Wolff ([1751] 1983, 180).
6. Leibniz attributes the "confusion" of perceptions to their resulting from "impressions that the whole [*sc.* infinitely complex] universe makes upon us" in §13 of "Principles of Nature and Grace", which was available to eighteenth-century readers (as well as in §33 of the "Discourse on Metaphysics", which was not.)
7. See the classic discussion of this passage by Cassirer (1953, vol. 1, 130–132). On the necessity of symbolization for thought, see also Pombo (1987, Ch. 2 of Pt. II).
8. Wolff might be thought similarly to confine perceptual experience to non-certainty-yielding functions (in mathematical reasoning) when he distinguishes "mathematical" from "mechanical" demonstration. In contrast to mathematical demonstration, mechanical demonstration involves the use of sensory evidence to estimate the size of lines and angles (Shabel 2003, 98–101). Kant (if not Wolff himself) is clear that the latter method cannot yield certain knowledge; see Shabel (2003, 103–104). However, Wolff seems to draw this distinction only with regard to geometrical reasoning; as far as I can tell, it does not figure in his general account of mathematical method. Hence, in this broader context, the involvement of the non-rational faculties does not have to be understood in terms of mechanical demonstration.
9. On Locke's view, demonstration consists in finding ideas to agree by a series of comparisons with "intermediate" ideas, such that at each stage two ideas are seen ("intuitively") to agree, and the ideas comprising the proposition to be demonstrated

are the first and last of the series. His example is that in order "to know the agreement or disagreement between the three angles of a triangle and two right ones", the mind "is fain to find out some other angles, to which the three angles of a triangle have an equality; and, finding those equal to two right ones, comes to know their equality to two right ones" (IV.ii.3; *cf. NE* IV.ii, 367). Here, the comparison appears to be with respect to observed size (of, in a "particular" demonstration, the angles contained in particular drawn figures). This reading fits with Locke's speaking of "comparing and measuring" ideas of angles at IV.xviii.4, and of the perceptibility of equality or difference of size at IV.ii.10. However, at IV.xi.6 Locke distinguishes demonstrating from "examining [figures] by sight", claiming that the latter gives "the evidence of our sight … a certainty approaching to that of demonstration itself". This suggests that the "agreement" with respect to which ideas are compared (in demonstration) is a more abstract relation, perhaps one grasped by the rational faculty. (Leibniz does not comment specifically on this paragraph, but remarks, seemingly approvingly, with respect to IV.vi that the "connection" of sensible things that constitutes truth "depends on intellectual truths grounded in reason" (*NE* IV.xi, 444).)

10. See also "Letter to Queen Sophie Charlotte of Prussia, On What is Independent of Sense and Matter": "the mathematical sciences would not be demonstrative and would consist only in simple induction and observation if something higher, something that intelligence alone can provide, did not come to the aid of *imagination* and *senses*" (1989, 188). To my knowledge, this text was not available to Wolff or his contemporaries.

11. A reason to discount the *Nouveaux Essais*, on the other hand, is given by Robert McRae, who suggests that "it is probably for tactical reasons that Leibniz, in making his own case against Locke, is silent about his own conception of the essential connection between mathematics and the imagination" (1995, 184).

12. Giorgio Tonelli argues persuasively that in general, "the works of Leibniz published prior to 1765 almost exclusively represented the general *metaphysical* point of view" that knowledge originates inside the soul, at the expense of his epistemological and psychological views of how truths of various kinds are known (1974, 438). This left Wolff, in particular, free to hold that epistemologically and psychologically speaking, experience plays an important role in generating knowledge (445).

13. Here Wolff is explaining how symbolic [*figürlich*] cognition can attain clarity and distinctness. He contrasts symbolic cognition, in which something is represented by means of "words or other signs", with intuitive cognition, which is representation of a thing itself or an image thereof ([1751] 1983, §316, 173). The contrast makes it appear that "before the eyes" is not meant literally, for Wolff associates intuitive cognition with sensory representation (§319, 77; *cf.* Ungeheuer (1983, 93)). Moreover, Wolff claims that by means of the "symbolic art", concepts "brought to mere signs" are thereby "completely abstracted from all images of the senses and imagination" (§324, 180). But Wolff also holds that by means of *ars characteristica combinatoria*, cognition can be "turned around" [*convertitur*] from symbolic to as-it-were [*quasi*] intuitive (*Psychologia rationalis* §312, 226). A natural way to understand this is that relations between things can be as it were seen in, or read off from, the formulae of a sufficiently apt system of signs (*cf.* Pimpinella (2001, 281–282). Clement Schwaiger holds that Wolff departs from Leibniz precisely by allowing that a whole can be grasped intuitively (as when we grasp by means of symbols how elements are related) even when we lack distinct cognition of its various parts (as when we do not know the symbols' denotation) (2001, 1180).

14. Wolff's fullest account of this procedure is in a minor work of 1709. See Arndt (1965).

15. See Anderson (2005).

16. I explain more fully how Wolff intends it to be useful in Dunlop, (m.s.).
17. See the *Mathematisches Lexicon* entry "*Numerus polygonus*" ([1716] 1965, 958–960).
18. For an account of how Leibniz represents predicates and their negations by positive and negative numbers, see Mittelstrass (1979, 609–610).
19. The *ars characteristica combinatoria* was to consist of the "universal characteristic", by means of which each primitive idea is designated by a symbol, and a combinatorial syntax (whose rules are powerful enough to generate all admissible strings).
20. (1754) 1968, I.18, 131–132. I follow the anonymous eighteenth-century translation *Logic, or Rational Thoughts on the Powers of the Human Understanding* (London: for L. Hawes, W. Clarke, and R. Collins, 1770, reprint Hildesheim: Georg Olms, 2003). *Cf.* Tonelli (1959, 53–54).
21. For instance, by Capozzi (2011, 317) and Sutherland (2010, §4).
22. Shabel quotes an eighteenth-century translation of Wolff ([1742] 1968) in which negative quantities are described as "absurd", "wanting reality" and "not real" (1998, 600 n. 17).
23. Geometrical representation of negative numbers was treated as a problem in the eighteenth century (Boyer 1956, ch. 7), and graphical representation of complex numbers was introduced (by Gauss) only in the nineteenth century.
24. The noted scholar Daniel Sutherland, for instance, writes (in an essay focusing on Kant's relationship to Wolff) that the *ars combinatoria* "is conspicuously absent" from Wolff's work (2010, 168).
25. Charles Corr argues that because Wolff regards the *ars characteristica combinatoria* as part of a larger *ars inveniendi*, he "pays little attention to [it] as a theoretical system". It is introduced "only to supplement the basic logic in the critical instance of scientific discovery" (1972, 333).
26. As Engfer also observes (1982, 198).
27. As in the essay "Primary Truths" (1989, 30–34) and the letter to Herman Conring of March 19, 1678 (1969, 186–187).
28. *Cf.* Engfer (1983, 54).
29. Kant's denial that a mathematical concept has any "significance" apart from that conferred by its definition (2:291) can be taken to express the same view. Kant typically uses "significance" [*Bedeutung*] to mean relation to an object (as at A239-41/B298-300), but it can also mean the object to which a representation relates.
30. In the first *Critique*, Kant "grants and presupposes" the definition of truth as correspondence with an object (A58/B82; *cf.* A820/B848). But he explicitly calls this definition "nominal", meaning that it does not reveal the underlying nature of its object. He can reasonably be supposed to have taken this attitude (even) in the pre-Critical period, when he was sympathetic to the Leibnizian view that truth consists in the determination of a predicate in a subject (1:392).
31. Rechter brings out the striking force of Kant's claim that correct use of rules in a system of arithmetical notation suffices to ensure that cognition established by the method cannot be false (2006, 25).
32. Parsons appears to endorse this reading when he says the Prize Essay "suggests" a view not "compatible" with Kant's critical position, namely that "operation with signs according to the rules, without attention to what they signify, is" itself sufficient to guarantee correctness (1969, 138).
33. Thus, Paul Guyer's account of Kant's change in view appears oversimplified. On Guyer's reading, the Prize Essay's claim that in mathematics "the definitions are the first thought which I can have of the thing to be explained" [*erklärt*] (2:281) forestalls, or sets aside, questions about the "truth or correspondence" of concepts "constructed" by arbitrary synthesis (1991, 37). The "supposition of 1762 that the

starting-points of mathematics are merely arbitrary constructions" is then "rejected" in the *Critique* (44). Yet, the claims that mathematical concepts are based on "arbitrary synthesis" and are "first given through" definitions recur at A729-731/B757-9.

34. *Cf.* Weir (2011).
35. Detlefsen calls such advocacy "perhaps the most distinctive component" of the formalist "framework" (2005, 237; *cf.* p. 263).
36. As Rechter notes (2006, 25).
37. Among Rechter's reasons for denying that "the contribution of [the] 'sensible' character" of mathematical signs involves their "perceptibility" is that Kant claims *words* "are not 'sensible' in their capacity as signs" (33). But I take Kant's point to be that the structure perceptible in a word does not correspond to that of the concept it signifies, in contrast to the mathematical case. See Section 2.4.
38. Rechter further observes that Kant does not provide an account of the notion of cardinal number or its application, which would be needed to explain how the reference of signs is verified by counting numeral tokens.
39. It must be kept in mind that by "number" Kant means the natural or "counting" numbers. See Sutherland (2006).
40. A reason to doubt that Kant relates arithmetic to geometry in this way in the Prize Essay is that he does not in the Critical period. In notes written in 1790 (responding to attacks published in J.A. Eberhard's *Philosophisches Magazin*), Kant states explicitly that "when the geometer says that a line could always be extended no matter how far one has drawn it, then this does not mean what is said in arithmetic about numbers, namely that one can always, and endlessly, increase them through the addition of other units or numbers" (20:419–420, quoted in the translation of Onof and Schulting [2014]). But in the context of the Prize Essay (§2 of the First Reflection), Kant seems clearly to want to provide a unified account of the use of signs in both branches of mathematics.
41. That we could not see the adequacy of what Rechter calls a "formal symbolic alternative to the representation of [philosophical] concepts in natural language" complements her point that we would need to already "have access to the distinct concept even in order to recover" or correct the defects of natural-language designations (2006, 28).
42. Tonelli notes (1959, 66) that the view that philosophy studies qualities, which are infinite in number, was widely held among Kant's contemporaries.
43. See Carson 1999 and Guyer (1991, 36–37).
44. "The procedure" of "mathematical and geometrical construction" is that "by means of which I put together in a pure intuition, just as in an empirical one, the manifold that belongs to the schema of a triangle in general and thus to its concept" (A718/B746).
45. See Dunlop (2012, §4).
46. In (2012), I argue specifically that the close connection between the schemata of mathematical concepts and our precise knowledge of their content (as expressed by their definitions) shows that the concepts are not arbitrary in the strong sense (of Section 2.1), as many of Kant's remarks would suggest.
47. I elaborate this contrast in my (2012, 94–96).
48. Capozzi (2011, 13) and Engfer (1983, 50) also see this as the outstanding difficulty for the view of the Prize Essay.
49. *Cf.* Parsons (2009, 165–167).

## References

*Note.* Kant's and Descartes's works are cited according to standard conventions and in the translations published by Cambridge University Press (except as noted). Locke's

*Essay Concerning Human Understanding* is cited by chapter and paragraph number. Leibniz's *Nouveaux Essais sur l'Entendement Humaine* is cited by chapter and paragraph of Locke's essay and page number of the 1765 edition (Amsterdam and Paris: Eric Raspe) and in the translation of Jonathan Bennett and Peter Remnant (Cambridge: Cambridge U.P., 1996).

Anderson, R. Lanier. 2005. "The Wolffian Paradigm and its Discontents." *Archiv für Geschichte der Philosophie* 87: 22–74.

Arndt, H. W. 1965. "Christian Wolffs Stellung zur 'Ars Characteristica Combinatoria'." *Studi e ricerche di storia della filosofia* (Torina) 71: 1–12.

Basso, Paola. 2011. "Le mythe de la démonstrabilité résiste-t-il encore?" *Astérion* 9 (n.p.).

Boyer, Carl. 1956. *History of Analytic Geometry*. Reprint New York: Dover, 2004.

Capozzi, Mirella. 1981. "Kant on Mathematical Definition." In *Italian Studies in the Philosophy of Science*. Boston Studies in the Philosophy and History of Science, edited by M. L. Dalla Chiara. Vol. 47, 423–452. Dordrecht: Reidel.

Capozzi, Mirella. 2011. "Philosophy and Writing." *Quaestio* 11: 307–350.

Carson, Emily. 1999. "Kant on the Method of Mathematics." *Journal of the History of Philosophy* 37: 629–652.

Corr, Charles. 1972. "Christian Wolff's Treatment of Scientific Discovery." *Journal of the History of Philosophy* 10: 323–334.

Desmaizeaux, Pierre, ed. 1720. *Recueil de diverse pieces ... par Mrs. Leibniz, Clarke, Newton, et autres Auteurs célèbres*. Amsterdam: Duillard et Changuion.

Detlefsen, Michael. 2005. "Formalism." In *The Oxford Handbook of Philosophy of Mathematics and Logic*, edited by S. Shapiro, 236–317. Oxford: Oxford University Press.

Dunlop, Katherine. 2012. "Kant and Strawson on the Content of Geometrical Concepts." *Nous* 46: 86–126.

Dunlop, Katherine, "Definitions and Empirical Justification in Christian Wolff's Theory of Science." Unpublished Manuscript.

Engfer, H.-J. 1982. *Philosophie als Analyse. Studien zur Entwicklung philosophische Analysiskonzeption*. Stuttgart: Frommann-Holzboog.

Engfer, H.-J. 1983. "Zur Methodendiskussion bei Wolff und Kant." In Schneiders, 1983 48–65.

Guyer, Paul. 1991. "Mendelssohn and Kant." *Philosophical Topics* 19: 119–152. Cited as reprinted in *Kant on Freedom, Law, and Happiness*. Cambridge: Cambridge University Press, 2000.

Hintikka, Jaakko. 1965. "Kant's 'New Method' of Thought and his Theory of Mathematics." *Ajatus* 27: 37–47.

Hintikka, Jaakko. 1967. "Kant on the Mathematical Method." *The Monist* 51: 351–372.

Hintikka, Jaakko. 1974. *Knowledge and the Known*. Dordrecht and Boston: D. Reidel Pub. Co.

Körner, Stefan. 1968. *The Philosophy of Mathematics*. Reprint New York: Dover, 1986.

Leibniz, G. W. 1969. *Philosophical Papers and Letters*. Translated and edited by L. Loemker. Dordrecht: Reidel.

Leibniz, G. W. 1989. *Philosophical Essays*. Translated and edited by R. Ariew and D. Garber. Indianapolis: Hackett.

Loemker, Leroy. 1966. "Leibniz's Conception of Philosophical Method." First published in *Zeitschrift für Philosophische Forschung*, Bd. 20 Heft 3-4. Cited as reprinted in Ivor LeClerc, ed., *The Philosophy of Leibniz and the Modern World*. Nashville: Vanderbilt University Press, 1973.

McRae, Robert. 1995. "The Theory of Knowledge." In *The Cambridge Companion to Leibniz*, edited by Nicholas Jolley, 176–198. Cambridge: Cambridge University Press.

Mittelstrass, Jürgen. 1979. "The Philosopher's Conception of *Mathesis Universalis* from Descartes to Leibniz." *Annals of Science* 36: 593–610.

Onof, Christian, and Dennis Schulting (transls.). 2014. "On Kästner's Treatises." *Kantian Review* 19: 305–313.

Parsons, Charles. 1969. *Kant's Philosophy of Arithmetic*. Cited as reprinted in Mathematics in Philosophy. Ithaca, NY: Cornell University Press, 1983.

Parsons, Charles. 2009. *Mathematical Thought and its Objects*. Cambridge: Cambridge University Press.

Pimpinella, Pietro. 2001. "*Cognitio intuitiva* bei Wolff und Baumgarten." In *Vernunftkritik und Aufklärung*, edited by M. Oberhausen, et al., 265–294. Stuttgart: Frommann-Holzboog.

Pombo, Olga. 1987. *Leibniz and the Problem of a Universal Language*. Materialien zur Geschichte der Sprachwissenschaft und der Semiotik vol. 3. Münster: Nodus Publikationen.

Rechter, Ofra. 2006. "The View from 1763." In *Intuition and the Axiomatic Method*, edited by E. Carson and R. Huber. The Netherlands: Springer.

Schwaiger, Clement. 2001. "Symbolische und intuitive Erkenntnis bei Leibniz, Wolff, und Baumgarten." In *Nihil sine ratione. Mensch, Natur und Technik im Wirken von G.W. Leibniz* (VII. Internationaler Leibniz-Kongress), edited by H. Poser. Hanover: Gottfried-Wilhelm-Leibniz-Gesellschaft.

Sepper, Dennis. 2012. "Spinoza, Leibniz, and the Rationalist Reconceptions of Imagination." In *A Companion to Rationalism*, edited by Alan Nelon. Chichester: Wiley-Blackwell.

Shabel, Lisa. 1998. "Kant on the 'Symbolic Construction' of Mathematical Concepts." *Studies in History and Philosophy of Science* 29: 589–621.

Shabel, Lisa. 2003. *Mathematics in Kant's Critical Philosophy*. New York: Routledge.

Shabel, Lisa. 2006. "Kant's Philosophy of Mathematics." In *The Cambridge Companion to Kant and Modern Philosophy*, edited by P. Guyer, 94–128. Cambridge: Cambridge University Press.

Sutherland, Daniel. 2004. "Kant's Philosophy of Mathematics and the Greek Mathematical Tradition." *Philosophical Review* 113: 157–201.

Sutherland, Daniel. 2006. "Kant on Arithmetic, Algebra, and the Theory of Proportions." *Journal of the History of Philosophy* 44: 533–558.

Sutherland, Daniel. 2010. "Philosophy, Geometry, and Logic in Leibniz, Wolff, and the Early Kant." In *Discourse on a New Method*, edited by M. Dickson and M. Domski, 155–192. Chicago and LaSalle: Open Court.

Tonelli, Giorgio. 1959. "Der Streit über die mathematische Methode in der Philosophie in der ersten Hälfte des 18.Jahrhunderts." *Archiv für Philosophie* 9: 37–66.

Ungeheuer, Gerold. 1983. "Sprache und symbolische Erkenntnis bei Wolff." In *Christian Wolff (1679–1754)*, edited by W. Schneiders, 9–112. Hamburg: Meiner Verlag.

Weir, Alan. 2011. "Formalism in the Philosophy of Mathematics." In *The Stanford Encyclopedia of Philosophy* (Fall 2011 edition), edited by E. Zalta.

Wolff, Christian. (1716) 1965. *Mathematisches Lexicon*. edited by J. E. Hoffmann. Hildesheim: Georg Olms.

Wolff, Christian. (1737) 1981. *Herrn Christian Wolffs Gesammelte kleine philosophische Schrifften, Dritter Theil*. Hildesheim: Georg Olms.

Wolff, Christian. (1738) 1968. *Psychologia Empirica*, edited by École Jean. Hildesheim: Georg Olms.

Wolff, Christian. (1740) 1983. *Philosophia Rationalis sive Logica*, edited by École Jean. Hildesheim: Georg Olms.

Wolff, Christian. (1742) 1968. *Elementa Matheseos Universae*. Hildesheim: Georg Olms.

Wolff, Christian. (1750) 1999. *Anfangsgründe aller mathematischen Wissenschaften, Erster Theil*. Hildesheim: Georg Olms.

Wolff, Christian. (1751) 1983. In *Vernünftige Gedanken von Gott, der Welt und der Seele des Menschens, auch allen Dingen überhaupt*, edited by C. A. Corr. Hildesheim: Georg Olms.

Wolff, Christian. (1754) 1965. *Vernünftige Gedanken von den Kräften des menschlichen Verstandes*, edited by H. W. Arndt. Hildesheim: Georg Olms.

Wolff, Christian. 1963. *Preliminary Discourse Concerning Philosophy in General*, edited by R. J. Blackwell. Indianapolis: Bobbs-Merrill.

Zac, Sylvain. 1974. "Le prix et la mention." *Revue de métaphysique et de morale* 79: 473–498.

# Kant on the construction and composition of motion in the Phoronomy

Daniel Sutherland

*Department of Philosophy, University of Illinois at Chicago, Chicago, USA*

This paper examines the role of Kant's theory of mathematical cognition in his phoronomy, his pure doctrine of motion. I argue that Kant's account of how we can construct the composition of motion rests on the construction of extended intervals of space and time, and the representation of the identity of the part–whole relations the construction of these intervals allow. Furthermore, the construction of instantaneous velocities and their composition also rests on the representation of extended intervals of space and time, reflecting the general approach to instantaneous velocity in the eighteenth century.

## 1. Construction and applied mathematics in the *MFNS*

A central aim of Kant's theoretical philosophy is to explain the possibility of mathematical cognition and the applicability of mathematics to experience, and in particular the possibility of the mathematical physics found in Newtonian natural philosophy.[1] Kant explains the possibility of mathematical cognition and the applicability of mathematics in the *Critique of Pure Reason* (*CPR*), which provides a grounding for and shapes Kant's account of Newtonian mathematical physics in the *Metaphysical Foundations of Natural Science* (*MFNS*). Kant argues in the *CPR* that mathematical cognition is distinguished from philosophical cognition by its method, the heart of which is construction, that is, the "*a priori* exhibition of an intuition corresponding to a concept." Furthermore, the applicability of mathematics to appearances depends on the fact that a special synthesis, the synthesis of composition, is the same synthesis underlying both mathematical cognition and our apprehension of appearances. It is therefore no surprise that construction and composition are central to explaining the possibility of mathematical natural science in the *MFNS*. This paper will examine the role of construction and composition in the first chapter of the *MFNS* in order to draw out the influence of Kant's theory of mathematical cognition on his account of Newtonian natural philosophy.

Natural science, according to Kant, contains principles of the doctrine of body, which takes as its basis a complete analysis of the concept of matter (Kant 2004, 4: 471–472).[2] Kant also maintains that the understanding traces back the predicates of matter to motion, so that natural science is "either a pure or applied doctrine of motion" (Kant 2004, 4: 476). Since the completeness of a metaphysical system rests on the pure concepts of the understanding adumbrated in the table of categories (Kant 2004, 4: 473), the *MFNS* are divided into four chapters, each of which considers motion in relation to the four classes of categories. The first chapter, the Phoronomy, considers motion in relation to the categories of quantity – Unity, Plurality and Allness – and treats motion insofar as it is a magnitude, that is, a *quantum*. More specifically, it "considers *motion* as a pure *quantum* in accordance with its composition, without any quality of the moveable" (Kant 2004, 4: 476–477). One of the aims of the Phoronomy is to explain how the composition of motion can be constructed, thereby establishing a crucial necessary condition of the application of mathematics to motion. That condition, together with further mathematical resources, above all geometry, will be sufficient to allow the mathematization of motion in natural philosophy. Additional appeal to the mechanical laws of motion and empirical data will then permit successive approximations to a privileged frame of reference in which the laws of mechanics hold and to a true uniform measure of time, and to the law of universal gravitation.

Any account of the Phoronomy treatment of the construction of motion and its composition must begin with Kant's understanding of motion and speed. Section 2 argues that Kant explicates motion and speed in the Phoronomy by appealing to extended intervals of space and time, and it is these that are mathematically constructed. Instantaneous speeds, velocities and accelerations play a crucial role in the mathematical treatment of motion in Newton's natural philosophy and a central role in most of the quickly developing applications of mathematics to nature in the eighteenth century. I will argue that the Phoronomy as a pure science of motion lays the foundation for these instantaneous properties by explicating the instantaneous speed or velocity of a point at a place in space or at an instant in time on the basis of ever-diminishing extended intervals of space and time.[3] This explication reflects the general approach to instantaneous properties in the eighteenth century.

Kant's understanding of motion and speed lays the foundation for an analysis of Kant's account of the construction of the composition of motion. Section 3 explains Kant's argument against using vector composition in the construction of motion. I argue that his argument depends on the representation of the identity of extended parts and wholes in composition, and is not motivated by the need to directly construct instantaneous speeds. I then argue that Kant's positive account of how we can construct the composition of motion rests on the construction of extended intervals of space and time, and the representation of the identity of the part–whole relations the construction of these intervals allow.

This reading of the Phoronomy assigns a priority to extended intervals of space and time over instantaneous speeds and times, which contrasts with the interpretation offered by Michael Friedman, who argues that the aim of the Phoronomy is to explain the constructability of instantaneous motion and speed without reliance on the construction of extended spaces and times. Friedman (2013) raises a set of philosophical, textual, historical and systematic reasons to think that Kant held that motion is, and must be, represented as instantaneous, and that a major point of the Phoronomy is to show how instantaneous speeds can be constructed. I explain those reasons in Section 4. The interpretation they support is compelling and reinforces larger themes in Kant's reformation of Newton's natural philosophy. In Section 5, I respond to each of these reasons, and end with raising a further sort of reason Friedman gives for Kant's Phoronomy to reject an account of instantaneous motion in terms of ever-diminishing intervals of space and time. It takes into account Kant's larger project of the *MFNS* and Kant's reform of the Newtonian appeal to absolute space and time. I briefly outline a response, which argues that explaining instantaneous motion in terms of ever-diminishing intervals of space and time is consistent with Kant's larger project.

There is a great deal more to the Phoronomy than will be considered here. For example, a good deal of the Phoronomy concerns and depends on an account of relative motion.[4] We will simply take the relativity of motion to a frame of reference as given. We will also not have time to discuss the purity or apriority of phoronomy as a pure doctrine of motion, or its connection to the apriority of mathematical construction. We will keep focus on the construction of the composition of motion, and as we shall see, even on this issue there is more to be said.

## 2. Kant's phoronomical conceptions of motion and speed

### 2.1. Kant's explication of motion and definition of speed

In the second explication of the Phoronomy, Kant contrasts the explication of motion of a point with motion of a thing: the motion of a point is change in place, while the motion of a thing is the change of its outer relations in space. The latter is more general and includes the former as a special case. Kant's motive, as his first remark to the explication makes clear, is that by "thing" he means to include a general conception of matter, one that would include bodies, which are extended. Since they are extended, a body can change its relation to other things without changing its position (defined by reference to some point within the body, such as its centre of mass) by means of rotation.

Nevertheless, the Phoronomy abstracts from all inner constitution of a body and from its magnitude, by which Kant means its extension. Matter as the subject of motion can therefore be treated as a point (Kant 2004, 4: 480). Kant explicates motion of a thing in order not to rule out a later treatment of rotations, while throughout the Phoronomy, Kant only considers the motion of a point, or more carefully, the motion of a body considered as a point. Kant says that although he will use the term "body," he will only use it in anticipation of the application of phoronomy to bodies (Kant 2004, 4: 480), and for the purposes of the Phoronomy itself, he states: "I could allow the common explication of motion as a change of place to be used" (Kant 2004, 4: 482). In what follows, I will only consider this explication of motion, and will refer to the motion of a point, even when Kant uses "body."

Kant's explication of motion as a change of place entails that, at the very least, motion requires being at one place in space at one instant of time and being at another place in space at another instant in time. Given properties of space that Kant would accept, there is some extended interval of space between these places, and hence Kant's explanation of motion entails, at a minimum, that when a point has changed its place, there is an extended interval of space between them. Kant holds that space and time and extended intervals of space and time are continuous. Together with rather minimal assumptions about the identity conditions of a point, references to the motion of one and the same point at least suggest that the point has occupied some continuous succession of places at successive times between the two places. In fact, in the remark to Explication 3, Kant states that a body in motion is at every point that it traverses. (Kant 2004, 4: 485); moreover, one of the aims of the remark is to argue that an attempt to represent discontinuous motion leads to contradictions. Furthermore, the third remark to the second explication, Kant refers to the "continuously [kontinuerlich]" changing direction of a body moving in a circle (Kant 2004, 4: 483). Thus, in his explication of motion as a change of place, Kant appears to be committed to a point or a body traversing a continuous extended interval of space between its starting and ending places.

At the close of the third remark of the second explication, Kant states: "In phoronomy we use the word 'speed' purely in a spatial meaning $C = S/T$," that is, *Celeritas est Spatium per Temporum*. (Kant 2004, 4: 484, 20 fn. 13 of Friedman's translation). At the beginning of this remark, Kant states that, if one abstracts from all other properties of the moveable, direction and speed are the two moments for considering motion (Kant 2004, 4: 483). He adds that he presupposes the usual definition [*gewöhnliche Definition*] of both, and it is that definition which he gives at the close of the remark. Thus, Kant's reference to the meaning of the word 'speed' should be taken as a definition. The definition seems to require that for any motion, an extended interval of space is traversed over an extended interval of time.

## 2.2. *Motion and speed: intervals and their endpoints*

Kant's explication of motion and definition of speed and their appeal to extended intervals of space and time appear to be fairly straightforward if one starts by imagining a simple case of uniform rectilinear motion or if one thinks of a motion in which one abstracts from a determination of either uniformity or non-uniformity and hence does not consider the latter. One cannot, however, simply take the endpoints of an interval as sufficient to determine either the motion or speed of a point over the interval, as becomes clear if one considers Kant's account of motion more carefully.

If a point or object is at different locations at the beginning and end of an interval, then by Kant's explication it has moved, but the converse cannot be asserted; if the point or object is at the same location at the beginning and end of an interval, that does not mean it has *not* been in motion, and hence has been at rest over that interval. Vibrations, oscillations and closed orbits (and more generally any closed paths) are all cases in which a point or object moves while returning to a place, and hence if an interval of time were chosen with such a place as endpoints, and we took Kant's explication as a necessary condition of motion, then it would entail that the object is not moving, which is clearly not the case.[5] Judged just by the endpoints of an interval, this would only tell you whether there was "net" motion over that interval. Thus, whether a point or object is moving does not merely depend on the places occupied at the end points of some interval of time, but what occurs within the interval.

Alternatively, and more reasonably, we can understand Kant's use of "change" in a fuller sense, in which one considers not just the place occupied at endpoints of some interval of time, but whether there was any change of place at *any* time during the interval. On this reading, a point that left and returned to the same place would count as having changed its place. This would raise change of place to a necessary as well as sufficient condition of motion; it would also be more in line with what Kant refers to as the "common" explication and is therefore most likely what Kant had in mind. It is significant, however, that defining change of place in this way implicitly refers to *any* sub-interval within an interval, of which there are indefinitely many.

The same issue arises for speed: one needs to take into account not just the endpoints of an interval, but what happens within an interval. Any difference of place is sufficient to show that a point has *some* speed over an interval of time, but if we only judged speed by the endpoints of the interval, we would at most arrive at the "net" speed of the object over that interval. How much speed the point has depends on the possibly indirect path, and hence the distance, the point travels between them, an issue independent of whether the endpoints are the same place or not. But speed raises further issues as well. Even if one takes into account the distance travelled, the speed determined by the extended interval of space travelled over an interval of time is limited in an obvious and well-known way: it will only give you the average speed over the interval, so that it would assign the same speed to a point that traversed a path uniformly and a point that traversed the same path in a herky-jerky fashion in the same interval of time.

This brings out a fundamental feature of motion and speed explicated and defined in relation to intervals. On the one hand, motions seem to be individuated by what occurs within an interval, so that a herky-jerky motion over an interval is a different motion than one that is uniform over that interval. While there is a respect in which the herky-jerky motion and a uniform motion are the same – namely, with respect to the whole interval of space traversed over the whole interval of time – they are nevertheless different motions. Furthermore, the multiplicity of possible motions holds for *any* interval, including any sub-interval as small as you please within a given interval. Thus, any specification of motion or speed simply in terms of the endpoints of an interval, even over the same path, does not completely determine the motion or speed of the point over the interval; there are multiple possible motions consistent with any such specification. On the other hand, it seems that motion is a fully determinate property of a point; that is, a point in motion has to have one possible motion for any interval and any sub-interval, no matter how small. This fact was reflected in the medieval theory of motions beginning in the early fourteenth century; at least some of these natural philosophers characterized uniform motion as that which traverses equal distances in *any* equal time intervals (Grant 1996, 100). It is noteworthy that an account of motion and speed of a point in terms of intervals drives one towards consideration of the indefinitely many sub-intervals within an interval, quite independently of the need for a clear account of instantaneous motion and speed for the application of mathematics in natural philosophy in the seventeenth and eighteenth centuries.

Returning to Kant, we find that he addresses motions that return to the same place – in particular, vibrations, oscillations and orbits – in his second remark to Explication 2. He states that they are motions limited to a given space and return on themselves [in sich zurückkehrende], in contrast to rectilinear motions or those curvilinear motions that enlarge the space and do not return on themselves [in sich nicht zurückkehrende Bewegungen].[6] Kant's concern is to rule out a not uncommon use of "speed" to refer to the frequency or period of rotation, which might lead one to say, for example, that a child on a playground spinner is revolving at greater speed than someone on a Ferris wheel, even if the latter has greater linear speed. Kant does not address the issue that, on the narrowest reading of his explication of motion as mere change of place, returning to the same place could lead one to conclude that the point is not in motion. The fact that Kant is explicitly considering closed paths and does not consider this is further evidence that Kant explicates motion over and interval as *any* change of place during that interval.

The second remark culminates in his rejecting the use of "speed" to mean merely the frequency of a vibration or the period of an oscillation or an orbit, and it is in this context that he states that in phoronomy, we will use speed in a purely spatial meaning as $C = S/T$. On this approach, the speed of a point moving in a circle would be measured not simply by the time it takes to travel the circumference, but the length of the circumference divided by the time. Thus, not only the definition of speed as $S/T$ but also the specific examples he has in mind make it clear that, in the Phoronomy, speed is a function of intervals of space traversed over intervals of time.

In summary, Kant's explication of motion of a body or a point as change of place is best understood as the claim that at least somewhere over an interval of time, it occupies two distinct places and that it traversed some continuous path between those places. Similarly, the definition of speed using the formula $C = S/T$ presupposes that an extended interval of space has been traversed over an extended interval of time. For the explication of motion to be adequate, we need to consider any interval of time within a given interval, not just its endpoints. Furthermore, the speed of a point determined by the endpoints of an interval does not completely determine the motion of a point over that interval, while a complete determination of the motion of a point or object would require taking into account indefinitely many sub-intervals between the endpoints.

This feature of explaining motion and speed in terms of intervals might drive one to think that any such account will ultimately prove inadequate. This thought might be reinforced by, or independently motivated by, the fact that instantaneous velocities are absolutely central to Newtonian physics. But what is required to make the intervals approach work is some explanation of motion and speed of a body or point at a particular point in space or a particular instant in time in terms of intervals of space and time. Kant provides that in the following explication, significantly deepening his account.

## 2.3. Rest and motion at points of space and instants of time

In Explication 3, Kant rejects a conception of rest as simply a lack of motion. As noted above and as Friedman (2013, 48) has emphasized, Kant wishes to rule out discontinuous motion, by which Kant has in mind an instantaneous change in speed or direction, that is, a single moment in time. Kant also wishes to explain the sense in which a body that is continuously and uniformly changing its velocity so that it reverses direction, such as a ball thrown up into the air, can be correctly said to be at rest at the instant in time (or the point in space) its direction reverses. This immediately requires addressing the relationship between motion and speed defined over intervals and the motion and velocity of a point at instants of time or points in space. Kant begins working out this relation in a passage I mentioned earlier, at the very start of his remark to Explication 3:

A body in motion is at every point of the line that it traverses for a moment [Augenblick]. The question is now whether it rests there or moves. Without a doubt one will say the latter; for it is present at this point only insofar as it moves. (Kant 2004, 4: 485)

Kant immediately dismisses the thought one might have if one thinks that motion is explicated, and speed defined, in a way that requires that intervals of space and time be traversed: the thought is that an object considered at a point of space is not changing its location at that point, so is not in motion. Kant's immediate reason for dismissing it is that a body is only present [gegenwartig] at a point insofar as it moves, which might seem to simply beg the question against someone genuinely troubled by this application of the concept of motion,

174

understood in terms of intervals to an instant or a point. But the trouble presupposes that no coherent account of motion at an instant or point can be given, so we will focus on Kant's positive account.

Kant takes as data two claims to which he holds everyone will agree. The first is the claim just discussed: at every point of a line that a body in motion traverses, the body is in motion. The second datum is that in cases like a ball thrown up in the air in which a body is undergoing a continuous and uniform change in velocity so that the direction of motion reverses, the body at the instant of reversal can be thought of as at rest.

These two claims are in apparent tension, for it seems a body uniformly changing velocity to reverse direction is traversing every point of the line it ascends and then descends, so that at the turn-around point the body is both in motion and at rest. But Kant may mean by traverse [durchläuft] a motion in which a body travels *through* a point and hence appears on both sides of it. Regardless, when it comes time to make room for rest, Kant uses a claim restricted to uniform motion:

> ... no body at a point of its uniform motion at a given speed can be thought of as at rest.

(Kant 2004, 4: 846)

A body tossed in the air is not moving uniformly, only its *change* of motion is uniform, and we are free to say without contradiction that the body is not in motion at the turn-around point.

But how does Kant explain why it is appropriate to describe the body at the turn-around point as at rest? More generally, how does he explain the relation between motion explicated and speed defined, in terms of extended intervals of space and time, on the one hand, and motion and rest at points of space and moments of time, on the other? He does so in a manner quite familiar since at least the seventeenth century: by appeal to indefinitely or infinitely small intervals of space and time.

In a particularly important passage worth quoting in full, Kant states that the body at the turn-around point is at rest because:

> ... the latter motion [i.e. motion of a body rising and falling back again] is not thought of as uniform at a given speed, but rather first as uniformly decelerated and thereafter uniformly accelerated, so that the speed at point B [the turn-around point] [is] not completely [diminished], but only to a degree that is smaller than any given speed. With this [speed], therefore, the body, if it were to be viewed always as still rising, so that instead of falling back the line of its fall BA were to be erected in the direction Ba, [the body] would uniformly traverse, with a mere moment of speed (the resistance of gravity here being set aside), a space smaller than any given space in any given time, however large, and thus would in no way change its place (for any possible experience) in all eternity. (Kant 2004, 4:486)

Note first that Kant states that the speed at point B is not completely diminished but only diminished to a degree smaller than any given speed. This sounds at first paradoxical; how could the speed change uniformly and continuously, from

10 ft/sec to − 10 ft/sec for example, without passing through the speed of zero, and how could that be thought of in any way as a speed that has been completely diminished? While there are paradoxes and problems in treatments of indefinitely small quantities and continuity that were only satisfactorily understood and solved in the late nineteenth century, this is a clear reference to the general treatment of instantaneous speeds that blossomed in the seventeenth and eighteenth centuries, a treatment in terms of a sequence of ever-diminishing intervals of space and time. Kant is employing the definition of speed as an extended interval of space over an extended interval of time with a fixed endpoint of the intervals being the turn-around point. No matter how much one diminished the still always extended intervals, the ratio of $S/T$, and hence the speed, would not be completely diminished to zero. In other words, there is no last ratio in the sequence of ratios of ever-diminishing but still always extended intervals of space and time.

Because Kant states that the speed is not completely diminished, the speed at B is not the limit, as we would say, of this diminishing interval, if the limit is thought of as the unique result of completely diminishing the interval. Nor is it an infinitesimal, if an infinitesimal is thought of as a completely diminished yet nevertheless extended magnitude. Kant's position reflects the fact that he does not think we are capable of cognizing a completed infinite sequence or series; treatments of the infinite are limited to representations of the potentially infinite.

However, the speed at point B as Kant conceives it shares a property standardly attributed to infinitesimals, for he states that the speed is smaller than any given speed. This appears contradictory: for any ratio in the sequence of ratios approaching the turn-around point, there are smaller speeds, so "the speed" cannot refer to any ratio in the sequence, and since the diminishing is never completed, it cannot refer to the result of actually completing the potentially infinite number of ratios in the sequence.

Nevertheless, Kant is clear that he is referring to the speed in two different ways: as the speed of the body as it rises and falls back conceived in terms of intervals and as the motion of that body *at point B* and hence at the instant it occupies that point.[7] Recall that his opening statement of Explication 3 states that at every point of a line that a body in motion traverses, the body is in motion. Kant is explicating a necessary connection between the instantaneous speed of the body at a point and the speed of the body as it traverses a line conceived in terms of intervals of space and time. Kant thereby attributes two properties to the instantaneous speed at B: it is both ever-diminishing and smaller than any other speed. This is the sort of apparent contradiction which the foundations of analysis struggled until the late nineteenth century.[8]

Kant then attributes a third property to the instantaneous speed on the basis of the second. Even if the speed at point B is not regarded as completely diminished, the fact that it is smaller than any given speed means for him that it shares an important property with zero speed, for the speed he attributes to the point B is such that, if after the body reached B it continued with the speed at point B, it would in fact not change its place "in all eternity." That is, the indefinitely small speed, characterized by $S/T$ on always extended but indefinitely small intervals of space and time, could not be multiplied by even an indefinitely large interval of time to yield a change in place, and hence motion. And it is therefore at rest. This third property is described in terms of what finite extended interval of space the point or body would traverse in a finite extended interval of time, providing another necessary connection between extended intervals of space and time and instantaneous speed.

It is important to see that the account does not just apply to instantaneous rest, but generalizes to any attribution of instantaneous motion and speed. If we chose a different point on the trajectory of the body thrown upward and consider the ratios of ever-diminishing intervals with that point as endpoint, and if we make assumptions corresponding to a pre-nineteenth century view of continuity, then the sequence of ratios of ever-diminishing intervals will approach a non-zero ratio.[9]

Note that Kant has formulated two bridges between the ratio of extended intervals of space and time on the one hand and instantaneous speed at a point of space or instant of time on the other. The first characterizes the speed at a point or instant on the basis of diminishing extended intervals of space and time taking the point or instant as an endpoint. The second connects this instantaneous speed attributed to a point or an instant with a hypothetical motion – how the point or object would move were it to continue over an extended interval of time with the speed attributed to it at that instant or point. Both bridges play an important role in Kant's analysis of instantaneous rest, motion and speed.

Kant's explanation of instantaneous rest allows him to draw a distinction between "perduring in a state [in einem Zustand beharren]" and "being in a perduring state [beharrlichen Zustand]." Perduring in a state requires that a thing remain in that state over an interval of time. In contrast, being in a perduring state only requires that the state of a thing at an instant in time be that of perduring, even if at any time before and after that instant, it is not in that state. In his treatment of motion, the sense in which something could be in a state at an instant is characterized by what it would do if that state were to persist over an interval of time. So the speed of an object at an instant in time is here too characterized in terms of the speed of the extent of space the object would traverse over an interval of time. In the particular case Kant considers, as we have seen, if the object were to continue to move with the speed it has at the instant of turn-around, the object would traverse no distance even in an eternity, and for this reason we say it is in a perduring state of rest, even if at no other instant it is at rest.

In the special case of an object actually perduring in a state of rest, that is, actually remaining at rest relative to a frame of reference over a period of time, the instantaneous speed at any instant during that period would be zero. The spatial magnitude in the sequence of ratios of $S/T$ would simply be zero for each diminishing interval of time. Thus, the account of instantaneous speed in terms of ever-diminishing intervals allows for the special case in which the spatial interval is simply zero.

In the passage above, Kant refers to the speed of the object at the point B as "a mere moment [einem bloßen Moment] of speed." Kant uses "moment" here in the sense that corresponds to the way it was employed in natural philosophy in Kant's time, that is, for an incremental change of a time-varying quantity that occurs in an interval of time smaller than any other time. Strictly speaking, the moment of speed at point B should be the change of speed, that is, the acceleration at point B, and hence the instantaneous acceleration. Kant may instead mean change of place, so that by the moment of speed he means how much the object would change its place in an interval of time smaller than any other time.

Regardless, instantaneous accelerations can also be explained in terms of the ratios of ever-diminishing magnitudes. In this case, it is the ratio of diminishing differences in speed over ever-diminishing intervals of time. Since Kant holds, as we shall see below, that speed is an intensive magnitude, in this case, the first magnitude in the ratio (the numerator) need not be an extensive magnitude. What is important for the account of instantaneous rest, motion, speed or acceleration is using ever-diminishing intervals of time with the instant as an endpoint.

As I mentioned, the general approach Kant takes to explaining instantaneous speed in terms of ever-diminishing intervals had been familiar since at least the seventeenth century. It was crucial to the development of the calculus, and found expression in Leibniz's appeal to infinitesimals and Newton's fluents and fluxions as well as his theory of limits, about which there was still debate in the eighteenth century. I will not attempt to settle what Kant thought of infinitesimals, fluents or fluxions, or ulitmate ratios, or the extent to which he may have favoured one approach over another.[10] It is sufficient for our purposes to recognize that Kant's discussion of rest reveals that his phoronomical treatment of motion includes some form of the view that the motion, rest and speed of a point at an instant or a point of space is characterized in terms of ratios of ever-diminishing extended intervals of space and time and the extended interval of space a body or point would traverse over a finite extended interval of time were it to continue to move with that instantaneous motion or speed. Thus, both speed over an interval and instantaneous speed are explained in terms of extended intervals of space and time, just as his definition of speed by means of $C = S/T$ suggests.

## 3. Kant on the construction of motion and composite motion

### 3.1. The construction of motion

We can now turn from the explication and definition of rest, motion and speed to their construction. Kant does not discuss how net rest and motion or average speed of a point over an interval can be constructed. On the interpretation offered here, that is because it is straightforward: through the representation of the interval of space traversed over an interval of time. The crucial issue is the construction of instantaneous speeds and motions. And as Kant makes clear immediately after drawing the distinction between perduring in a state and being in a perduring state, this has important implications for the applicability of mathematics to experience:

> To be in a *perduring state* and to *perdure in this state* (if nothing else displaces it) are two different, although not incompatible, concepts. Thus rest cannot be explicated as a lack of motion, which, as $a = 0$, can in no way be constructed, but must rather be explicated as perduring presence in the same place, since this concept can also be constructed through the representation of a motion with infinitely small speed [unendlich klein Geschwindichkeit] throughout a finite time, and can therefore be used for the ensuing application of mathematics to natural science. (Kant 2004, 4: 486)

When Kant states that rest as a mere lack of motion cannot be constructed, he suggests that it is because "$a = 0$," but Kant is not saying that zero speed cannot be exhibited *a priori* in intuition. Rather, Kant's claim here concerns an inability to construct a bare denial of motion without reference to infinitely small extended intervals of time, coupled with the claim that were the point to move with the speed it has at that instant for a finite time, it would "in no way change its place."[11] Thus, what makes the representation constructible are the two bridges between motion explicated by appeal to intervals of space and time and motion at an instant or point of space.

When Kant claims that instantaneous rest is constructible, he appeals to the second bridge between intervals of space and time and instantaneous motion and rest, rather than the first bridge: what is constructible is an infinitely small speed [unendlich klein Geschwindigkeit] through a finite time. This is because the key property of the speed is that it would not traverse an extended interval of space in a finite interval of time; it is in a perduring state of rest at the instant because if it were to continue with the speed at that instant it would perdure in the state of rest.

We have seen that the construction of net rest and motion and average speed over an interval is straightforward on the interpretation offered here: it is through the *a priori* representation in intuition of intervals of space over intervals of time. These constructions are then employed in the construction of instantaneous rest, motion and speed. They are constructed by appeal to indefinitely small intervals of space and time preceding the instant of time or point of space, and by appeal to the space that would be traversed if a point or body were to continue for a finite interval of time with that instantaneous speed. Having settled at least to this extent Kant's views on motion and speed and their construction, we are now in a position to examine Kant's view of construction of a composite motion and its implications for the applicability of mathematics.

### 3.2. From motion to composition of motion: a requirement of applied mathematics

In the fourth explication, Kant does not expand further on the constructability of rest, motion and speed, but instead shifts to a discussion of construction of a *composite* motion. Kant states:

> To **construct** the concept of a *composite motion* means to present a motion a priori in intuition, insofar as it arises from two or more given motions united in one movable. (Kant 2004, 4: 486)

It is this construction that is required for applied mathematics:

> ... so much must be constituted wholly a priori, and indeed intuitively, on behalf of applied mathematics. For the rules for the connection of motions by means of physical causes, that is forces, can never be rigorously expounded, until the principles of their composition in general have been previously laid down, purely mathematically, as a basis. (Kant 2004, 4: 487)

Kant has in mind that we need a demonstration that motions can be composed with each other in a way that allows mathematics to be applied to them (taking into account direction as well as speed). Without such a demonstration, we will not be justified in concluding, for example, that if an object is moving 10 ft/sec, then adding 10 ft/sec to its speed in the same direction entails that it is going 20 ft/sec, or that adding 4 ft/sec in the opposite direction entails that is going 6 ft/sec, or that adding 10 ft/sec at a right angle to its motion entails that it is going $10\sqrt{2}$ ft/sec on the diagonal.

Kant does not state that the construction of composition is a sufficient condition of applied mathematics, though it is clearly at least a necessary condition, and a crucial one. While the concrete examples of composition I just gave are mathematically rather straightforward, a great deal more besides a construction of composition is required to apply mathematics to the phenomena even in these very simple cases. Kant's sole aim at this point in the Phoronomy is to demonstrate that a motion is the sort of thing that has a manifold of parts within it that when composed together constitute a whole composite motion. Kant makes this clear in Explication 5, where he says:

> The *composition of motion* is the representation of the motion of a point as the same as two or more motions of [this point] combined together. (Kant 2004, 4: 489)

Kant claims that the representation of composition requires the representation of the identity of the composed parts with the whole they compose, and this claim plays a crucial role in the arguments that follow. Note that the composition in question only makes reference to the relationship between parts and wholes. I want to emphasize this point. Kant has not said anything about an equality relation that may obtain among parts, and in my view, nothing he has said implies an equality relation. One might think that the notion of magnitude in general itself already implies that an equality relation obtains, or that the notion of the composition of magnitudes in general presupposes this relation, but I think that this is a mistake. [12] On the other hand, Kant takes the proposition that the whole is greater than a (proper) part to be analytic, and he uses this to define greater than and less than in terms of the part–whole relation, and he applies it to magnitudes; hence a whole magnitude is greater than any of its parts. But the greater than and less than relation among parts is prior to and independent of any relation of equality.

The Phoronomy and the *MFNS* as a whole aims to explain the possibility of the mathematization of motion, and the construction of the composition of motion establishes a crucial necessary condition. If we assume that we have geometry in place, which includes an equality relation on spatial magnitudes, we are on our way to establishing sufficient conditions for the applicability of mathematics to motion, and eventually, by means of the mechanical laws of nature, successive approximations to a true uniform measure of time. But this crucial necessary condition argued for in the Phoronomy only concerns the identity of parts of a motion with the whole composite motion they constitute, not equality relations among those parts. Kant emphasizes the focus on identity in his remark to Explication 5:

> ... phoronomy is the doctrine of the composition of the motions of one and the same point [eben desselben Punkts] in accordance with its speed and direction, that is, the representation of a single motion as one that contains two or more motions at the same time, or two motions of precisely the same [eben dasselben] point at the same time, insofar as they together constitute one motion *together*, that is, they are the same [*einerlei*] with the latter. (Kant 2004, 4: 489)

Kant's clear emphasis in this passage is on the representation of the identity of composed parts with the whole they compose.

Kant is articulating a constructability condition for the applicability of mathematics of any kind of magnitude. In the case of spaces, the composition can be readily constructed; one can exhibit *a priori* in intuition that two parts of a line segment, for example, are identical to the whole line segment they compose. In fact, it is not possible to *not* represent this identity, and hence one cannot but represent the composition relation on spaces, since space is an extensive magnitude, which means that the representation of the whole presupposes a representation of the parts with which it is identical. In the case of motion, it is not quite so easy, which brings us to one of the main points Kant wishes to make in the Phoronomy.

### 3.3. How composition of motion cannot be constructed

The familiarity and success of using line segments with arrows to represent the speed and direction of the motion of points might tempt one to think that one can simply employ what we call vector addition to represent the combination of motions. But Kant argues that this fails the condition of representing the identity of the composed motion with the motions composing it. All one thereby represents is the composition of line segments, and hence spatial magnitudes. In taking mere directed line segments to represent motions (or any other magnitudes for that matter), one is simply assuming that the arrows can represent motions in combination; that is, one is simply presupposing that the motions can combine with each other in just the way that directed line segments can be combined.

Vector composition is standardly used not just to represent the composition of motions but also and crucially the composition of forces, and Kant claims that we may be misled if we think of the motions as combined as cause and effect. Kant's insight is that however we think of the vector composition, we cannot simply *assume* that it accurately represents the composition relation of motions (or any other magnitudes); we need to *demonstrate* that the needed composition relation holds.

But what, *exactly*, is the problem with using spatial vectors to represent the composition of motions? Note first that a mere spatial vector cannot be taken to represent motion on its own, without imparting some temporal aspect to it. We usually do this by assuming some common unit of measure, such as a second, and using the vectors to represent the distance that would be covered in that unit of time. We then simply compose the spatial vectors. As Friedman (2013, 68–76) has emphasized, this simply assumes that the purely spatial composition adequately represents the underlying motions, and hence that the motions are composable. But geometry is not phoronomy. What is required is a representation of motion itself, which requires a representation of intervals of space over intervals of time.

The demand to represent intervals of space over intervals of time, however, uncovers deeper issues. One might attempt to do so by explicitly representing some standard unit of time, such as a second, and representing not just the spatial vectors but the point moving along the vector during that interval. This immediately raises the issue of how we represent some standard interval of time, which presupposes a measure of time. And that appears problematic, for, unlike space, we do not have a direct representation of time. In fact, as Friedman explains, part of Kant's larger project in the *MFNS* depends on using the mechanical laws of motion and our experiences to determine the proper measure of time. But then, it seems that the measure of time must wait upon the Mechanics, and we cannot appeal to the representation of intervals of time to represent motions. And if that is the case, one might conclude that one cannot appeal to the representation of intervals of space and intervals of time at all to represent the composition of motion. One would instead need to find a way to directly construct the composition of instantaneous motions. This is the interpretation endorsed by Friedman (2013, 61–66).

While I think it is true that Kant appeals to the mechanical laws of motion to determine the proper measure of time, I do not think that this is what is at issue here. I think Kant is getting at a different, even more fundamental issue. As I have been emphasizing, Kant's conception of construction of composition concerns the representation of the identity of parts of magnitude with the whole magnitude they compose, and the problem is with vector composition, not with vector addition, if addition is taken to assume particular measures of space and time, and hence to assume that the relation of equality obtains for both space and time. Even if one were to represent a point moving along a vector to represent motion over an interval of space during an interval of time, without taking into account either the spatial or temporal measures of either, we cannot construct the identity of the parts of motion with the whole motion they compose.

Reflection on vector composition bears Kant out. Suppose I represent a motion not by a mere spatial vector, but by the motion of a point from the tail to the tip of the vector. Call this a "moving vector." Now suppose that I wish to represent the composition of two motions (for simplicity, two motions in the same direction), and in the standard way place these vectors tip to tail. Although I can represent the mere spatial vectors as identical to the whole merely spatial vector they compose, I cannot thereby represent the composition of the two moving vectors; I cannot represent the two *motions* as identical to the whole motion they compose. The two motions would have to be of one and the same point moving at one and the same time. But each vector represents a *different* point travelling from tip to tail at the same time. If, on the other hand, I imagine one and the same point travelling along the first vector and then along the second, then I will have represented the same point travelling along the two vectors at successive, but different intervals of time. While the result yields a composition of the total spatial distance covered, it does not represent a composition of the motions themselves.

The same holds for composition of motions in opposite directions or in otherwise differing directions. In the case of composition of motions in opposite directions the composition does not result in a whole motion greater than each of the motions composed (and in the case of composition of motion in otherwise differing directions the composition may not do so). Nevertheless, the unrepresentability of the identity of the composed motions with the motion they compose still holds.[13]

It is crucial to note that this argument does not depend on the introduction of a measure into either extended intervals of space or time. It instead rests entirely on reflection on the nature of part–whole composition, and in particular on the representation of the identity of parts with the whole they compose. Nevertheless, we are so familiar with introducing measures into space and time that the underlying problem can be made more salient if we allow ourselves the representation of particular measures. Suppose I represent two motions in the same direction as a motion of 1 inch/sec by a line 1 inch long, and represent to myself a point travelling the length of each arrow in one second. If I now place them tip to tail, I can represent the two motions at the same time only if I represent the motion of two different points. And if I represent the point as first moving along the first vector and then moving along the second vector, then there is only one point, but it has not traversed the vectors in one and the same second, but in two successive seconds. Furthermore, the speed is still 1 inch/sec, just considered over a 2-inch interval of space and two-second interval of time. This concrete example with measures makes the problem particularly vivid, but the underlying problem to which Kant is drawing attention does not depend on the introduced measures of space and time.

It is particularly important to note that on this reading, Kant's objection to vector composition is not that we cannot represent motion as what I have called "moving vectors" because we do not yet have a measure of time available in the Phoronomy that will allow us to represent intervals of time. We could allow representing motion in terms of particular measures of intervals of space and time, and that will *still* not allow us to represent the required identity between the parts of a motion and the whole composite motion they compose.

### 3.4. How the composition of motion can be constructed

Having ruled out appeal to standard vector representation, Kant next shows that one can construct the composition of a motion in space by use of a moving frame of reference. If we represent the motion of a point as, for example, moving 2 ft/ sec relative to a frame of reference, and also represent that frame of reference as moving 2 ft/sec in the same direction, then we will thereby represent the motion of one and the same point as identical to the motion of two different motions of one and the same point at one and the same time.

Let us reflect more carefully how this representation of composition works. Consider a point at a particular location relative to our frame of reference at the beginning of its motion. To make it vivid, imagine a frame determined by three axes at right angles to each other and let us say it is at the position $(1,1,1)$ as measured from the origin on a scale of feet, and over the course of one second, the point moves to position $(3,1,1)$. Thus, according to the definition of speed, it did so with a speed of 2 ft/sec. Now imagine the entire frame of reference moving relative to a further frame of reference at rest with respect to us. (Call the first "the moving frame" and the second the "the rest frame.") To simplify, imagine each frame determined by three axes at right angles to each other, that the axes of each frame coincide at the beginning of the second, and that the moving frame moves at 2 ft/sec in the same direction relative to the rest frame as the motion represented as moving within it. If we consider the moving point relative to what is now the moving frame, it is moving at 2 ft/sec. If we consider the movement of this *very same point* relative to the rest frame, it is moving at 4 ft/sec. Furthermore, the movement of the moving frame is at 2 ft/sec. So we have thereby represented the composition of two motions in one and the same moving point.[14]

The construction of the composition of instantaneous motions follows quite straightforwardly. Because the speed at an instant is represented by the ratios of ever-diminishing intervals of space and time with that instant as an end point, the composition of two instantaneous speeds can be represented by the construction of the composition of the motions at each of the diminishing intervals of time using a moving frame of reference. Furthermore, we can construct the composition of two instantaneous motions by constructing the composition of the motions a point or body would have if it moved over a finite interval of time with the speed it had at the instant.[15]

I want to pause here to clarify and then draw on the results of Sections 3.2 and 3.3. First, as noted above, Kant is arguing that the construction of the composition of motion is a necessary condition of a mathematical treatment of motion, not a sufficient condition. Kant's focus is on a quite basic property of motion, and the demonstration of composability does not require much. In my example, I chose the composition of 2 ft/sec and 2 ft/sec, but the argument does not depend on assigning particular speeds to the motions; I did so just to make the example as concrete and vivid as possible. Kant himself does not assign particular speeds to the motions. Moreover, I chose an example in which the two speeds are equal so that the resulting speed is double each of its components. While Kant does not assign particular speeds according to some specific metric, he does refer in a general way to adding speeds that are equal to each other and hence doubling the speed. He thereby assumes an equality relation on speeds. But the point about the representation of composition Kant makes against the standard appeal to vector composition and for an appeal to moving frames of reference does not depend on equality relations, only on the conditions for representing the identity relation between parts and wholes. On my interpretation, then, Kant refers to composing to equal speeds into a double speed simply to make the point more salient.[16] All one needs to demonstrate that motions compose in the right sort of way is that parts are identical with the whole they compose. In fact, I need not even represent the motions as uniform. All I need to represent is a point moving relative to a frame of reference, which is in turn moving.

This clarifies what we need not represent. What we *do* need to represent, however, is motion as a change in location, that is, the traversing of an extended interval of space over an extended interval of time. And as noted in the Section 2.3, the representation of a motion at a particular point in space or particular instant appeals to the representation of extended intervals of space and time by means of the two bridges between intervals of space and time and instantaneous rest, motion and speed. As noted, rest can also be attributed to an instant in the special case in which it is simply not moving at all with respect to the frame of reference. In that case, the spatial interval traversed in each smaller interval of time with that instant as an endpoint is always zero, and were the body or point to move with the speed attributed to it at that instant for some finite interval of time, it would not change its place. Nevertheless, the attribution of rest at an instant is in terms of the diminishing intervals of time leading up to the instant.

In contrast, Friedman claims that Kant wishes to *directly* represent the composition of instantaneous speeds at a given point:

> For Kant, the problem of conceptualizing speed or velocity as a magnitude therefore involves the construction or exhibition of an addition operation directly on the set of *instantaneous* speeds defined at *a given spatio-temporal point*. (Friedman 2013, 61)

Friedman explains how one can conceive of the construction of an addition operation directly on instantaneous speeds. (See KCN, p. 61 fn. 42). It seems to me, however, that this does not fit as well with what Kant says as the interpretation outlined above, in which what is constructed are motions represented by means of extended intervals of space and time. Kant's examples of the composition of motions that we are to represent to ourselves is by means of a point moving over an extended interval of space relative to a reference frame over an extended interval of time, and likewise for the reference frame itself.

I have presented an interpretation of Kant on the construction of motion and instantaneous motion in terms of intervals of spaces and times. This fits with Kant's explication of motion and his definition of speed. It also explains his references to ever-diminishing speeds and to how a point would move if it continued with the motion it had at an instant for a finite time. Furthermore, it does so in a way that has Kant agreeing with the general framework for explaining instantaneous velocities in the seventeenth and eighteenth centuries. It also provides us with a reading of Kant's arguments concerning the construction of the composition of motion, which highlights the importance of representing the identity of parts of magnitude with the whole magnitude they compose. There are, however, a number of quite serious challenges to the interpretation I am putting forward, to which we will now turn.

## 4. Problems for the account

There are good grounds for thinking that Kant did not think of instantaneous motion and rest in the way common in the seventeenth and eighteenth centuries, and that in Kant's view, the representation or attribution of instantaneous motion and rest is independent of extended intervals of space and time. Friedman (2013, 52–67) describes those reasons and makes a convincing case. I already briefly discussed one of those reasons above: that before introducing a measure of time in the Mechanics, Kant does not have a well-defined notion of an interval of time. I will return to this objection in Section 5.5. In this section, I will focus on more direct reasons, not turning on measurement, to believe that Kant would not account for instantaneous rest, motion and speed in terms of intervals of space and time.

### 4.1. Reducing speed to extensive magnitude

As we have seen, Kant defines speed as $C = S/T$ and I have argued that by $S$ and $T$, Kant has in mind extended intervals of space and time. I have also argued that Kant accounts for the speed of a body or point at an instant in time in terms of two bridges, that is, in terms of the diminishing extended intervals of space and time leading up to the instant, and to the motion the body or point would have if it were to continue with that instantaneous speed for a finite time after the instant. Friedman calls the first bridge the "standard procedure," since it was prevalent in Kant's time. But intervals of space and time are extensive magnitudes, and defining instantaneous speed in this way seems to make speed itself an extensive magnitude. As Friedman puts it:

> On this standard procedure, then, we would, in effect, have reduced the intensive, instantaneous quantity of speed or velocity to an extensive magnitude (more precisely, to an infinite sequence of extensive magnitudes) .... (Friedman 2013, 61)

On the other hand, Kant is quite clear that speed is an intensive magnitude. The problem can be summarized by three incompatible claims:

1. Kant explicitly holds that intervals of space and time are extensive magnitudes.

2. The standard procedure reduces instantaneous speed to an extensive magnitude.
3. Kant explicitly states that speed is an intensive magnitude.

One need not appeal to Kant's definitions of extensive and intensive magnitude to appreciate the difficulty; we seem compelled to reject the standard procedure of explicating instantaneous speed in terms of an unending sequence of ratios of extended magnitudes.

## 4.2. Instantaneous apprehension

The difficulty is straightforward as it stands, but even a cursory consideration of Kant's account of intensive magnitudes increases the pressure considerably, for Kant claims that intensive magnitudes can only be apprehended in an instant (Kant 1998, A167/B209). If this is true, and speed is an intensive magnitude, Kant is apparently committed to the claim that we somehow or other represent speed instantaneously, for the very strong reason that we can *only* apprehend it in an instant, a feature of Kant's view that Friedman rightly emphasizes (Friedman 2013, 59). Thus, an account of instantaneous motion and speed in terms of ever-diminishing intervals appears to get Kant's account backwards. We must begin with instantaneous speeds and represent speeds over intervals in terms of them.

## 4.3. Instantaneous motion in medieval philosophy

Placing Kant's account of extensive and intensive magnitudes in historical context further compounds the challenge. As Friedman rightly points out, Kant's distinction between extensive and intensive magnitudes draws directly from the medieval theory of the intension and remission of forms first developed in the early fourteenth century at Merton College, Oxford, and then in Paris, above all by Nicole Oresme. This theory opened up a mathematical treatment of motion by treating it as a quality or form of a body that could vary in intensity and hence as a quality having an intensive magnitude. With this revision to the place of motion in metaphysics, they were able to begin an analysis of different kinds of motion: uniform and non-uniform motion, that is, motion with a constant and changing speed, and in particular non-uniform but uniformly changing motion, that is, motion under constant acceleration. The high point of their work was the famous mean speed rule, according to which a uniformly accelerating movement starting from rest covers the same interval of space as a uniform movement moving at a speed equal to the speed of the first movement at the middle instant of the time of acceleration (Clagett 1959, 205).

It was crucial to the theory of the intension and remission of forms that motion be treated as a quality that could vary from instant to instant, and hence was a property possessed by an object at an instant. From a metaphysical standpoint the quality of motion was fundamentally instantaneous, and was not reduced to a ratio of extended intervals of space and time. At the same time, the instantaneous quality of a thing nevertheless stood in a necessary relation to motion over extended intervals of space and time, and was even characterized by it. Both Heytesbury and Swineshead, for example, stated that the motion of a point at an instant is considered or measured by the path that *would* be described by the point if, in a period of time, it were moved uniformly at the speed which it is moved at that instant (Clagett 1959, 210, 236, 243, 244, 364). Despite this necessary connection between instantaneous motion and motion over an interval, however, motion as an instantaneous quality was regarded as fundamental. This priority is central to their views. The path that would be described in the hypothetical case was called the quantity of motion, measured by the space covered, and was considered the effect of the quality at the instant.

Kant agrees with many of the most fundamental features of this medieval theory. As we have seen, he holds that speed is an intensive magnitude. Moreover, Kant locates his treatment of intensive magnitudes under the categories of quality, implying that speed is a quality of a thing. And what I have called Kant's second bridge is precisely the way the medieval theorists considered the motion at an instant, and they did so while still maintaining that the instantaneous motion was more fundamental. Finally, as we saw in the previous section, Kant insists in his account of intensive magnitudes that we apprehend them in an instant. It therefore seems overwhelmingly likely that Kant agrees with the medieval theory in holding that speed is fundamentally instantaneous and cannot be reduced to a definition of speed as intervals of space traversed over intervals of time.

### 4.4. *Instantaneous motion from the medieval theory to Newton*

One final consideration would seem to seal the case that Kant thought that speed is fundamentally instantaneous: an appreciation of the importance of treating speed as instantaneous for the development of natural philosophy from the medieval period through Galileo and in Newton's natural science. The medieval theory and its treatment of velocity as an instantaneous property provided Galileo with some of the conceptual tools he needed to describe the uniform and identical acceleration of all bodies falling under the influence of terrestrial gravity. Furthermore, the mathematical treatment of the motion of an object or the action of a force on object at an instant in time was crucial for the application of the calculus to natural philosophy, and without it, Newton's achievements would not have been possible. As Friedman (2013) explains in detail, a good deal of Kant's *Metaphysical Foundations* is an attempt to account for our mathematization of momentum and mass through the laws of mechanics, and this requires that we consider the manifestation of forces through the motions they induce – the instantaneous velocities in particular. Since Kant is providing a metaphysical foundation for Newtonian natural philosophy, it is not at all surprising that the representation of instantaneous motions play a central role.

## 4.5. Summary

Taken together, the various considerations outlined here – from the nature of intensive magnitudes and Kant's views on their apprehension, Kant's relation to the medieval theory of the intension and remission of forms, and his appreciation of the centrality of instantaneous velocity for the mathematical treatment of natural science – provide a powerful argument that Kant thought of speed not as explained as the ratio of extended intervals of space and time, but as fundamentally instantaneous. And this in turn suggests that an aim of the Phoronomy is to directly construct instantaneous velocities, thereby legitimating their mathematical treatment.

On the other hand, these conclusions seem to contradict the results of Sections 1, 2 and 3, which pointed out Kant's explication of motion, his explicit definition of speed as $C = S/T$, and his appeal to never completely diminished speed in his explanation of instantaneous motion and rest, which strongly suggest the standard procedure for treating instantaneous velocities in terms of extended spaces and times. It also contradicts the account of Kant's arguments concerning the conditions of the construction of motion in the Phoronomy, which appeals to extended spaces and times. Finally, it has Kant rejecting the general framework for explaining instantaneous speeds and velocities in the seventeenth and eighteenth centuries, which threatens to make him an anachronism in his own time. In the next section, I will show that Kant's views of intensive magnitudes and their apprehension, the influence of the medieval theory of motion, and Newtonian science are all compatible with the interpretation I outlined above.

## 5. Response to the challenges

### 5.1. Speed as an intensive magnitude

The first objection – that explicating instantaneous speed in terms of the ratio of ever-diminishing intervals of space and time reduces speed to an extensive magnitude, while Kant explicitly holds that speed is an intensive magnitude – is *prima facie* the most compelling. It is indisputable that Kant counts speed an intensive magnitude. But is also indisputable that in the Phoronomy, Kant defines speed, in what he says is "the customary way," using the formula $C = S/T$, and from our analysis it certainly seems he has in mind the ratio of extended intervals of space and time. I will argue that it is the very intuitive second claim that fails. Kant does *not* hold that defining a magnitude as a ratio of extensive magnitudes entails that the resulting magnitude is itself extensive.

This seems counter-intuitive, but it is not if one carefully considers Kant's definitions of extensive and intensive magnitude. As Friedman (2013, 59) rightly emphasizes, Kant's distinction between extensive and intensive magnitudes does not correspond to our own. According to Kant, a magnitude is extensive when the representation of the whole presupposes the representation of the parts. Space is the paradigm example of an extensive magnitude and illustrates Kant's meaning;

one cannot represent a line extended in space without thereby representing the parts of the line. (Kant sometimes links this feature of spatial magnitudes with the claim that the parts are "outside" of each other.) The line is composed of its parts in such a way that it is impossible to represent the line without thereby directly and immediately representing the parts.

For Kant's account of intensive magnitude, it is helpful to keep in mind the intensity of a light, which can serve as a paradigm, even though the intensity of a force, along with the intensity of a speed, will be most important for the *MFNS*. In the *CPR*, Kant calls a magnitude intensive if it can only be apprehended as a unity and in which its multiplicity can only be represented through approximation to negation $= 0$ (Kant 1998, A165/B207ff). Kant wishes to make several points, one about representation parallel to the claim about extensive magnitudes, and one about apprehension. I will postpone a discussion of apprehension until Section 5.2 and focus here on representation. In contrast to extensive magnitudes, the representation of an intensive magnitude does not presuppose the representation of its parts. In fact, an intensive magnitude is always represented as a unity, and I cannot even represent its manifold at all unless I represent it as diminishing as I approximate zero or as increasing from zero up to a given value (Kant 1998, A165–168/B208–210). My representation of the intensity of a light does not presuppose the representation of smaller intensities of light in the way that the representation of a space presupposes the representation of its parts, and it is this difference Kant is describing.

With these definitions in place, let us return to speed. In the Greek theory of magnitudes developed by Eudoxus and found in Book V of Euclid's *Elements*, only magnitudes of the same kind could stand in ratios, so that lines could stand in ratios with lines, areas with areas, and so on, but a line could not stand in a ratio with an area. Magnitudes that were of the same kind were called homogeneous to indicate that that they allowed of standing in this fundamental mathematical relation to each other. The Eudoxian theory of magnitudes had a tremendous influence that persisted into the eighteenth century, and also greatly influenced Kant. But the understanding of magnitudes and their treatment also changed in important and even fundamental ways. In particular, since about the time of Galileo, ratios of magnitudes that the Greeks considered inhomogeneous were admitted into the theory of magnitudes, and were themselves treated as magnitudes (Friedman 2013, 58 fn. 7). Speed as the ratio of an interval of space and an interval of time is perhaps the most important example. It is nevertheless a different kind magnitude from either space or time, and does not necessarily inherit all the properties of space and time. In particular, the fact that space and time are extensive magnitudes may or not have implications for whether speed is an extensive magnitude in Kant's sense.

So the first question is whether the definition of speed as $C = S/T$, where $S$ and $T$ are extended intervals, entails that speed has corresponding extended parts. The answer is "no." The representation of speed does presuppose the representation of an extended interval of space traversed in an extended interval of time, and the representation of any interval of space or time presupposes the representation of its parts. But there are no parts of speed corresponding to either the parts of an interval of space or the parts of an interval of time. To see this, consider a point moving with uniform speed at 2 ft/sec for one second. Its speed over the whole interval is not composed of the speed of the first half of the space traversed added to the speed of the second half of the speed traversed (and likewise for the two halves of the time interval). This is the same the point made in Section 3.3 about attempting to construct the composition of motion with two "moving" vectors tip to tail. For a motion with uniform speed, for example, the speed of the first and second halves of the space traversed are the same as the speed over the whole space traversed. In short, neither the parts of space nor the parts of time, nor even the ratios of intervals of spaces and times, correspond to parts of the speed. Thus, defining speed as $C = S/T$, where $S$ and $T$ are taken to be extended intervals of space and time, does not entail that a speed has corresponding extensive parts that would make it an extensive magnitude.

Even if one acknowledges this disconnect between the extensive parts of space and time and even the speed over those parts, on the one hand, and the parts of a speed, on the other, one might still ask whether speed defined in terms of extended spaces and times is nevertheless extensive. Does the representation of the whole speed presuppose the representation of smaller speeds out of which it is composed? The distinction between what is necessarily presupposed in a representation and what is explicitly represented complicates matters. Nevertheless, when I represent a straight line 10 feet long, there is a clear sense in which I thereby necessarily represent the first through the fifth 2-foot segments, even if I am not explicitly thinking of the line as so composed. But do I, in representing a point moving at 10 ft/sec thereby represent five 2 ft/sec speeds, even if I am not explicitly thinking of them as so composed? It does not seem so, and certainly Kant insists that we do not. In the second remark to the proposition of the Phoronomy, Kant challenges the assumption that one can simply explicate a doubled speed as one in which a doubled distance is covered in the same time. He states:

> ... it is not clear in itself that a given speed consists of smaller speeds, and a rapidity of slownesses, in precisely the same way that a space consists of smaller spaces. For the parts of speed are not external to one another like the parts of space, and if the former is to be considered a quantity, then the concept of its quantity, since this is *intensive*, must be constructed in a different way from that of the *extensive* quantity of space. (Kant 2004, 4: 493)

The different way in which speeds and their composition need to be constructed is, of course, through moving frames of reference. And here again, we can ask, does *this* representation of a whole speed relative to a frame of reference presuppose the representation of all the speeds of which it is composed? If Kant is right, then the parts whose representation are presupposed would be all the moving frames of reference corresponding to the representation of the whole. Does the representation of a point moving 10 ft/sec relative to a frame of reference presuppose the representation a sequence of five frames of reference each moving 2 ft/sec relative to its predecessor? It does not seem so, and certainly Kant does not think it to be so.

Contrary to what one might expect, the fact that speed is defined as the ratio of extended spaces and times does not entail that speed is reduced to an extensive magnitude in Kant's sense of extensive magnitude, and Kant can consistently claim that speed is an intensive magnitude.

## 5.2. *Instantaneous apprehension of intensive magnitude*

The previous section focused on Kant's claims about the difference between extensive and intensive magnitudes with respect to the *representation* of their parts. But Kant also thinks that this point about representation is reflected in our *apprehension* of extensive and intensive magnitudes. In the above discussion of intensive magnitude, we found Kant connecting his claims about the relation between the representations of parts and their wholes with apprehension. He holds that our apprehension of an extensive magnitude is based on a successive synthesis proceeding from the parts to the whole representation. In contrast, Kant claims, intensive magnitudes can only be apprehended in an instant (Kant 1998, A167/B209).

The claim that we apprehend intensive magnitudes in an instant might seem plausible with phenomena such as the intensity of light or an impressed force on our body, where there is a degree of influence on our sense at each instant. But this simply seems false for our apprehension of motion of the sort discussed in the Phoronomy, the motion of a point or object in space. We apprehend the motion of objects – with respect to both speed and direction – by apprehending their change in place over time. If I were given only an instantaneous snapshot of the motion of a body passing in front of me, I would not apprehend any motion or rest at all. Thus, even if one thought that Kant holds that we in some way or in some contexts, for example the context of mathematical physics, directly *represent* speeds to ourselves as instantaneous, it does not seem that we *apprehend* a motion all at once at instant.

The issue we are facing is complicated by the fact that Kant holds that the senses can only be affected through motion (Kant 2004, 4: 476). Thus, any influence on our senses is ultimately a result of some sort of motion. This looks like a claim about how our senses work. If one takes motion here to be local motion, it looks like an empirical claim about the operation of our senses; for example, that our sensation of light is the result of a vibration communicated to our retina. As Friedman (2013, 83–90) points out, this seems quite out of place in an *a priori* foundation of natural science. If one instead takes motion here in a broader sense (going back to Aristotle) that comprises not just local motion but any change, then perhaps it could be taken as an *a priori* claim about the nature of affection and its dependence on some sort of change. Or perhaps, as Friedman (2013, 43–46) suggests, Kant's point is an interpenetration of the mathematical and empirical concepts of motion. We need not settle this for our present purposes. However we understand the claim exactly, motion as an affection of the senses would result in a degree of intensity of the sensation that corresponds to a degree of reality in what is represented by means of the sensation. This fits well with Kant view of intensive magnitudes set out in the *CPR* when applied to the intensity of a light: motion (in some sense of motion) affects the senses and results in an intensity of sensation of light which corresponds to the perceived intensity of light. It also fits well with his claim that the intensity of a light is apprehended in an instant, for the degree of influence corresponding to the intensity operates at each instant.

On the other hand, it is unclear how this account should work in the case of the particular sort of local motion at issue in the Phoronomy, the change of place of a point or body. In our apprehension of an intensive magnitude, there is a direct correlation between the degree of influence on the senses and the intensity that is simply not present in our apprehension of the change of place of a point or object. Consider an object crossing your field of view. The motion of the object results in changes in the visual sensations I am having, and according to Kant's claim about the relation between affection and motion, those visual sensations are the result of motions induced in my eye by the light reflecting from the object into my eye. But the change of place of the object is *not* itself directly apprehended as a degree of influence on the sense, that is, as a sensation with a degree corresponding to the speed of the point or object. This is clear if one thinks of apprehending an object moving past in a plane perpendicular to the line of sight of a viewer, first at one speed and then at double that speed. The degree of influence on my eye caused by the reflected light, the intensity of the light, does not double. Rather, the rate of change in the succession of sensations of the object at different places has doubled. And we do not apprehend that change of place in an instant.

There seems, however, to be a way out. Kant qualifies his claim about apprehension in an important way:

> Apprehension, merely by means of sensation, fills only an instant (if I do not take into consideration the succession of many sensations). (Kant 1998, A167/B209)

Kant's parenthetical remark could be taken to suggest that although apprehension by means of sensation is properly speaking instantaneous, there is a broader sense of apprehension according to which we apprehend something in apprehending a succession of sensations corresponding to it. If so, then I can be said to apprehend the local motion of an object by means of the succession of sensations of the positions of the object.

In fact, there is some reason apart from the present problem to think that Kant allows a broader view of what can be sensed over an interval of time. Shortly after the above passage, Kant says that "every sensation is capable of diminution, so that *it* can decrease and gradually disappear" (Kant 1998, A168/B209–210, my italics). Thus, one and the same sensation can decrease over an interval of time. On the other hand, in the very next sentence he concludes: "Hence between reality in an appearance and negation there is a continuous nexus of many possible intermediate sensations ..." so there is a different sensation corresponding to each instant. The identity conditions of sensations are not what is important to Kant; what is important is that the apprehension of a sensation that diminishes over time contains and is based on apprehension of a succession of sensations each of which fills only an instant. Similarly, my apprehension of a motion contains and is based on the apprehension of a succession of sensations, that is, the sensations of the positions of the object, each of which fills only an instant. But that is not to claim that when we are apprehending a moving object by means of sensations of the object at successive instants, we are apprehending the instantaneous motion of the object at that instant. In short, the instantaneousness of the apprehension of the sensations in a succession of sensations that allow us to apprehend motion does not entail that we are apprehending instantaneous motion.

This interpretation of our apprehension might drive one back in the direction of thinking that motion of an object or point is an extensive magnitude after all. If we apprehend motion by apprehending a change over an extended interval of space over an interval of time, then are we not apprehending it as itself extensive? Recall, however, that we have defused this natural and intuitive thought in the previous section through a careful consideration of Kant's definitions of extensive and intensive magnitudes. Even if we apprehend motion through the apprehension of successive positions over intervals of space and time, and even if the apprehension of the intervals of space and time presuppose an apprehension of their parts, it does not follow that the representation of the whole motion presupposes a representation of its parts. Motion is still an intensive magnitude.

## 5.3. *Kant and the intension and remission of forms*

As mentioned above, Kant's theory of extensive and intensive magnitudes is deeply influenced by the medieval theory of the intension and remission of forms, according to which motion is a quality with an intensive magnitude and is fundamentally instantaneous. Despite the many points of agreement, however, I believe Kant departs from the medieval doctrine precisely on the explanation of instantaneous velocity. To appreciate how, we need to consider the medieval theory a bit more closely.

As noted above, the medieval school articulated a necessary connection between an instantaneous speed and extended spaces and times, and even used this to characterize the instantaneous speed. The instantaneous speed is considered as the path it would traverse in some time if it were to continue with the speed it had at that instant, and this connects the quality of a motion with its effect, the quantity of motion.

On might think that this opened the way for reducing instantaneous velocity to the ratio of extended intervals of space and time, but there were two fundamental limitations of their theory that prevented this. First, as I mentioned above, prior to Galileo, it was assumed that only mathematically homogeneous magnitudes could stand in ratios, so that only lines could stand in a ratio to a line, an area to an area, a time to a time, and so on. Thus, the "ratio" of an interval of space to an interval of time was not itself of magnitude and did not even express a ratio. Since the ratio between an interval of space and an interval time was therefore ruled out, so was the definition of speed in terms of such a ratio. As a result, discussions of the relation between the intensive magnitude of an instantaneous motion and the quantity of motion as its effect focused on either distance or time. For example, the quantity of one motion was twice that of another if the interval of space covered in an equal amount of time was twice as large. This presupposes that we are considering the two intervals of space covered in an equal time; nevertheless it is the ratios between the spaces that are used to express the relation between the two quantities of motion. Thus, the intensive magnitude of a motion at an instant was necessarily connected to the extensive magnitudes of space and time, but only as its effect. There was simply no possibility of reducing the instantaneous speed to a ratio of extended intervals of space and time.

There was a second limitation as well. The correlation between instantaneous velocity of a thing and the speed it would have over an interval of time only permitted quantifying the instantaneous velocity in special cases. If the motion is uniform, then the speed over *any* interval is the same. In fact, uniform motion was defined by appeal to this property. Thus, the same intensive magnitude could be attributed to each instant of the interval. If the motion is uniformly difform (i.e., constantly accelerating) over an interval beginning at rest, then the mean speed rule establishes that the intensity of the instantaneous velocity at the midpoint of the interval is half of the average velocity over the interval. Despite these special cases, there is no general way to attribute instantaneous velocity at a point on the basis of its effects over extended intervals of space and time. What is required is precisely a characterization of the velocity at an instant in terms of the ratios of ever-diminishing extended magnitudes with that instant as an endpoint, which is conspicuously absent from the medieval treatment.

Kant, in contrast, was not hindered by either of these limitations; ratios of intervals of space over intervals of time were permitted, and were themselves considered magnitudes. And as we have seen, Kant appeals to the ratio of ever-diminishing intervals of space and time to characterize motion and rest at an instant. This means that a much more intimate relationship is possible between speed, including instantaneous speed, and extended intervals of space and time for Kant.

This seems to invite Kant not just to define and construct, but to *reduce* instantaneous speed to the ratio of extended intervals of space and time. But he does not do so, for although Kant thinks that motion is a magnitude, he thinks that the kind of magnitude that results from taking the ratio of an extended space and an extended time differs in the nature of its part–whole relations, because the representation of this magnitude does not presuppose a representation of the parts.

In a nutshell, Kant does indeed explicate motion in terms of change of place, define speed in terms of the ratio of extended intervals of space and time, and characterize instantaneous speed in terms of a sequence of ever-diminishing intervals of space and time. But that does not amount to *reducing* speed to that sequence, in the sense of simply replacing instantaneous speed with that sequence.

### 5.4. Instantaneous velocities in Newtonian physics

The previous section explained how Kant could accept the framework of the medieval theory of intension and remission of forms and many of its most characteristic claims while still defining speed as the ratio of extended intervals of space and time. But as has been made abundantly clear from recent work, Kant possessed a thorough understanding of Newton's *Principia* and remarkable insight into the conceptual framework on which it rests. The treatment of instantaneous velocities in Newtonian physics plays a pivotal role not only in Newton's derivation of the law of universal gravitation, but in natural philosophy in the century after Newton published the *Principia*. Thus, one can expect that whatever Kant's understanding of the metaphysical status of intensive magnitudes and their apprehension, his account will give a treatment of instantaneous velocity adequate for its role in Newton's physics. And that would seem to elevate instantaneous velocity over velocity defined in terms of extended intervals of space and time. I believe, however, that Kant's account of instantaneous velocity in terms of a sequence of ratios of ever-diminishing extended intervals of space and time is precisely an attempt to accommodate the eighteenth century understanding of how to mathematically treat instantaneous velocities.

### 5.5. The definition of speed and the problem of the determination of time

I hope to have shown how Kant can explain motion and define speed in terms of extended intervals of space and time in a way consistent with his commitment to motion being an intensive magnitude and with the need to represent the speed of a point at an instant as required by the natural science of Kant's day. There is, however, a very different sort of reason Friedman gives for thinking that Kant could not define speed as $C = S/T$ understood as the ratio of intervals of space and time. As described in Section 3.3, it is based on Kant's larger aim to reform Newtonian physics by freeing it of a reliance on absolute space and time. He does so by appeal to relative frames of reference and measures of time successively singled out using his mechanical laws of motion, a procedure Kant bases on Newton's own method of refining frames of reference. As a result, Kant can no longer define motion in terms of intervals of absolute space and time. According to Friedman, however, he also cannot define it in the Phoronomy by appeal to relative spaces and times, since prior to the mechanical laws of motion described in the Mechanics, the idea of a mathematically determinate temporal magnitude of duration is not well-defined (Friedman 2013, 65, 66). Thus, Kant must reject the standard procedure, and attempt to directly construct instantaneous speeds.

Unfortunately, I do not have the space to include a full treatment of this issue here, which requires a thorough account of the relation between magnitudes and their measurement and the relation of both to mathematical cognition. I would nevertheless like to indicate the direction that treatment takes. I agree with Friedman's interpretation of how Kant replaces the role of absolute space and time in Newtonian natural philosophy and that it is a central aim of the *MFNS*, but I believe that defining speed in terms of extended intervals of space and time in the Phoronomy is consistent with it. In a nutshell, there is an important distinction between, on the one hand, having representations of extended intervals of spaces and times sufficient to explicate motion, define speed and demonstrate the constructability of motion in the Phoronomy, and on the other, determining a measure of time through application of the analogies of experience and Kant's version of the mechanical laws of nature.

In Section 3.3 I noted that, as I understand it, neither Kant's conception of magnitude nor the notion of composing magnitudes presupposes the relation of equality, and that Kant's arguments about how the composition of motions can and cannot be constructed turn on the constructability of the identity between parts of motion and the whole motion they compose. No measure of space or time is needed in order to construct the composition of motions, and that construction establishes a crucial necessary condition for the applicability of mathematics. At the same time, one kind of magnitude, namely space, does allow of the relation of equality: following Euclid, Kant holds that space allows the relation of equality, and that together with the composability of space and the fact that space is an extensive magnitude is sufficient to generate a measure of space. This measure can then be leveraged to introduce a measure into other kinds of magnitudes, including motion. A determination of true motion will also require a true measure of time, and, as Friedman explains, the full story of the determination of the "true" uniform measure of time does depend on Kant's mechanical laws of motion. Nevertheless – and this is the key claim – there are representations of intervals of space and time that make no appeal to their true measure and are sufficient for the Phoronomy.

Making good on these and other supporting claims requires, among other things, distinguishing levels of determination within the System of Principles, and showing how the determination of space and time described in the Axioms of Intuition allows for the representation of extended intervals of space and time without thereby fully determining their mathematical properties, and in particular, introducing a measure. It also requires a careful consideration of the interrelations among the representations of motion, spaces and times at this level of determination, and explaining how these representations relate to the use of motion in measuring time. But a full defence must wait another occasion.

## 6. Summary

I have outlined a reading of the Phoronomy that explains how the construction of motion and the composition of motion rest on the construction of extended intervals of space and time. It explains how Kant combines elements of the medieval view that motion is an intensive magnitude with the seventeenth- and eighteenth-century approaches to explaining instantaneous speeds in terms of the ratio of ever-diminishing intervals of space and time. I believe it makes most sense of the text and arguments concerning construction, while showing how Kant's phoronomical treatment of motion provides the foundation on which the treatment of motion in the later chapters can build. In particular, it is consistent with Kant's attempt to reform Newtonian natural philosophy by ridding it of an appeal to absolute space and time, and with Kant's procedure for successively approximating the true measure through the mechanical laws of motion.

This reconstruction of Kant shows how the *CPR* account of mathematical cognition and magnitudes, and especially the synthesis of composition and the doctrine of construction, both sets the framework for and is designed to account for the possibility of mathematical physics. In doing so, it highlights a feature of Kant's doctrine of construction that might be underappreciated: the importance of constructing the composition of magnitudes to demonstrate the identity of a whole magnitude with its parts in order to establish the applicability of mathematics to them.

### Acknowledgements

I thank the participants at a Workshop on Kant's Philosophy of mathematics and science at the Max Planck Institute for the History of Science in Dahlem, Germany, on 28 July 2014. I thank Vincenzo De Risi for organizing the workshop, and De Risi, Katherine Dunlop and Tal Glezer for especially helpful comments. I also thank Friedman for extended discussions on the role of mathematics in Kant's philosophy and for comments on the penultimate draft of this paper.

### Funding

This work was supported by a University of Illinois at Chicago Institute for the Humanities Faculty Fellowship for 2013–2014, for which I am most grateful.

### Notes

1. This is made abundantly clear by a great deal of work on Kant's philosophy of science in the last two decades. Friedman (2013) has been particularly helpful to me in understanding the depth and sophistication of Kant's engagement with Newtonian science; this article is throughout indebted to it and would not have been possible without it.
2. All references to Kant will be to the Akademie edition pagination by volume and page number, except references to Kant's *Kritik der reinen Vernunft*, which will follow the standard A/B form of reference. I follow the Michael Friedman's

translation of the *MFNS*, and Paul Guyer and Allen Wood's translation of the *CPR* unless otherwise indicated.

3. Natural philosophers did not always clearly express the distinction between speed as a scalar magnitude and velocity as a vector comprising both scalar magnitude and direction. "Velocity" is sometimes used for speed, or for the scalar magnitude of a velocity, though sometimes direction is also meant. My paper will focus almost exclusively on speed, though I too will sometimes refer to velocity.

4. See Friedman (2013) for a clear, full treatment of the Phoronomy that takes into account its role in the *MFNS* as a whole.

5. Thus, Kant would have articulated a sufficient but not a necessary condition of motion, and we cannot infer a lack of motion from a point being in the same place at two different moments. Because Kant is only giving an explication of motion and not a definition, this issue does not undermine his account, though it does underscore its limitations. The same could not be said about speed, however, which Kant classifies as a definition.

6. Kant takes the two properties to be co-extensive, which indicates that he has in mind typical vibrations, oscillations and orbits that follow the same paths, and not closed paths in general, for which it is easy to imagine the properties coming apart. This reflects the fact that he is here concerned only with those motions for which speed is often used to describe frequency or period.

7. Kant's example focuses on the instantaneous rest of the body at the turn-around point, and for that reason the ever-diminishing intervals take that spatial point as an endpoint. But a similar construction could be made for the speed of a body taking a particular instant of time as an endpoint. For simplicity I will sometimes refer below to the instantaneous speed at an instant, but I have both cases in mind.

8. I say "apparent" since Abraham Robinson (1966) rehabilitated in finitesimals in his development of non-standard analysis, and showed how one can think of them in a non-contradictory way.

9. The generalization also follows from the relativity of motion: if we consider the same body tossed in the air relative to a frame of reference that is moving downward 5 ft/sec relative to the original frame of reference, then the ratios of ever-diminishing intervals will approach an upward speed of 5 ft/sec relative to the new frame of reference. See Friedman (2013, 47). Of course, this must be the result if Kant's account is going be adequate; however, this way of establishing the relativity of instantaneous motion presupposes that motions can be composed, and Kant only demonstrates the composability of motions later in the Phoronomy, as we shall see. Thus, a generalization by considering instantaneous speed at other than the turn-around point within one reference frame is preferable. See note 15 below for further discussion.

10. See Friedman (2013, 49 fn. 22 and fn 23, 51 fn 25) for discussion of this matter. If Newton's theory of fluxions gives priority to instantaneous velocities and velocities over intervals are secondary, Kant's approach on the interpretation I am offering would rule it out. Kant refers to infinitely small [unendlich klein] speed. (See the next footnote.) I do not think that Kant is committing himself to a speed that is simultaneously actually infinitely small and yet thought of as defined over an interval; he may simply mean indefinitely small. But I cannot delve further into this issue here. I would like to thank Friedman for helpful comments on this point.

11. Note that Kant refers to a motion with "infinitely small [unendlich klein]" speed. See the previous note.

12. I would like to thank Michael Friedman for pressing me on this issue and for continued discussion. I will return to it again in Section 5.4.

13. I would like to thank Michael Friedman for prompting me to make this clear.

14. There is a very direct sense in which we have considered the motion of one and the same point moving 2 ft/sec relative to the moving frame and 4 ft/sec relative to the rest frame. This seems to be sufficient to demonstrate the composability of motion. Nevertheless, we do not represent the motion of the moving frame of reference as a motion of one and the same point quite as directly. Is that a problem? Every point of space relative to the moving reference frame is moving at 2 ft/sec, so the path traversed by the moving point is moving at 2 ft/sec. We represent any motion by representing it as traversing a path, and this seems sufficient.

15. As I mentioned in footnote 9, one might argue that Kant's characterization of instantaneous motion generalizes to any instantaneous motion simply by appealing to the relativity of motion to a frame of reference; the instantaneous motion at a point in one frame of reference will be the same as instantaneous rest of the point in a frame of reference that moves uniformly with the velocity of that instantaneous motion. But that way of generalizing the construction of instantaneous rest assumes the composability of motions, which Kant has not yet demonstrated. Accounting for the composition of instantaneous motions in the way described here avoids this problem.

16. Recall from Section 2.2 that Kant sometimes uses "body" rather than "point" in the Phoronomy, but only to anticipate later portions of the *Metaphysical Foundations* and to make the exposition "less abstract and more comprehensible" (Kant 2004, 4: 480). In my view, the same sort of thing is going on here: a concrete example that appeals to doubling a speed is much more vivid and comprehensible, but the point Kant is making does not depend on the relation of equality thereby presupposed.

## References

Clagett, Marshall. 1959. *The Science of Mechanics in the Middle Ages*. Madison: University of Wisconsin Press.

Friedman, Michael. 2013. *Kant's Construction of Nature: A Reading of the Metaphysical Foundations of Natural Science*. Cambridge: Cambridge University Press.

Grant, Edward. 1996. *The Foundations of Modern Science in the Middle Ages*. Cambridge: Cambridge University Press.

Kant, Immanuel. 1998. *Critique of Pure Reason*. Translated and edited by Paul Guyer and Allen Wood. Cambridge: Cambridge University Press.

Kant, Immanuel. 2004. *Metaphysical Foundations of Natural Science*. Translated and edited by Michael Friedman. Cambridge: Cambridge University Press.

Robinson, Abraham. 1966. *Non-standard Analysis*. Princeton: Princeton University Press.

# Kant on conic sections

Alison Laywine

*Department of Philosophy, McGill University, Montreal, Canada*

This paper tries to make sense of Kant's scattered remarks about conic sections to see what light they shed on his philosophy of mathematics. It proceeds by confronting his remarks with the source that seems to have informed his thinking about conic sections: the *Conica* of Apollonius. The paper raises questions about Kant's attitude towards mathematics and the way he understood the cognitive resources available to us to do mathematics.

*For Charles Laywine*

Kant may be understood to have developed a philosophy of mathematics. He had to do so, just to the extent that he wanted to understand why pure reason sometimes achieves brilliant results (as in mathematics, natural science and logic) and sometimes not (as in traditional metaphysics). But he never wrote a single, focused philosophical treatise on mathematics. His remarks on the subject, scattered here and there, are often unhelpfully abstract. When he does offer concrete examples, they are often so elementary or so underdeveloped that they raise as many questions as they settle. Kant should have had something to say about any and all mathematics on the go at his time. His philosophy of mathematics would be inadequate otherwise. As it turns out, he does have something to say about conic sections, in at least four different passages in works spanning a period of almost 30 years. The trouble is what to make of them. They do not contain throwaway remarks. On the contrary, they are rich and take up the same themes, though admittedly from different angles and with different emphases. But they are also highly condensed, which makes them enigmatic. So I am going to try to tease out some of what is going on in them. My strategy is first of all to use these passages to state a problem associated with Kant's account of conic sections. The second part of my strategy is to use the statement of this problem to put into perspective what Kant took to be the cognitive resources available to us for doing mathematics. I do not claim originality for everything

I have to say about this. But I think that my strategy can help bring to light puzzles – sometimes overlooked or underplayed – associated with Kant's understanding of mathematics as such and the relation between reason, imagination and the power of judgement (*Urteilskraft*). I will try to bring these puzzles and difficulties into focus by confronting Kant's account of the mathematician's cognitive resources with the historical sources that apparently informed his own understanding of conic sections.

Before setting out in earnest, let me add that I am in many ways indebted to the paper by Friedman (1992) that forms Chapter Four of his book *Kant and the Exact Sciences*. I offer the reflections in my paper not as a challenge, but as a complement to his. Friedman dealt with only one of the four passages of interest to me. I propose to look at all four of them together. To that extent, I will be doing something more than he tried to do. But I will also be doing something less. For the emphasis of my paper will be strictly on what Kant has to say about the mathematics of conic sections, whereas Friedman was concerned with Kant's views about the relationship between mathematics and physics.

## 1. The four passages

The almost 30-year period spanned by the four passages of interest to me was from 1763 to 1790. That was not just a long time. It was a period of very significant change in Kant's thinking about theoretical philosophy in general and metaphysics and mathematics in particular. The first passage comes from *Der einzig mögliche Beweisgrund zur Demonstration des Daseins Gottes* (2.95.19–31), an attempt from 1763 to mount a proof for the existence of God. While imaginative and idiosyncratic, this early work moves squarely within traditional metaphysics, which Kant viewed at the time as a discipline in need of reform, but not without prospects of future success. I discuss the *Beweisgrund* passage in Section 2.2 of my paper. The second passage comes from the so-called *Prolegomena* of 1783, a work in which Kant tried to present to a lay audience his new doubts that traditional metaphysics could count as a science on its own terms and his project for undertaking a critique of pure reason to determine systematically what reason can and cannot hope to accomplish. The passage of interest to me appears in Section 38 of the *Prolegomena*. This is the passage of interest to Friedman and the object of discussion in Section 2.2 and Section 3 in my paper. The larger context of *Prolegomena* Section 38 is the question how mathematics and natural philosophy are possible as sciences (while metaphysics is not). Part of what distinguishes this passage from the one in the *Beweisgrund* is Kant's appeal, as part of an answer to this question, to a new conception of space (and time) that he had first begun to work out in the 1770's. By 1770, Kant had come to believe that space is not a thing in itself, as perhaps Newton thought, nor a system of relations among things in themselves, as Leibniz seems to have thought, but some kind of subjective condition at the basis of human sensibility and hence at the basis of all human knowing: in the first instance, mathematics and natural philosophy. Kant speaks of space in the passage of interest to me from the *Beweisgrund* in terms that suggest certain aspects of his account of space in the *Prolegomena*. But the idea that space is something subjective in the relevant sense is not in evidence in the *Beweisgrund* or other works from the 1760s. The third passage of interest to me, and the object of discussion in Section 5 of my paper, comes from the Appendix to the Transcendental Dialectic of the *Critique of Pure Reason* (B690–696): it underwent no special revisions and thus appears the same, word for word, in both editions (1781/1787). The interest of this passage is, first of all, that it comes on the heels of an extended argument in the Transcendental Dialectic designed to show that the *constitutive use* of pure reason in traditional metaphysics cannot fail to lead us into impasse and, second of all, that it (the passage of interest) tries nevertheless to allow for a fundamental contribution by the *regulative use* of pure reason in mathematics and natural science: indeed, the mathematics of conic sections is supposed to give us an instructive example of this. A third interest of this passage is that it characterizes the distinction between the constitutive use of pure reason and its regulative use in terms that we can recognize, with the benefit of hindsight, as indicating a need for the distinction drawn in the *Critique of Judgement* of 1790 between determining and reflective judgement. The *Critique of Judgement* is itself the source of the fourth and last of the passages I will be considering (in Section 5 of my paper). This work is best known as the mature statement of Kant's aesthetics. But the passage of interest to me in Section 62 (5.362.6–366.23) comes from the rather more neglected part of this work devoted to natural teleology, and I should say that it takes up issues about teleology in mathematics that were already at work in the *Beweisgrund*.

If you are interested, as I am, in understanding the motivations behind Kant's development in theoretical philosophy, the four passages I just mentioned are important, precisely because they mirror and illuminate Kant's evolving views on some of the most central aspects of his philosophy over this significant period from 1763 to 1790. But what is more important for our purposes right now is something that remains constant in all four of them. That is a remark Kant can be understood to make about conic sections. Let me begin with a brief paraphrase of what seems to be the thrust of the remark. We will have occasion to look at Kant's own statements of it later.

Kant will typically say that every kind of conic section exhibits a certain 'necessary unity amidst the greatest manifoldness' (*Einheit bei der größten Mannigfaltigkeit* – 2.95.19–20). Sometimes, instead of unity, he will use the words 'order' (*Ordnung*), 'harmony' (*Harmonie*), connection (*Zusammenhang*) or 'agreement' (*Wohlgereimtheit*). These terms seem to be closely associated in his mind with laws: it is by virtue of some law that unity and order are exhibited in an infinite 'manifold'. Kant begins by invoking the circle, which he treats as a conic section in its own right.[1] The interest of the circle is always that it exhibits the relevant unity in the simplest way possible. Kant often illustrates this claim by appealing to a favourite Euclidean theorem: Proposition 35 of *Elements*, Book

Three. This proposition says that, given any pair of intersecting chords in a given circle, the rectangle on the two segments of the one chord is equal to that on the two segments of the other. The manifold in question comprises the infinitely many pairs of intersecting chords in the given circle. The manifold exhibits necessary unity, because it obeys the law stated by Proposition 35 of *Elements*, Book Three, and because there are no exceptions to this law. Kant will then claim that each of the other conic sections exhibits some kind of unity like that, or related to that, of the circle and thereby implies or directly claims that conic sections collectively exhibit some kind of unity. Perhaps, it would not misrepresent his view to speak of a higher order unity in this case: that expressed by whatever law, principle or common ground unites the conic sections into a special class of planimetric figures. As it turns out, this will mean something concrete for Kant: the characteristic property of the circle stated by Proposition 35 in *Elements*, Book Three, can be extended to conic sections, with certain qualifications. I will discuss this idea at greater length in Section 3 of my paper.

For now, let me add that the claim or set of claims I just stated on Kant's behalf fits a pattern distinctive of his thinking throughout his life. He will say in the passages under consideration and elsewhere that our higher cognitive faculty – sometimes he calls this faculty 'understanding' (*Verstand*), sometimes he calls it 'reason' (*Vernunft*) – is not satisfied to seek out laws or principles in whatever discipline it may be. It ultimately tries to understand how the laws or principles it may have discovered relate to one another. This it does by trying to understand these laws or principles as expressions, special cases or consequences of higher ones. Kant drew one of his showcase examples of this tendency from physics. He pointed out numerous times that Newton's law of gravitation enjoys a privileged status for reason or understanding. This is because it is the ground or principle that 'unites' (as he would say) many other natural laws, notably Galileo's law of falling bodies, but also the motion of the comets, that of the ocean tides, the flattening of the earth at the poles, among other things. In Section 39 of the *Prolegomena*, the section of this work right after the passage of interest to us, Kant gives a characterization of this tendency or preference of our higher cognitive faculty in perhaps its most pronounced form, namely as it is found among philosophers. The passage, as I translate it, reads as follows[2]:

> Nothing can be more desirable for a philosopher than if he can derive the manifold of concepts or principles that had previously presented themselves to him in an isolated way, through the concrete use he made of them, from a principle *a priori* and *in this way to unite everything in one cognition* [my emphasis – AL]. Before doing so, he believed that everything remaining after a certain abstraction, and seeming to constitute a special sort of cognitions through the comparing of them with one another, had been completely collected. But it was only an *aggregate*. Now he knows that precisely so many [items] – no more, no less – can constitute the relevant sort of cognition, and he grasped [*sah ein*] the necessity of his division, which is an understanding [*Begreifen*], and only now does he have a *system*. (4.322.22–32)

It turns out that what is immediately at issue here is Aristotle's list of categories. The complaint, to start out with, is that Aristotle drew up this list in an ad hoc way: if some notion seemed to him to be more fundamental than others, he would put it on the list. But in the absence of a defensible criterion, the choices were never philosophically defensible. Now that complaint, all on its own, is not yet distinctively Kantian: you do not have to be Kant to notice that there are no arguments in Aristotle's *Categories*. The distinctively Kantian twist is the idea of a principle that will 'unite everything in one cognition'. If you can find that principle, you can understand what a category is supposed to be and why some concepts belong among the categories and others do not. Just as important, you will understand how the genuine categories relate to one another. That means you can explain why you ended up with precisely these categories, no more, no less. Thus, you will have a true 'system' that unites the categories as a very special class of concepts, rather than just a shopping list, as Aristotle did.

A parenthetical remark is in order here. The word in play now is 'system', but I trust it is clear that, for Kant, having a system and having 'necessary unity amidst the greatest manifoldness' go hand in hand. You will have a system of categories if you have succeeded in 'uniting' all the category contenders 'in one cognition'.

Kant's predilection for systems and his claim to the systematicity of his own table of categories have often been the source of merriment among his detractors: he is seen as a sort of philosophical neat freak. But Kant was surely right to think that Newton sought the elaboration of some sort of system in the *Principia*. At the beginning of *Principia*, Book Three, Newton says 'It remains that we now lay forth, from the same principles [sc. the mathematical principles of Book One], the constitution of the system of the world' (1972, 549). The interest in systems is explicitly announced in the very title of Book Three as a whole: *De mundi systemate*. It is in Book Three, namely in Proposition 4 and its scholium, that Newton establishes the connection between the inverse-square law and Galileo's law of falling bodies: a connection that seems, as I say, to exemplify Kant's understanding of systematicity, at least in physics or mechanics. Our higher faculty of cognition, as exemplified in Newton, seeks out systematicity in physics; exemplified in others, it seeks it out in philosophy as well. There is no field of enquiry in which reason does not seek it out. The question is what the system-seeking of reason looks like in the mathematical case – or, to keep the question manageable – what it looks like, according to Kant, in the case of conic sections.

Here a natural thought may suggest itself: the system-seeking that Kant ascribes to the mathematician would be best satisfied for conic sections by an algebraic treatment based on the equations for the different curves familiar to us from analytic geometry. This thought derives its plausibility from another thought suggested by the very notion of 'systematic unity': we would naturally expect that the higher the level of systematic unity, the more powerful the techniques available to us. The algebraic treatment of conic sections is indeed more powerful than the classical geometric treatment, as presented to us in Apollonius' *Conica*. Consider, for example, the programme of *Conica*, Book Four: to show in how many different ways conic sections may meet one another and in how many ways they may meet circle circumferences and circle arcs (Apollonius: Rashed 2008, 252–253; Heiberg 1891, 4). Apollonius works his way painstakingly, case by case, through all the different configurations of conic sections and circles, distinguishing the case when the relevant sections cut each other and when they touch, and seeming – to the modern reader – to prove the same thing over and over again. This is no doubt what explains Heath's deflating assessment in volume two of the *History of Greek Mathematics*: 'Book IV is on the whole dull' (1981, 157). He was wrong about that. Fried and Unguru argue energetically and convincingly that Book Four is very interesting indeed. For the details of their account and their argument, I refer the reader to their book, *Apollonius of Perga's Conica: Text, Context, Subtext* (Fried & Unguru 2001, 117–146). The thing that matters here is that the obstacle they have to overcome to make their case (and the reason for Heath's assessment) is, as they point out, that the algebraic techniques developed by Descartes in *La Géométrie*, and that have inevitably coloured our perception, make *Conica*, Book Four, completely dispensable (Fried & Unguru 2001, 134–135). If, then, the idea of systematic unity, as Kant understands it, may be taken to imply a greater capacity to handle together as a single case cases that would otherwise have to be handled separately, then, indeed, the algebraic treatment of conic sections is the (or one of the) most natural way(s) to cash out Kant's claim in the four passages of interest that theory of conic sections exhibits a high level of systematic unity.

## 2. The problem

That thought, however natural, is what leads me to the problem I mentioned at the beginning of this paper. The later Kant does try to account for algebra: in the *Critique of Pure Reason* and elsewhere. But this account – if that is what we may call it – is obscure and fragmentary: more so, indeed, than any other part of his mature 'philosophy of mathematics'. It too is supposed to rest on intuition and the formal conditions of our sensibility. One might wonder whether and how such an account would work. The natural thing to think, on the face of it, is that algebra evacuates intuition from mathematics. The algebraic equations that state both the necessary and sufficient conditions for a curve to count as either a circle, ellipse, parabola or hyperbola seem to remove the conic sections from our mind's eye: we need no longer imagine them as plane figures, much less as plane figures generated as sections of a cone. Thus, one might suppose that, however Kant may have tried to account for algebra, the algebraic treatment of conic sections just was not on his horizon. Now this supposition proves to be false, as I will show in Section 2.1 of my paper. But as I will argue, Kant did not use algebra as a point of reference for thinking about conic sections (and the philosophical questions they raise about mathematics more generally). The problem of my paper, then, is to determine what he took to be the source of their 'systematicity' or 'over-arching unity'. One could also wonder whether Kant was aware of projective geometry – I will return to this question briefly at the end of Section 3 of my paper.

Let me now make two observations designed to bring the problem more sharply into focus.

## 2.1. The first of the two observations

First of all, throughout his life, i.e. even before developing the ideas distinctive of the *Critique of Pure Reason* and the *Prolegomena*, Kant's understanding of mathematics was informed by ancient Greek geometry – in this one respect at least: he took it to depend largely on techniques of construction, deployed in the interest of solving constructive problems (but see also Sutherland [2006] for another way in which ancient Greek mathematics was significant for him). I do not mean to suggest it could have escaped his notice that we can undertake to solve algebraic problems too. I mean only that the business of solving constructive problems in geometry loomed large in his understanding of mathematics as a discipline. That is part of what gives rise to my problem: it is not obvious how a *general* theory of conic sections is possible if your resources come from constructive problem solving in geometry. I do not mean to suggest that geometrical problem solving can never achieve generality; Apollonius' example shows us otherwise (and it is indeed his example that seems to have guided Kant, as we will see in due course). But it does so incrementally, even (perhaps especially) in the order of reasons, over a long, difficult and sometimes tedious haul. The algebraic techniques are indeed more powerful, and therefore they do seem more fit to produce the 'overarching unity', as Kant might put it, required for the 'system' or general theory familiar to us today from analytic geometry.

Here I think it is important to document my observation and the questions it raises. This means turning to the short work Kant published in 1790 in Königsberg under the title: *Über eine Entdeckung, nach der alle neue Critik der reinen Vernunft durch eine ältere entbehrlich gemacht werden soll.*[3] The purpose of this work was to reply to a series of attacks against the *Critique of Pure Reason* launched by Johann August Eberhard in a serial publication called *Philosophisches Magazin* published by Gebauer in Halle starting in 1788. At the beginning of the 'Reply to Eberhard', as I will now refer to this work, Kant mentions the construction Apollonius uses in the *Conica* to establish, among many other things, the principal property of the parabola in Proposition 11 of *Conica*, Book One. He (Kant) is guided in this by the question how geometers establish the 'objective reality' of their results:

> Apollonius first constructs the concept of a cone, i.e., he displays it [*darstellt*] *a priori* in intuition (this is now the first procedure [*Handlung*] whereby the geometer initially establishes the objective reality of his concept). He then cuts the cone according to a certain rule, e.g., parallel to a side of the triangle that passes through the summit and cuts the base of the cone (a right cone) at right angles. He proves, in

the *a priori* intuition, the properties of the curve that is produced by this cut on the surface of the cone and thus brings out a concept of the ratio [*Verhältnis*] in which the ordinates stand to the parameter. This concept, namely (in this case) that of the parabola, is given *a priori* in intuition and its objective reality, i.e., the possibility that there can be a thing having the aforesaid properties, is proved in no other way than this: *one bases it upon the corresponding intuition.* (8.191.7–21)

We learn from this passage that, by 'objective reality', Kant means the possibility that there is a thing having the distinctive properties – say – of a parabola. To prove the objective reality of the concept of parabola, the geometer must first use a certain procedure (described below) to construct a cone in intuition and then use a certain sectioning technique (also described below) to generate the desired curve. If we succeed in carrying out both steps of the construction, we will have succeeded in proving (at least in a preliminary way) the objective reality of the parabola, because, having actually constructed a parabola in intuition according to universal rules, we will have shown, in effect, that there really can be such things as parabolas. We will also have succeeded in showing that we understand (in a preliminary way) what a parabola is (that we really are in possession of its concept), because we could not have deliberately set out to construct one, and then succeeded, unless we possessed this understanding and were guided by it. This will not yet characterize the parabola in terms of its principal property (σύμπτωμα), i.e. specify the conditions that all points of the curve must satisfy if they are to lie on the perimeter of a parabola. Kant indicates in the passage I just quoted that constructing the parabola and characterizing it are two different things. Judging from the end of the passage, he thinks that, until the parabola has been characterized, its objective reality (that of its concept) has not been fully secured. His reason is no doubt that, until then, we do not know what property, beyond the mode of its construction, is distinctive of all parabolas. But he also indicates that characterizing the parabola in terms of its principal property presupposes the relevant construction in intuition: that of the cone and that of the relevant section.

In the passage from the 'Reply to Eberhard' quoted above, Kant only mentions the procedure for constructing the cone. It is fully described by Apollonius at the beginning of the definitions at the beginning of *Conica*, Book One:

> If, by a straight line, there be joined a certain point and the circumference of a circle, and if the point and the circle circumference do not lie on the same plane, then if the straight line joining them be extended in both directions and if the point be fixed, then if the straight line be carried around the circumference of the circle until it returns to the place from whence it started, then I call each of the two surfaces traced by the line as it comes around – the two surfaces being opposite the one to the other and able to be extended indefinitely inasmuch as the straight line [that generates them] is extended indefinitely – a *conic surface*. (Apollonius: Rashed 2008, 253–255)

The procedure described here only yields the skin of a cone, or rather that of a double-napped cone. The fixed point is the vertex of the two opposite conic surfaces; the line joining that point to the centre of the circle (and extended in both directions) is their axis, which we can think of as a sort of tent pole. The (finite) cone itself is just whatever is contained by the circle and this skin. Given the construction so described, we can call into question the possibility of cones no more than we can call into question the possibility of joining a point to any other point by drawing a straight line or the possibility of fixing one end of a line segment and moving the free end of the line segment around the circumference of a circle. If we call these things into question, we might just as well call into question Euclid's first and third postulates at the beginning of the *Elements*. That would be silly. As a result, Kant would say that the objective reality of the cone has been secured by construction (in intuition).

The objective reality of parabolas can be secured too. Now that the cone has been generated, it is first of all a matter of generating the relevant curve. This is done by passing one plane through the axis of the cone, from the summit of the cone to the centre of its circular base: this gives us the axial triangle. Then, we pass a second plane through the cone parallel to one of the sides of the axial triangle: the parabola is the curve traced by this plane on the conic surface. The possibility of passing two intersecting planes through a cone according to the 'universal rule' just indicated ensures the possibility of the parabola and the corresponding concept. It now remains to characterize the parabola in terms of its principal property, i.e. to specify the condition that must be satisfied by all points lying on the curve traced out on the conic surface by the second cutting plane. As we have seen, this cannot be settled by the generative procedure all by itself. But it is easy in the case Kant is explicitly considering, which is that of a right cone, i.e. one whose axis is perpendicular to the base. For it will then turn out that the principal property of the parabola is determined by the proportion of the squares on the ordinates to the rectangles formed on the abscissae and the latus rectum, where the latus rectum ('parameter', as Kant refers to it in the quoted passage) turns out to be twice the line from the vertex of the cone to the vertex of the conic section on the side of the axial triangle.[4] This can be shown from the symmetry of the right cone, Euclid III.35, the theory of parallel lines and the proportionality of similar triangles. (The reader may consult Dijksterhuis [1987, 58] for the details of the proof, stated explicitly for the case where the vertical angle of the axial triangle is a right angle, but applicable even when it is other than a right angle.) All of this is ultimately assured, Kant would presumably say, by construction in intuition. With the generative procedure for the parabola and the characterization of its principal property in hand, we can say with confidence that there are parabolas and which property is distinctive of them, i.e. we can take their objective reality to be fully secured. Kant believes that construction in intuition is what makes this possible.

Now Eberhard takes Apollonius to have believed that we can advance from truth to truth in geometry without troubling about the 'reality of the object' and hence the 'reality' of the corresponding theory (Eberhard 1789, 158). The example he cites in support of his claim is this:

> Apollonius and his interpreters built up the whole theory of conic sections without anywhere explaining how the ordinates are applied to the diameters of these curves, and yet the reality of the whole theory rests on this. If this application were not

possible, it would be impossible to carry out the construction of conic sections. It would be doubtful whether there be a subject to which belong the properties that are proven of it in all the lovely problems of the theory. (Eberhard 1789, 158)

This comment is so peculiar that one might wonder whether Eberhard ever read Apollonius for himself; Kant himself explicitly asserts that he (Eberhard) did not know Apollonius directly (8.191.27–28). I think he (Kant) must be right.

Given any cone (right or oblique), Apollonius explicitly defines the ordinates of a conic section in terms of their diameter: they are all those lines, parallel to one another (and to another given line that extends to the sides of the section), that are bisected by the diameter. See the first set of definitions at the beginning of *Conica*, Book One (Apollonius: Rashed 2008, 255, 12–17; Heiberg 1891, 6.23–29). Proposition 7 of Book One shows that the line of intersection ZΛH between the axial triangle ABΓ and the second plane cutting the surface of the cone (but not the summit) along the curved line ΘZM is the diameter of a conic section.[5] For ease of comprehension, I have reproduced two of Heiberg's diagrams below (Heiberg 1891, 24). The ordinates are just those parallel lines bisected by the diameter ZΛH, i.e. the line MΛΘ and all lines parallel to it. Their direction is defined by the line of intersection ΔE between the second cutting plane and the circular base of the cone: that line of intersection is a chord of the circular base set at right angles to the diameter of the circular base BΓ (itself the base of the axial triangle)[6]; all the ordinates of the conic section are parallel to this circle chord. There is no doubt that they exist, because they, like the diameter ZΛH with respect to which they are taken as ordinates, are artefacts of the sectioning of the cone. Nor is there any doubt that they are bisected by ZΛH. That is precisely what is established by Proposition 7, notably by appeal to Proposition 6.

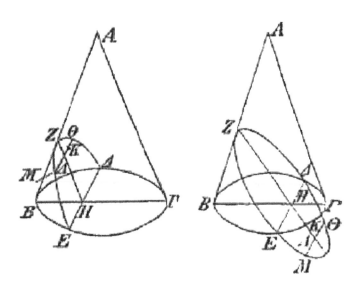

If we are satisfied that Proposition 7 establishes the existence of a conic section's diameter, we will be satisfied that the existence of the ordinates to that diameter has been established too. Apollonius goes on in Propositions 15 and 16 to introduce, for the ellipse and the double-branched hyperbola, the notion of the diameter *conjugate* to the (first) diameter of these sections – the existence of which diameter is established in Proposition 7. He extends his purview to *any* diameter of any conic section, starting in Proposition 41 and continuing beyond.[7] But again, ordinates are relative to diameter, whichever one it may be. If we are satisfied that we can construct the relevant diameter, we will not be in doubt about the ordinates.[8]

Eberhard's comment is – at best – uninformed, and it provokes an angry response from Kant who then attacks Eberhard on what he, Kant, takes to be the central issue between the two of them: whether or not Apollonius leaves the 'reality' of conic sections or their corresponding theory up for grabs. Kant is surely right to focus on this issue: Eberhard seems to think (perhaps because he has been informed by an algebraic presentation of conic sections)[9] that Apollonius proceeds hypothetically, and indeed without ever discharging the hypothesis. He is clearly mistaken about this. But in the course of arguing against Eberhard, Kant does not reflect on the relationship between ordinates and diameters and their joint dependence on the relevant sectioning of the cone, as we might have expected. Instead, he (Kant) presents the construction of the parabola and the right cone in the passage I quoted earlier and argues thus: that this construction and the relevant characterization are possible ensures the 'objective reality' of Apollonius' theory (at least for the parabola and, no doubt, for the other conic sections as well, following the same line of argument). I take it that Kant's rhetorical strategy is driven by Eberhard's remark, quoted above, that 'the reality of the whole theory rests on this [sc. knowing how to apply the ordinates to the diameter]. *If this application were not possible, it would be impossible to carry out the construction of conic sections* [my emphasis]'. Eberhard apparently imagines that, because Apollonius has not (on Eberhard's view) discharged the relevant hypothesis, we may still doubt whether conic sections can be constructed at all. Kant can be understood to reply thus: the construction of conic sections is in no doubt whatsoever; on the contrary, here – in rough outline – is the way to carry it out (for the case of the parabola). Eberhard should be moved by this. But I think Kant's point would have been rhetorically more effective if he had explicitly said that the existence of the ordinates relative to the diameter at issue in Proposition 7 in *Conica*, Book One, is itself assured by the sectioning of the cone. In other words, he could (and should) have pointed out that the order of dependence is the reverse of what Eberhard says it is. Still, one thing emerges very clearly from this odd passage, namely Kant's deep commitment to geometrical constructions in the theory of conic sections. And this commitment is what matters for our purposes.

For it raises a number of questions, the most important of which is this. Geometrical constructions can yield general results. But it just is not obvious how they yield overarching, systematic unity of the kind Kant himself says is the ideal of our higher cognitive faculties in all fields of enquiry. One can feel the force of this question in light of Apollonius' programme in *Conica*, Book One.

Before Apollonius, in the treatises of Aristaeus and Euclid, conic sections were always generated as sections of a right cone, itself generated by carrying a right triangle around one of its sides (see Definition 18 in Euclid's *Elements*, Book Eleven). The cutting plane would pass through the cone, always meeting one side of the axial triangle at right angles. This meant that conic sections had to be classified in terms of the vertical angle of the axial triangle. The ellipse was generated by passing a cutting plane at right angles to one side of the axial triangle of an acute-angle cone. That is why it was traditionally referred to as 'section of an acute-angle cone' – in Greek: ὀξυγωνίου κώνου τομή. The (single-branched) hyperbola was referred to as the ἀμβλυγωνίου κώνου τομή: 'section of an obtuse-angle cone'. The parabola was referred to as the ὀρθογωνίου κώνου τομή: 'section of a right-angle cone'.[10] The constructions used by Aristaeus and Euclid can be taken to have raised the following questions. Can all the conic sections be constructed from one and the same cone? Can they be constructed from oblique cones? Is a more general theory of conic sections possible than that of Aristaeus and Euclid: is it possible to derive the properties of conics no matter what kind of cone they be generated from? If so, how? These seem to be the questions Apollonius addresses – at least in Book One of the *Conica*. To that extent, we can take him to have been moved by a concern for 'systematic unity' in something like Kant's sense.

Now it might be said – in a seemingly Kantian spirit – that Apollonius surpassed the earlier theory by replacing the construction of Aristaeus and Euclid with one having greater potential for 'systematic unity'. The construction he describes in his definitions of 'cone' and 'conic surface', which I quoted earlier, is indeed more promising in this respect. Kant's discussion of it in the 'Reply to Eberhard' only takes into account its potential for generating a right cone – for the same reason, no doubt, that Galileo imposes the same restriction on his presentation, in Day Four of the *Discorsi*, of Apollonius' demonstration of the principal property of the parabola and its tangency conditions, namely as a matter of convenience and simplicity. To be sure, Apollonius' construction will yield a right cone so long as the axis of the cone is perpendicular to the circular base. But I said earlier that we may regard the axis of a cone as a tent pole that holds up the conic surface. Tent poles can list. When the axis of the cone lists, we get an oblique cone. Thus, we can now get cones of any and all kinds, as we could not with Aristaeus and Euclid: for their construction was a solid of revolution, which is symmetrical, while the oblique cone is not. Apollonius then generates each of the conic sections from any cone whatsoever.

To get the parabola, he passes the second cutting plane through a conic surface, generated in the way indicated, so that it (the plane) is *parallel* to one side of the axial triangle: this gets him a parabola whether the cone is right or oblique, and if it is a right cone, the vertical angle of the axial triangle need not be a right angle, as it was for Aristaeus and Euclid. To get the hyperbola, he passes the cutting plane through a conic surface so that it is *not parallel* to one side of the axial triangle, but cuts *one* of its sides, when the latter is extended, above the summit of the cone. The cone could be right or oblique, and if it is right, the vertical angle need not be obtuse, as it was for Aristaeus and Euclid. Because Apollonius has a handy way of generating a double-napped cone, he can handily generate, and then study, the double-branched hyperbola. Then, last of all, to get the ellipse, he passes the cutting plane through a conic surface so that, again, it is not parallel to one side of the axial triangle, but meets *both* of its sides below the summit of the cone. Again, the cone could be oblique, but if it is a right cone, the vertical angle of the axial triangle need not be an acute angle, as it was for Aristaeus and Euclid. Apollonius can now demonstrate the principal property of each of the conic sections in a much more general way than his predecessors.

We can see this, for example, in the case of the parabola. For Aristaeus and Euclid, having generated the parabola from a right-angled right cone, the square on any given ordinate would always be equal to the rectangle formed on the abscissa and twice the length of the line from the summit of the cone to the vertex of the parabola on the side of the cone. In Apollonian terms, we can understand that length to be the latus rectum. But, in the case at hand, it is determined by the symmetry of the right cone. Since oblique cones are not symmetrical, the latus rectum of a parabola will not always be equal to this length.

Apollonius shows us how to establish the principal property for all parabolas, whether generated from a right cone or an oblique cone, in Proposition 11 of *Conica*, Book One, namely as a rectangle applied to a certain length called the 'latus rectum' (ὀρθεία πλευρά) and whose width is equal to the abscissa. He characterizes the hyperbola and the ellipse, in Propositions 12 and 13, respectively, along the same lines.[11] The principal property of the hyperbola is determined by a rectangle applied to the latus rectum and whose width is equal to the abscissa, *plus* a rectangle similar to a given rectangle. That of the ellipse is determined by a rectangle applied to the latus rectum whose width is equal to the abscissa, *less* a rectangle similar to a given rectangle. Thus, Apollonius is able to characterize the three conic sections in general terms, but also in such a way as to show, in general terms, how they relate to one another – thereby achieving still greater systematic unity. This seems to be reflected in the new names that he bestowed on them. These are the names we use today.

'Parabola' comes from the verb παραβάλλω, which is the technical term for applying a figure to a straight line segment. It designates the section characterized by a rectangle applied *exactly* to the latus rectum. 'Hyperbola' comes from ὑπερβάλλω, which means to exceed. It designates the section characterized by a rectangle applied to the latus rectum with an *excess*. 'Ellipse' comes from the verb ἐλλείπω, which means, in the passive, to be wanting or defective. It designates the section characterized by a rectangle applied to the latus rectum that *comes up short*. A Goldilocks theme is emerging here. But we will have no opportunity to retell the story of the Three Bears as a geometrically Fractured

Fairy Tale (and hence no use for Apollonius' new, systematically unifying names), if we generate the three conic sections by cutting three different right cones at right angles to one side of the axial triangle in each of the three cases. For we will then have no way to compare how close or how far they are from being 'just right', i.e. how exactly or inexactly the relevant rectangle applies to the relevant line. (It is amusing to note that 'the section that is just right' is literally the term used in the Arabic translation of the *Conica* for 'parabola': 'القطع المكافئ'.)

These considerations suggest, as I say, that if geometrical construction can be a brake on systematic unity, as in the case of Aristaeus and Euclid, then it can also be an engine of systematic unity: provided that we pick or design the appropriate construction, as Apollonius did in the case of conic sections. If this is the right way of thinking about the matter, Kant's commitment to geometrical construction and his commitment to systematic unity are not necessarily at odds with each other. But it really cannot be as simple as that, as the example of Apollonius shows us.

Apollonius achieves greater systematic unity still: for he not only associates a latus rectum with the (first) diameter of any section whose existence is proved by Book One, Proposition 7, but also associates a latus rectum with *any* diameter of the section whatsoever. That cannot be achieved just by staring at the relevant constructions. It takes Apollonius well into the forties of Book One: Proposition 49 for the parabola, Proposition 50 for the hyperbola and the ellipse and Proposition 51 for the hyperbola of two branches. Kant must have understood this, since he understood that the construction of a cone and a conic section does not yield all by itself the characterization of the relevant section in terms of its principal property. The characterization of the section depends on its construction, but steps beyond the construction must be taken. These steps may themselves depend on other constructions and arguments from them. But systematic unity will be the fruit not of some single construction, but rather of the way a number of specially picked or designed constructions are managed. The nature of this management requires discussion from Kant.

This point can be sharpened by the following observation. *Conica*, Book One, concludes with eight constructive problems: to construct, in a given plane, with a given diameter, given the latus rectum associated with this diameter and the angle of inclination of the ordinates to this diameter: a parabola (Propositions 52 and 53), a single-branched hyperbola (given also the vertex to be constructed – Propositions 54 and 55), an ellipse (Propositions 56–58), a double-branched hyperbola (Proposition 59) and conjugate hyperbolas (Proposition 60). The angle of inclination of the ordinates to the given diameter is taken respectively to be a right angle or other than a right angle. The solution to these problems consists in constructing the cone from which the sought after section is generated. The cone is always a solid of revolution and hence a right cone, which is initially surprising just because of the effort we have noted to embrace sections generated from oblique cones. But there is at least one good reason for this. The constructions carried out in these final propositions might be thought to ensure and display the 'objective reality' – in Kant's sense – of the programme, as a whole, of *Conica*, Book One. Apollonius explicitly uses his distinctive characterizations of the conic sections and his propositions for switching diameters, established in earlier propositions. Thus, he can be understood to argue as follows: my results can be deliberately generated from the same cone used to get the traditional results; there can be no cavilling from traditionally minded sceptics that my results apply only to funny cases – they are indeed perfectly universal. This suggests that the relationship between geometrical construction and 'objective reality', on the one hand, and systematic unity, on the other, is often the reverse of that which we were considering earlier. Whereas the thought earlier was perhaps geometrical construction is the engine of systematic unity, the solution of the constructive problems at the end of *Conica*, Book One, suggests that it is systematic unity that drives geometrical construction and thus yields objective reality. It is the new level of systematic unity reached by the earlier propositions of Book One that permits the display of objective reality, by means of geometrical construction, in the final propositions of this book.

Let me add finally that we learn from the same passage of the 'Reply to Eberhard' we have been considering that the algebraic treatment of conic sections was indeed on Kant's horizon. For he explicitly states the equation $y^2 = ax$ as defining the parabola (8.192.14). He must take this equation to stand in some relation to Apollonius' geometrical characterization of the parabola in terms of its principal property. For one thing, he himself explicitly alluded to Apollonius' characterization of the parabola a few lines earlier. For another thing, he has deliberately picked the example of the parabola to make a certain comparative point about geometry and algebra (see below). For another thing still, there is indeed a relation between the two.

To start with the last point, let it suffice to say just this. We can take the $y$-value in the algebraic equation for the parabola to express the length of the ordinates, the $a$-value to express the length of the latus rectum and the $x$-value to express the length of the abscissae. We can take the equation as such to state the necessary and sufficient conditions satisfied by all the points that lie on the curve we call a parabola. This does not imply that Apollonius' geometrical characterization of the parabola is itself an equation[12] or an instance of what people used to like to call, following Zeuthen's cue, the 'geometrical algebra of the ancients'.[13] It does mean, however, that we can understand the equation to translate geometrical insights into the language of algebra. Exploring what this might mean would take as afield; the question for us is what Kant himself thought it meant. Alas, he does not say directly. But he does briefly address the question whether those who work on algebra are more susceptible to the point made by Eberhard about Apollonius than Apollonius himself. Thus, Kant himself wants to address the question whether the 'more modern' mathematicians, engaged in algebra, fail to secure the objective reality of their object (8.192.7–13). He pointedly denies that they do.

I take Kant to be making the following claim: unlike the historical Apollonius, the modern algebraic mathematicians proceed hypothetically, but, unlike Eberhard's Apollonius, they (can and) will discharge their hypotheses – at least in their treatment of conic sections – if they are challenged to do so. The question for us is how Kant thinks the algebraic mathematicians would do this. I think that the evidence in the passage under discussion is pretty clear: if pressed, they will invoke the historical Apollonius' construction. For Kant says explicitly that the algebraic mathematicians are always aware of the ('pure, schematic') construction that underwrites what they are doing (8.192.11–12 – cf. the footnote running from 191 to 192). In context, it is hard to see how he could intend any other construction than that of Apollonius. Here is why.

As I indicated, Kant cites the algebraic equation for the parabola. We may well wonder what, according to Kant, the algebraic mathematician could have in mind when this equation is invoked. One simple thing will likely occur to us to start out with. Kant must be supposing, at a minimum, that the algebraic mathematician knows, for example, that the y-value in this equation expresses the length of the ordinates, and that the other values express the length of other lines, and also which lines these are. Last and no less important, she presumably understands that the algebraic operations carried out in the equation on the a-, x- and y-values can be interpreted as yielding a relation of equality between two areas. In other words, the algebraic mathematician knows how to back translate the relevant equation into the language of the geometers, and hence she always knows what she is talking about. This does not mean that the 'modern' algebraic mathematicians are above reproach. Kant says that they 'conceive [of parabolic curves by means of the relevant equation] in an arbitrary way and do not, following the example of the ancient geometers, first show that they [sc. the relevant curves] are given as sections of the cone ...' (8.192.13–16). I take this, in the first instance, to be a complaint about presentation. The algebraic presentation does not make room for the beginning of the theory: it neglects to show how the cone is constructed and how the curves called 'conic sections' are generated from the cone. The presentation is thus incomplete. Kant himself says that it lacks 'the elegance' of the geometrical treatment.[14] But, on reflection, the complaint about presentation surely touches on the larger theme of the passage as a whole, which is objective reality. For the algebraic mathematician's awareness that, say, the equation for the parabola can be back translated into the language of the geometers is not all by itself a proof that the equation has 'objective reality', as Kant understands it: the equation as such does not establish that there are any such things as parabolas having the relevant principal property. This can only mean that, on Kant's view, the algebraic mathematicians proceed hypothetically as long as they are only solving equations. It will mean too that, on Kant's view, they will have to do something to reassure those of us in need of reassurance that conic sections are not a matter of mere speculation. The only reassurance they can give is to appeal to Apollonius' techniques for constructing and sectioning the cone. If Kant has a complaint about the algebraic mathematicians, it is that they do not voluntarily do this.

It is obvious that the point at issue here is very local: it concerns only the theory of conic sections and the source of its 'objective reality'.[15] But that is all that we need for our purposes. The source of the objective reality for the geometric treatment and the algebraic treatment of conic sections seems to be the same: construction in intuition – ultimately that of Apollonius. Just for that reason, the questions I was raising earlier about the relationship between objective reality and systematic unity, in Kant's sense, seem to be as pressing for the algebraic treatment as for its geometrical counterpart.

## 2.2. The second observation

I said earlier that I meant to make two observations. The first concerned the central role of geometrical construction in Kant's account of mathematics in general and conic sections in particular. The second is a point I would like to make about Kant's conception of space. It is an observation that turns on remarks in our *Beweisgrund* passage and *Prolegomena* Section 38 that are directly at odds with each other.

In both the *Beweisgrund* and *Prolegomena* Section 38, Kant seems to treat space – that which is of concern for geometry, in the first instance – as a limitless reservoir of anything we might possibly be given or need to carry out any conceivable constructive problem. I might perhaps decide, or be challenged, to solve the problem of dividing a given circle arc in half. But part of what makes that eventuality possible is that I never have to worry whether I will be denied any and all circle arcs. Much less do I have to worry that, once some circle arc is given to me, I will be denied the line segments I need to solve the problem.[16] Now an important difference in the *Beweisgrund* conception of space and that presupposed by *Prolegomena* Section 38, in addition to the one I mentioned earlier, turns on the question of whether space itself, construed in the reservoir way just indicated, is or is not the source of 'necessary unity' or overarching unity that Kant associates in both works with systematicity, the high level of power and generality that we would expect from a general theory of anything in mathematics. In the *Beweisgrund*, Kant says that this unity lies in the objects of mathematics and *hence in space itself*. Thus, he writes in a key passage:

> In order to discern *unity amidst the greatest manifoldness among the necessary properties of space* and connection in the one thing that otherwise seems to have a necessity completely cut off from the other, we have simply taken a look at the circle, which has still infinitely many [sc. necessary properties] of which only a few are known. [NB: Kant has just recalled *Elements* III.35 and observed in light of it that the circle possesses a far-ranging 'necessary unity' – AL.] There can be inferred from this example what an *immensity of harmonious relations must otherwise lie in space*, many of which are explained by higher geometry in the relations of kinship [*Verwandtschaften*] among the different species of curved lines .... (2.95.19–31, emphasis is mine)

The point here is clearly stated, however purple the prose: the necessary unity that mathematicians are especially eager to find is there to be found in space itself. This suggests that mathematics (anyway, geometry) is a study of space itself by some kind of excavation that little by little reveals an immense network of necessary unities embraced by overarching unities. This excavation would involve trying to discern the potential for future problem solving in each successful solution to a given geometrical problem. Kant does not say more in this passage about curves in general or conic sections in particular. But immediately before it, he points out that Euclid III.35, his all-time favourite, plays a role in Galileo's solution to the kinematic problem of showing that the trajectory of projectile bodies is parabolic. However he may have wanted to cash things out, Kant's general idea here is that the theory of conic sections succeeds in bringing to light the higher order unity that unites the ellipse, the parabola and hyperbola, by digging it up from space, where it had been lying concealed all along.

Let me now point out that, exactly 20 years later, Kant would directly contradict in Section 38 of the *Prolegomena* what he said in the passage I just quoted from the *Beweisgrund*.

As in the *Beweisgrund*, Kant opens *Prolegomena* Section 38 with the circle and his all-time favourite Euclidean theorem, *Elements* III.35. He then asks: 'Does this law [sc. *Elements* III.35] lie in the circle ...?' (4.320.29). And he answers flatly, No! He proceeds to discuss the conic sections, indicating in passing – but quite explicitly – that he regards the circle as a conic section in its own right. He says just as categorically that whatever is distinctive of each of the conic sections, and no doubt whatever is distinctive of them collectively as a class of curve, is also *not* to be found in space. He adds this remark to cap it all off: 'Space is something so uniform and in respect of all specific properties so indeterminate that indeed one will certainly seek out in it no treasure of natural laws' (4.321.33–36). Now natural laws like Newton's law of gravitation are what is immediately at issue in this remark. I should point out, though, that they were also at issue in our *Beweisgrund* passage. But that need not detain us here. The point Kant is making applies just as well to mathematical truths: they – no more than the natural laws – are to be found in space. This means that there is no use undertaking an archaeological excavation of space. Whatever form such a venture would take, it will not yield the overarching unity required for a theory of conic sections. Indeed, it will not yield anything at all, because there is nothing there to excavate. That, of course, is what sharpens our problem: if the unity sought by the mathematician is not discoverable in space, then where is it? Or perhaps it is better to put the question this way: how is it to be achieved? The *Beweisgrund* position had the advantage of giving us at least a suggestion (however vague). Now it just seems we are left with a mystery.

Here it may help a bit to consider some of what moved Kant to reject so unequivocally in the *Prolegomena* the position he had staked out in the *Beweisgrund*. One could think that his change of heart had something to do with his new idea that space is something subjective: it is neither a thing in itself, nor a system of things in themselves, but rather a pure intuition that serves as a fundamental condition of human sensibility. That is certainly part of the story. I tried to tell a fuller story in a paper in this journal that focussed on the state of Kant's thinking about sensibility and the understanding in the mid-1770s (Laywine 2003). My idea is not to revisit that paper here, but rather to complement it by looking ahead to evidence from the mid-1790s. I will begin by making the following point whose relevance will become clear, I hope, by the end of this paper.

Even before the *Prolegomena* in 1783, Kant had come to take very seriously the surely reasonable thought that all advances in human knowledge are the product of hard, intellectual work. No amount of such work will yield fruit, however, unless we are given something to work on. That something, according to Kant in the 1780s (and already in the 1770s), is sensibly given to us under the 'formal' conditions distinctive of our sensibility. But Kant also wants to say that nothing sensibility can give us, much less its formal conditions, can substitute for the work itself. Now even if you find this thought attractive, you might be tempted to reply that excavation is a kind of work in its own right. So why not allow space to keep its *Beweisgrund* unity and then tell us how mathematicians can dig it up? Kant's short answer to this objection might be taken to deny the basic premise: if excavation really is work, then it is not what it seems. That is, it is not really excavation or the discovery of something just lying around waiting to be noticed. Fruitful work is the building up in thought of the object of our enquiry under certain kinds of constraints that include, first of all, items of knowledge relevant to our enquiry that we believe secure from doubt and second (and more fundamental) the formal conditions of sensibility. This talk of 'building up in thought the object of our enquiry' is admittedly vague,[17] but it does have as a consequence a kind of philosophical immunization against woolly headed hopes of finding short cuts that will take the work out of our intellectual lives: if the conditions of our sensibility come pre-packaged with the unity or systematicity needed to advance our favourite disciplines, you could still hope that with the help of divine inspiration, the latest in high resolution imaging technology or yogic exercises designed to reveal what is locked up in our pure intuitions, we could gain a kind of direct vision into the truths of interest to us, including the truths of mathematics.

Kant was always on guard against any such romantic *Schwärmerei*. I think it is important that this be documented. So it can, notably in 'Von einem neuerdings erhobenen vornehmen Ton in der Philosophie'[18] published in the May issue of the *Berliner Monatsschrift* in 1796. The 'Lofty Tone', as I will now refer to this essay, is a scathing attack on J.G. Schlosser who had brought out a book the year before that included a German translation of Plato's letters along with notes and an introduction on the history of Syracuse and Plato's repeated and repeatedly failed efforts to convert its tyrant, Dionysius II, into a philosopher king.[19] The politics of the book and Kant's allergic reaction to it matter.

Schlosser was writing just as the Jacobin revolution was getting started in France. He took himself to be offering would-be revolutionaries and rabble rousers in the German states a wholesome lesson: retire from politics, as Plato did, resign yourself to your rulers (if they will not heed your advice) and take refuge in some kind of higher life of the mind. Such a life would require freedom from ordinary distractions and a receptiveness to truths and realities that can only be felt (but never expressed discursively) by higher souls of a noble disposition. Kant recognized in this appeal a sort of political snobbism that looks down on anybody who does not have the time or luxury to spare himself the toil of life in this vale of tears. 'In a word', says Kant, 'all reckon themselves lofty [*vornehm*] inasmuch as they believe that they do not have to work' (8.390.13–15). But he also recognized in Schlosser's appeal a closely related intellectual snobbism, no less objectionable for all that. Noble minds do not have to work at learning the truth: they just have to feel it. That, at least, is how Kant reads Schlosser[20]:

> The principle of wanting to philosophise through the influence of a higher feeling is the one most suited of all for the lofty tone. For who cares to dispute my feeling with me? If I can now make it believable that this feeling is not just subjective in me … but also objective and hence an item of knowledge, and thus not cooked up as a concept by a semblance of reasoning, but holds good as an intuition (the grasping of the object itself), then I am at a great advantage over all those who must first of all seek justification ere they vaunt themselves for the truth of their claims. (8.395.1–9)

Kant takes Schlosser to be both presumptuous and politically dangerous. He is presumptuous, because he imagines it is enough to strike a lofty tone in order to gain insight. He is dangerous, because he imagines that he and his ilk can offer sagacious advice to rulers, should there be any enlightened enough to consult with them, and thus direct the affairs of state from behind closed doors. This is *Schwärmerei* at its absolute worst. The only effective antidote is a critique of pure reason that will teach us the important lesson: there is no short cut in enquiry of any sort into any subject matter; you have to work for a living (intellectually and otherwise).

The attack on Schlosser is directly relevant to our understanding of Kant's mature conception of how and where systematic unity in mathematics and other fields of enquiry is to be sought. There is no short cut to systematic unity in mathematics. It is not just found, discovered lying in space, as Kant apparently takes Plato to have thought (8.391.1–29, plus footnote). Kant seems to have believed that his stripped-down, unity-free conception of sensibility and his strict division between sensibility and our higher faculties of cognition was the best way to bring home his point about the significance, value and unavoidability of intellectual labour.[21] This conception was to go hand in hand with an effort to discipline our higher faculties of cognition so that they exercise themselves incrementally on whatever is given to us under the conditions of sensibility – in mathematics by an ongoing effort to produce the appropriate constructions in intuition.

The problem now, of course, is whether this homily against short cuts, however wholesome, might be (in light of the questions raised in Section 2.1) a block to accounting for the 'unity' needed to give us a theory of conic sections. I am going to argue in the next sections, at least in a programmatic way, that it is certainly not intended as a block, and is – in any case – the key to the story. Kant says in the continuation of *Prolegomena* Section 38 that the relevant unity lies not in space or sensibility, but rather in the understanding, a faculty of cognition in us that is capable of doing intellectual work (unlike sensibility). What he means by this is that the sought after unity is an artefact of the understanding, i.e. something the understanding constructs (under the direction of reason). This thought will require some unpacking. It may help now to notice the specific piece of mathematics that seems to have informed Kant's thought about conic sections.

## 3. The specific example

In *Prolegomena* Section 38, after mentioning the circle and *Elements* III.35, but before turning the position of the *Beweisgrund* on its head, Kant explicitly refers to an important result or set of results. The passage reads as follows:

> If we now extend this concept [sc. that of the construction of the circle as the ground underlying *Elements* III.35 – AL] and pursue the unity of manifold properties of geometrical figures under common laws still further and consider the circle as a conic section that thus stands together with other conic sections under the same fundamental conditions of construction, we find that all chords that intersect inside the circle, the ellipse, the parabola and the hyperbola always do so such that the rectangles formed out of their segments, though not always equal [sc. as in the case of the circle – AL], nevertheless always stand to one another in the same ratios [*in gleichen Verhältnissen*]. (4.321.3–11)

Kant here states the 'common law' to which all the conic sections are subject as a special class of curve. The law of interest to him gets developed in a cluster of thematically unified propositions in the third book of the *Conica*, Propositions 16–29. These propositions can be understood to have at least two things in common. First, they are all concerned, one way or another, with the equality of a ratio between rectangles on certain lines and that of the squares on certain other lines: Kant alludes to this in the passage just quoted. Second of all, they can be understood to extend to conic sections the property of the circle established by Euclid in Propositions 35 and 36 of *Elements*, Book Three. Kant is calling attention to this too – and I will address a question raised by this at the end of the current section of my paper. Following Jakob Steiner, we have come to call the relevant property the 'power of a point with respect to a circle'. But since the issue at hand is the extension of this property, we must clarify the sense in which it may be appropriate (conceptually, if not historically) to speak of the 'power of a point with respect to a conic section' in the context of Apollonius.

*Elements* III.35 considers any two intersecting chords of a given circle. Steiner says in *Einige geometrische Betrachtungen* that the point whose 'power' is at issue in this theorem is that at which the two chords intersect: that point divides each chord in two segments such that the rectangle formed on the segments of the one chord is equal to that formed on the segments of the other. If *P* is that point and if QPR is any chord of the circle passing through *P* (thereby intersecting some other chord at *P*) and meeting the circle at *Q* and *R*, then the area of the rectangle on the lengths QP and PR is constant: the magnitude of that area is what defines the power of the point *P*. The point whose 'power' is at issue in *Elements* III.36 lies outside a given circle. Then, the circumference of the circle will divide any line extending from this point through the circle and cutting the circle at two points on its circumference. Let *P* be the point and let the line extending from *P* cut the circle first at *Q* and then again at *R*. If another line is now extended from *P* and is tangent to the circle at *T*, then the square on PT will be equal to the rectangle formed by PR and PQ. Hence, the area of the rectangle formed by PR and PQ is constant too, and the power of the point *P* is defined by its magnitude (Steiner 1881, 22–23). One important difference in the case of conic sections is that the rectangles formed on the relevant lines are not constant; rather they are in some constant ratio with each other. Kant himself points this out in the passage from *Prolegomena* Section 38 quoted above. But there is another difference that he does not mention.

Given that we are dealing with circles in the two Euclidean propositions, we do not have to define the directions of the divided lines, on whose segments the rectangles of interest are formed: according to *Elements* III.36, the angle between the lines PT and PQR can vary; the only constraint is that there be some angle between them such that one of the two lines meets the circle at one point and the other cuts the circle circumference at two. By contrast, the directions of the relevant lines must be defined in the case of conic sections. Let us consider just one example, namely that of Proposition 17 in *Conica*, Book Three, where we have, in any given conic section or circle, a pair of intersecting chords whose respective directions are defined inasmuch as they (the chords) are parallel to a pair of intersecting tangents.

Suppose we are given a conic section or a circle. Suppose *A* and *B* are any two points on the section or the circle. Let AΓ and BΓ be the tangents at *A* and *B*, and let Γ be their point of intersection. Take *E* and Δ, two other points on the arc AB. Then, let there be constructed the line EZK through *E* parallel to AΓ, and let there be constructed the line ΔZΘ through Δ parallel to BΓ. *K* and Θ are on the section or the circle. The lines EZK and ΔZΘ are chords that intersect at *Z*. Proposition 17 says that the rectangles on the segments of the two intersecting chords are in a certain constant ratio, namely the same ratio as the squares on the two intersecting tangents. That is rect. ZK,ZE:rect.ZΘ,ZΔ::sq.ΓA:sq.ΓB. For ease of comprehension, see Heiberg's diagrams below.

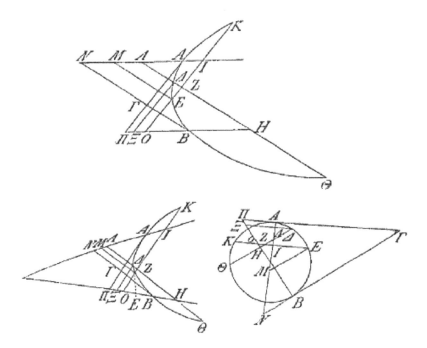

It is very easy to see that Proposition 17 is some kind of extension of *Elements* III.35 to conic sections.[22] For Proposition 17 explicitly applies to the circle. When we consider that case, we find that the tangents AΓ and BΓ are equal and hence that the rectangle contained by the two segments of EZK and the rectangle contained by the two segments of ΔZΘ are also equal. This is just to say that, in the case of the circle, we can define the directions of EZK and ΔZΘ by reference to the tangents AΓ and BΓ, if we wish, but there is no need to, precisely because the rectangle on the line segments of circle chords passing through Z is constant by reason of *Elements* III.35. For the three conic sections, the rectangle on the two segments of EZK and that on the two segments of ΔZΘ are equal in one case only, namely when the tangents AΓ and BΓ are equal: this happens only when Γ lies on what Apollonius calls the 'axis' of the section, i.e. the one diameter that bisects its ordinates at right angles (see the first set of definitions at the beginning of *Conica*, Book One). Hence, in the case of the three conic sections, the areas of the rectangles on the segments of the intersecting chords are not usually equal, but they stand in a constant ratio, under the condition of sameness of direction in the relevant sense. Under this condition, we may adapt Steiner's expression and say that Proposition 17 in *Conica*, Book Three, establishes the power of the point Z for a given section and for directions that are defined inasmuch as a pair of intersecting chords are parallel to a pair of intersecting tangents. If we take some point X other than Z, it will turn out that the power of Z with respect to the given section, for the directions defined relative to a pair of intersecting tangents AΓ and BΓ, will be to the power of $X$, with respect to the same section and the same directions, in the same ratio as the power of $Z$, for directions defined relative to tangents *other* than AΓ and BΓ, to the power of $X$ for these other directions.

I said earlier that Propositions 16–29 in *Conica*, Book Three, form a thematically unified group. This can be seen not just in light of their theme, as I have tried to characterize it, but also in the way that they relate to one another and develop the theme. Different propositions consider different cases. Some extend the results established for a certain case by an earlier proposition to some other case. For example, Proposition 19 extends the result of Proposition 17 to the double-branched hyperbola; Proposition 23 extends the result of Proposition 19 to the case of conjugate hyperbolas. No pair of propositions in the group can establish the 'power of the point with respect to a conic section' in the way that *Elements* III.35 and 36 do for the circle. I might add that the systematic unity of these propositions can also be seen in the work that they (or rather some of them) do at the very end of *Conica*, Book Three.

Apollonius had announced in the letter to Eudemus that serves as a general preface to the *Conica* as a whole that he was in possession of a solution to the three and four line locus problem (Apollonius: Rashed 2010, 252–253; Heiberg 1891, 4). Pappus gives a statement of the three line locus problem in Book Seven of the *Collectio* that may be paraphrased as follows (1876–1878, 678). Suppose that three straight lines are given in position. Suppose that from a single point straight lines are drawn to the given lines at given angles such that the ratio of the rectangle contained by two of the drawn lines to the square on the remaining drawn line is fixed. The problem is to find the locus of the three lines. The solution is that it is a conic section. In the case of the four-line locus problem, four lines are given in position and the condition is that the ratio of the rectangle on two of the drawn lines to the rectangle on the remaining two drawn lines is fixed. Now Apollonius does not give us his solution to the three and four line locus problem in the *Conica*. But he does say explicitly in the general preface that propositions in Book Three are needed to solve the problem (Apollonius: Rashed 2010, 253; Heiberg 1891, 4). The reader can see Knorr for a reconstruction of the solution and the relevance of Propositions 16–17 of *Conica*, Book Three (1986, 120–126). The last three propositions of *Conica*, Book Three – Propositions 53–56 – give solutions to the converse of the problem: to show that the central conic sections and the circle are loci of three or four lines. They depend on Propositions 16, 18 and 20.

By reason of their thematic unity, the way they extend to conic sections, in different cases, the power of a point with respect to a circle, and their role in solving the three and four line locus problem and its converse, Propositions 16–29 in *Conica*, Book Three, give us – if nothing else – a taste of the 'systematic unity' that Kant says is the ideal of mathematics and indeed all other sciences. I say a 'taste', because they may be understood to whet our intellectual appetite, without properly satisfying it. To establish the analogue for conic sections of the 'power of a point with respect to a circle', we have to trudge through 13 propositions and consider separately cases for which Kant would surely predict we would desire a single, overarching treatment – and this comes after a trudge, more dogged still, through the first 15 propositions of Book Three that establish the equality of areas of different triangles and quadrilaterals and that serve as lemmas for Propositions 16–29. Kant was surely right to excoriate Schlosser. It is worth reminding ourselves that we all have to work for a living and that inquiry is itself a form of work. But that bracing reminder should not be taken to imply that we have to make our work as hard as possible. It is not *Schwärmerei* to think that, if possible, we should spare ourselves the hard labour of – say – Apollonius' efforts to extend the characteristic property of the circle to conic sections in Propositions 16–29 of *Conica*, Book Three. In fact, Kant himself says that reason demands labour saving devices, inasmuch as it values systematic unity. Where a high degree of systematic unity has been achieved, we can expect a greater yield of results with fewer means. So perhaps the thing to say is this: the reading of Apollonius is peculiarly well suited, from a Kantian point of view, to spark in us a demand for systematic unity, because it gives us such a distinctive taste of what we are looking for. But by the time Kant came on the scene, it would also just as naturally have produced a sense of let-down in mathematically minded readers. That thought can be sharpened in light of the development of projective geometry.

It might be thought that projective geometry achieves the systematic unity that eluded Apollonius. For it shows us that circle and conic section are the same thing by finding a characteristic property of conic sections that is also a property of the circle. But it does this without the reservations, qualifications and consideration of special cases that we found in Propositions 16–29 of *Conica*, Book Three. Pascal's treatment of conic sections already achieves this level of systematic unity. It depends on the theorem that the points of intersection of the three opposite sides of a hexagon inscribed in any conic section are collinear and on the recognition that this property is found in the circle too. Pascal's theorem holds for the Euclidean plane. But, here, there will be exceptions that must be taken into account. For a pair of opposite sides of the hexagon can be parallel to each other. Indeed, all three pairs of opposite sides can be pairs of parallel lines. When that happens, there is no pair of opposite sides that intersect. This exception does not arise on the projective plane, since a line at infinity is added and the respective pairs of parallel opposite sides will meet at three points on this line. Pascal was deliberately aiming towards a high level of 'systematic unity', since he says, in the text on conic sections that has survived, that he will use his result to give a complete account of the 'elements of conic sections' on the basis of his theorem (1964, 232).

It is natural to wonder what Kant would have made of Pascal's theorem, its implications and projective geometry more generally. Having said that, it would not be surprising at all if Kant knew nothing about Pascal's contribution directly (and the work of Desargues that it depended on). Leibniz had seen Pascal's essay, but it was not in circulation. Still, Kant could and should have been aware of Philippe de la Hire, a student of Desargues who developed the theory of conic sections out of the notion of harmonic division (de la Hire 1685). It turns out that passages can be found in Kant's writings that at least suggest an awareness of projective geometry. I know of two.

One can be found at the beginning of Part One of the *Universal Theory and Natural History of the Heavens* of 1755. Kant had been arguing that the structure of our solar system can be found reiterated at greater and greater scales throughout the physical universe. Our solar system is a system of planets sweeping out concentric orbits in the same direction on the same plane around a central body, the sun. But ours is only one such system embedded in the larger system of our galaxy: the central bodies of all the smaller solar systems in our galaxy orbit in the same direction on the same plane around the central body of the galaxy as a whole, and this central body orbits, along with the central bodies of other galaxies, around the central body of a system even greater still. The universal plane of the greater system extends to (and beyond) the Milky Way. These ideas lead Kant to say that we should then expect to observe heavenly bodies that appear to us as single stars but are, in fact, whole systems of stars. The passage of interest reads as follows:

> If a system of fixed stars that relate in their positions to a common plane in the way we have sketched for the Milky Way is so far removed from us that it is impossible even with a telescope to discriminate the individual stars that make it up; if its distance from the stars of the Milky Way is in just the same ratio as the distance of these stars from our sun; in short, if such a world of fixed stars be intuited by the eye of an observer at such an immense distance, then that world would appear, under a small angle, as a small place [*ein Räumchen*] illuminated by faint light whose shape will be circular if its surface is presented directly to the eye and elliptic if it is regarded from the side. (1.253.23–254.3)

Kant now reports that such 'nebulous stars' have indeed been observed. But the interest of the passage, for our purposes, is precisely the thought expressed in it that the difference between circle and ellipse is, as it were, a matter of perspective. That thought might indicate an awareness of projective geometry. As a matter of curiosity, let me add that, in a footnote to a remark two paragraphs before the passage just quoted, Kant explicitly mentions a contribution of de la Hire – to observational astronomy.

The second passage I can think of relevant in this context appears in the Appendix to the Dialectic in the *Critique of Pure Reason*: the very passage in which Kant tries to argue that reason's demand for systematic unity can play a regulative role in scientific enquiry. He characterizes the 'unity of reason' in this passage as a '*focus imaginarius*'. His language is at least suggestive of the point at infinity. Thus, he writes:

> I say then that the transcendental ideas are never of constitutive use such that concepts of certain objects are given thereby. To the extent that they are taken to be such they are merely sophistic (dialectical) concepts [*vernünftelnde Begriffe*]. But,

> on the other hand, they have an excellent and indispensable regulative use, which is to direct the understanding towards a certain aim with respect to which the guide lines [*Richtungslinien*] of all its rules concur in a point. Although it is only an idea (*focus imaginarius*), from which the concepts of the understanding do not really proceed since it lies quite beyond the limits of possible experience, nevertheless it serves to provide them with the greatest unity together with the greatest expanse. (B672)

A page or so later, he explicitly characterizes 'systematic unity' in the sense we have been discussing as a 'projected unity' (B675 – '*eine projectierte Einheit*').

These two passages are certainly suggestive of central notions from projective geometry. But they are only suggestive and just as plausibly suggest a concern with more traditional problems of perspective. Thus, the '*focus imaginarius*' might be taken to be the vanishing point of a picture rather than the point at infinity in the projective plane. It would be natural for Kant to have been interested in perspective, and he may well have been stimulated by Lambert's works on the subject and the related section on 'phenomenology' in Lambert's *Neues Organon*. This presents us with a puzzle I can only state for now, but not solve, concerning Kant's attitude towards mathematics and the way it seems to have differed from his attitude towards the physical sciences.

The interest of Friedman's discussion of *Prolegomena* Section 38 in the early 1990s and now that of his new book on the *Metaphysical Foundations of Natural Science* (2013) is that they show, among other things, that Kant had a serious, life-long engagement with the physical sciences: he kept up; he was fluent with the technical details and he drew from them philosophically imaginative and rich conclusions about the constraints under which reason can make progress in enquiry. But, in the case of mathematics, things look different. I do not mean to say that Kant's conclusions about mathematics were philosophically unimaginative and impoverished. I mean just that he does not seem to have been immersed in mathematics as a living discipline. That makes what he has to say about mathematics a good deal less progressive. Here is an argument to support that claim.

Consider the two suggestive passages I just quoted. Suppose we decide, on balance, that they cannot be taken to indicate an interest in, and awareness of, projective geometry. Suppose it turns out that Kant was aware of de la Hire's work in observational astronomy, but not his extensive, original work on conic sections. Should we not then conclude that he missed an opportunity to investigate the way reason achieves higher and higher degrees of systematic unity in mathematics? Should we not conclude that Ernst Cassirer, as the philosopher of mathematics and geometry portrayed recently by Jeremy Heis, was a better Kantian than Kant himself? What we learn from Heis is precisely that Cassirer set out to understand, in the light of projective geometry and other innovations in mathematics of his time, how reason's demand for systematic unity drives progress in mathematics and unites it as a single, *living* discipline (Heis 2011). But suppose now that we decide, on balance, that the two suggestive passages I quoted above *do* indicate an interest in, and awareness of, projective geometry, or suppose that there are other passages I have overlooked that show this conclusively. Then, if anything, my questions become harder, because it will be all the harder to understand why Apollonius' results in Book Three of the *Conica* are the *privileged* example of systematic unity in mathematics in *Prolegomena* Section 38. (Let me stress: I am not saying that Apollonius is uninteresting: quite the contrary! I am saying only that his achievements do not look like the obvious case to pick, in the *eighteenth century*, as the privileged example of systematic unity.) To be sure, we can always point to Kant's relation to Newton. The *Principia* ostentatiously engages with the classical, ancient treatments of conic sections. Indeed, Proposition 17 from *Conica*, Book Three, is put to work, not surprisingly, in the relevant sections of *Principia*, Book One. Given that *Prolegomena* Section 38 is not only concerned with mathematics but also with a special kind of systematic unity achieved by reason when physics and mathematics are united in what Kant takes to be the right way, we should expect Newton's treatment of conic sections in the *Principia* to be an important point of reference there. But this important observation does not make it any less odd that we do not find a treatment, in works by Kant covering almost 30 years, of the distinctive and remarkable ways that reason, in the eighteenth century, was achieving higher degrees of systematic unity in the theory of conic sections.

The question I just stated I will have to leave open. The question I will take up in the final two sections of my paper is this: what cognitive resources are available to mathematics, on Kant's account, that have made possible such systematic unity as we find exhibited by Propositions 16–29 in *Conica*, Book Three? This question is pressing in view of the two observations I made in the previous section. First, Kant's point of reference in all this will be construction and no single construction will suffice all on its own. Second, Kant denies in *Prolegomena* Section 38 that the sought after systematic unity is to be found in space as such.

## 4. What are the cognitive resources available to mathematics for achieving systematic unity?

Let's start with the easy and obvious. We know from *Prolegomena* Section 38 that, by 1783, Kant would deny that space and thus the conditions of formal sensibility are the source of the overarching unity on display in Apollonius' conic sections. We also know that this unity is not something that mathematicians uncover, unearth or excavate, but rather something that they actively build up by some kind of work. We saw too that Kant believes that intellectual work is carried out, in the first instance, by a faculty distinct from sensibility. He calls that faculty the 'understanding' (*Verstand*) (though we must not forget that the understanding carries out this work under the supervision of reason – see below). It is very easy to spell out the nature of the work carried out by this faculty (the understanding). For Kant characterizes it in the *Critique of Pure Reason* as a faculty for thinking

(B74): all thinking consists in the use of general concepts in the formulation of judgements (B74, B93–94). A general concept is that by which our understanding represents what is common to things (B93, A106). The concept Cat represents what is common to a whole class of carnivores (some little, some big). So it is obvious that the work contributed by our understanding to the theory of conic sections is to specify what is common to the whole class of conic sections. Kant would thus say presumably that Apollonius Book Three, Proposition 17, helps put this concept on display for us.

But it is just as obvious that this part of the story does not take us very far, because it simply raises the further (and very difficult) question: how did the understanding succeed in putting together and then justifying the general concept of conic sections on display in this theorem and its demonstration? It did not just pull these things, like a rabbit out of its hat – witness the programme of Apollonius' *Conica*, Book One, as characterized in Section 2.1 of this paper.

Much could be said at this point.[23] But I will focus in a quite programmatic way on just one element of Kant's thinking that is relevant here, namely the imagination. I do not mean imagination, as Locke or Hume understood it, namely that capacity we have to freely associate ideas, because this capacity yields no 'unity' in the relevant sense of the word: freely associated ideas are just whatever comes into your mind when somebody suggests that you think about something. The imagination at issue for Kant should be understood, instead, as a capacity that can operate *a priori* in such a way as to help the understanding convert the deliverances of our sensibility into thoughts or universal concepts about, say, mathematical objects. To be more specific, it produces figures (in, say, a diagram) that graphically correspond to whatever the understanding is trying to think about. Without this assistance, the understanding could not think about anything determinate, nor would it have any determinate concepts. But the imagination can be of no assistance at all if it merely produces *images*. That is because images as such are never universal in scope: they are just the marks you happen to see on the page. So the imagination has to produce its figures in such a way that they can be recognized by the understanding as applying universally to a whole class of things. Kant calls the technique used by the imagination to produce figures in this way a 'schematism' (B179–180). There is a famous passage in the *Critique of Pure Reason* about this so-called schematism – a passage that seems obviously applicable to the diagrams and demonstration of Apollonius Book Three, Proposition 17. The passage reads as follows:

> In fact, it is not images [*Bilder*] of objects that underlie our pure sensible concepts [sc. those concepts at work in mathematics], but rather schemata. For no image would ever be adequate to the concept of a triangle in general. For it would not achieve the universality of the concept which makes this concept hold for triangles – right triangles as well as oblique, etc. –, since it [sc. the image] is always restricted to only one part of this sphere. The schema of the triangle can never exist anywhere else than in thought and refers [*bedeutet*] to a rule of the synthesis of the imagination in respect to pure figures [*Gestalten*] in space. (B180)

In short, the figures produced by the imagination, on the basis of a schematism, allow the understanding to preserve the universal scope of its mathematical concepts while focussing its thought on an image in such a way that it can actually follow the accompanying argument. It is what ensures that the statement of a theorem or problem in ancient Greek mathematics, often practically unintelligible on its own, has been properly cashed out in the part of the proof called the 'ekthesis' or the specific example following the statement (I gave the ekthesis of Apollonius III.17 in Section 3 of my paper). To that extent, it might also be said that the schematism is what allows us or the understanding to check or verify our grasp of what has to be proved and then whether it has been proved, after all is said and done.

Now something like a schematism of the imagination might be supposed to play an important role in Apollonius III.17. The diagram I reproduced above from Heiberg's edition helps us to a very general understanding of conic sections. But it does so to the extent that we are not misled by the image printed on the page, which looks like not one, but three different diagrams: one of a parabola, one of a hyperbola and one of a circle. Apollonius' theorem holds not only for these sections, but also for the ellipse (Heiberg's edition only has the three diagrams; Rashed's edition of the Arabic translation also only has three, but gives the ellipse in place of the circle). We certainly recognize this, but do so, according to Kant, only on the basis of a schematism of the imagination. Here, it would be interesting to know what Kant might have said about single proofs that feature multiple diagrams, as – for example – in the manuscripts and editions of Apollonius. Would Kant say that a separate schematism is required for us to recognize the universal import of each individual diagram: thus, in the case of Apollonius III.17 in Heiberg's edition, one to cover all parabolas, one to cover all circles and one to cover all hyperbolas? Or would he say that it is by virtue of a single schematism that we take in all three diagrams (making room for the ellipse too) and understand a certain result schematically as thereby applying universally to the three conic sections and the circle, taken together? He does not say.

I am sure that the schematism of the imagination is indeed playing an important role for Kant's story about conic sections. But the important thing to see is that an appeal to the schematism in the way I just suggested, even if the answer to the last question I asked is affirmative (and I think it may not be), cannot really do justice to the mathematical example that Kant himself picked, namely Apollonius III.17. That is because, first of all, Kant goes out of his way in *Prolegomena* Section 38 to call our attention to the relatively high level of generality achieved by this theorem and second, of all, because he would presumably call on the schematism of the imagination to tell the same story about theorems on conic sections at a much lower level of generality. Our question all along has been how Kant will account for the relatively high level of generality that the theory of conic sections ultimately achieves by the time it reaches Apollonius III.17. That question has not yet been answered. Not even programmatically. I think that this may have been overlooked by others working on Kant's philosophy of mathematics, and – as a result – I think something interesting and important may have been overlooked about the schematism of the imagination. Namely this.

It now seems natural to think that, for Kant, the schematism is not merely at the service of the understanding, but also of reason itself. Reason is our highest cognitive faculty. One way it differs from the understanding is that it is the capacity we have to set ends for ourselves, not just in the practical sphere, but in the theoretical sphere as well (Laywine 1998). Among the ends we can set for ourselves is the solving of geometrical problems, like that of dividing a given circle arc in half. So it is presumably reason that sets problems for mathematicians, not the understanding. The role of the understanding is just to direct the imagination in such a way as to solve, where that is possible, the problems that reason has set for us. But if the understanding requires a partner to produce graphic or figurative translations of its mathematical concepts, *a fortiori* reason requires a partner to produce figurative translations of its problems. For its favourite problem – the problem that directs all its problem-setting – is that of establishing unity as far and wide as possible. But that is not a specific problem, as can be seen from remarks I made earlier. It was one thing for Newton to establish unity in celestial mechanics, another thing for Kant himself to set up (as he believed) a unified table of categories, and still another thing for Apollonius to establish as much unity in the theory of conic sections as he did. Let's set aside the special problem of the table of the categories. The thing to see now is that reason, as embodied in Newton, could never have undertaken to establish unity in celestial mechanics without having some conception of what the problem itself actually required. Similarly for reason, as embodied in Apollonius, in the case of conic sections. For Apollonius what that problem would have required, to start out with at least, is showing that all the conic sections can be produced from the same cone and indeed from any cone. It seems natural to think that Kant would have had to say (and would have wanted to say) that reason, as embodied in Apollonius, could set its problem in a specific and an appropriate way, because it too, like the understanding, had found a willing partner in the imagination and a powerful resource in the schematism of the imagination.

The thought I just floated should be rounded out by another. Kant believed, of course, that reason is that higher cognitive faculty in us that carries out demonstrations. I take it that carrying out demonstrations and requiring that we establish the highest possible unity are two distinct, if complementary, activities that reason engages in. You could produce lots of highly rigorous, fruitful demonstrations about conic sections, and yet fail to achieve the level of unity on display in Apollonius III.17. This may well have been true of the treatises on conics by Euclid and Aristaeus before Apollonius, if Eutocius' report about them is true (it should be noted, however, that Archimedes states Apollonius III.17 as Proposition 3 of his *Conoids and Spheroids* with the same level of generality and says that it had already been proved in the 'elements of conics' – he presumably means those of Euclid, if not also those of Aristaeus). Suppose, for the sake of argument, that Eutocius' report may be accepted and that Euclid and Aristaeus proved lots of different things about conic sections without achieving the level of systematic unity in this theory that Apollonius would later. Then we could say that Apollonius had greater success than they did in coupling both activities of reason: its endless searching for maximal unity in the sciences and its efforts to supply rigorous demonstrations. It is tempting to think that Kant would have said (had he direct access to the work) that Apollonius could not have succeeded in this unless he had succeeded in putting the schematism of the imagination at reason's disposal. That's because Apollonius had to work out (the word 'work' should be stressed here) a complete programme of demonstrations that build on one another in the appropriate way so that we ultimately arrive at a successful demonstration of Proposition 17 in Book Three of the *Conics* (and, of course, everything that follows). Reason, as Kant conceives it, would not have been able to design this programme of demonstration if the imagination had not translated into figurative terms what the steps of the programme would require.

## 5. A new and final problem, and a possible solution

The final problem we must confront concerns the two complementary thoughts that brought the previous section to a close. As I said, they turn on the idea that reason must be helped somehow by the imagination and its schematism. But Kant himself says things in the Appendix to the Transcendental Dialectic of the *Critique of Pure Reason* that can be understood, in the worst case, to reject any such idea and, in the best case, to muddy the waters.

The Appendix to the Transcendental Dialectic is concerned to argue that, though reason is led into fallacy when we put it to constitutive use in metaphysics, its regulative use is essential to the sciences. We may not expect it to yield any knowledge *a priori* of any special object, but we may expect it to guide our use of the understanding: in particular, because reason always demands systematicity, generality and unity in the results we achieve through this faculty. Kant puts it this way:

> If we survey in its full extent the knowledge we have through the understanding, we find that that which is quite especially under the command of reason and which reason tries to bring about is the *systematic aspect* of knowledge, i.e., its connection due to a principle. (B673)

A few pages later, Kant cites the theory of conic sections as an example of knowledge that exhibits systematicity and hence shows the stamp of reason (B690–691). All this talk concerns the relation between reason and understanding. But it therefore predictably raises the question of the relation between reason and the schematism of the imagination. Here Kant seems to hedge:

> Understanding constitutes an object for reason just as sensibility does for the understanding. To make systematic the unity of all possible empirical operations of the understanding is a task of reason, just as the understanding joins the manifold of

> appearances through concepts and brings it under empirical laws. But operations of the understanding are indeterminate without schemata of sensibility; so too the unity of reason is in itself indeterminate even in respect to the conditions under which, and the degree to which, the understanding is supposed to combine its concepts systematically. (B692–693)

This passage says that reason, with its demand for unity, relates to the understanding in something like the way that the understanding relates to sensibility: whereas the understanding 'joins the manifold of appearances through concepts', reason 'makes systematic the unity of all possible operations of the understanding'. Without the schematism of the imagination, the operations of the understanding are indeterminate. Likewise, reason's demand for unity is indeterminate, at least if that demand is taken in itself, without any relation to the understanding. But one has to wonder whether the relation of reason to the understanding is such as to give reason access to the schematism. More to the point, is it such as to permit reason's demand for unity to be schematized? If not, we have a problem. For how can reason concretely set specific problems for the understanding to solve? How can it judge the solutions the understanding comes up with? Given Kant's remarks at B673 and all the things we heard him say in *Prolegomena* Section 38 and elsewhere, reason cannot be indifferent to the results – say – of Apollonius and his predecessors, assuming that Eutocius' report is true. All other things being equal, it should prefer Apollonius' *Conica*, just because that work achieves a higher level of 'systematic unity'. But if reason's demand for such unity remains indeterminate, as I think it must be if it cannot be schematized, how could it (reason) possibly judge between the two sets of results? We cannot ask the understanding to adjudicate, because, considered in itself, it does not care about systematic unity. This consideration leads one to expect that Kant will continue the passage from B692 to 693 with some thought or other that will give reason access to the schematism of the imagination. But he does not do that. Instead, he says that reason requires some kind of an 'analogue' – for want of a schematization of systematic unity.

The passage reads as follows:

> But although there can be found no schema in intuition for the thoroughgoing systematic unity of all concepts of the understanding [demanded by reason – AL], there must and can be given an analogue of such a schema, which is the idea of the maximum of division and unification of knowledge through the understanding in a principle. For the Greatest and the most Complete can be conceived in a determinate way. (B693)

The trouble is that the very last claim made in this passage seems false. How can 'the Greatest and the most Complete' be conceived in a determinate way? Kant tries to answer in what follows: 'For all restricting conditions that yield undetermined manifoldness are left out' (B693). However Kant imagines that this thought is to be understood, the implication seems to be that reason still cannot decide between Apollonius and his predecessors. For let us suppose that it can indeed conceive in a determinate way the idea of achieving maximum systematic unity from a single principle in all our knowledge: if it cannot schematize its demand for systematic unity in intuition (and Kant says explicitly in the first clause of the passage just quoted that it cannot or that it cannot do so in a thoroughgoing way), how can it assess concretely the level of systematic unity Apollonius and his predecessors respectively achieve? That sort of an assessment will require translating its demand for systematic unity into terms or cases that can be exhibited in sensibility. Surely such a translation is impossible without some kind of schematization.

By way of a conclusion, let me suggest that, if Kant has a way out of this dilemma, it may be by calling on the notion of reflective judgement in the third *Critique*. After all, reflective judgement is supposed to serve as some kind of bridge between reason and understanding. To the extent that reflective judgement is characterized in the third *Critique* as a capacity to seek the universal when the particular is given, it looks like the sort of faculty that inherently feels drawn to systematic unity (and hence to reason itself), while being – as the 'faculty of judgement' – closely associated with the understanding (and hence to the schematism of the imagination). It should not be any surprise that Kant devotes the opening section of his critique of teleological judgement to mathematics and that he there reprises his beloved conic sections. I will not quote the passage; I refer the reader to Section 62 and especially 5.364.26–365.36. Let me just add two last points by way of conclusion and in support of my thought.

The first is this. Readers of the *Critique of Judgement* will remember that Kant associates with this faculty the 'feeling of pleasure and displeasure'. One might think that feelings have no place in mathematics. But Kant would disagree. Surely he is right about that. Indeed, by the end of the passage at 5.364.26–365.36, he speaks of a special wonder or astonishment (*Bewunderung*) that systematic unity in mathematics inspires – a feeling he had already described in the parallel passage of the *Beweisgrund* I discussed for other reasons in the second section of this paper. For the sake of clarity and precision, it should be noted that this astonishment is to be distinguished from the feeling Kant says is associated with the sublime. For we experience that feeling only in the presence of magnitudes so great as to defeat our imagination. Such magnitudes can be encountered in the absence of systematic unity. As a result, the two feelings seem to be quite different. The feeling Kant associates with the sublime has a dash of terror mixed in. The feeling of astonishment he associates with systematic unity is precisely that feeling we experience when we read Apollonius against the historical backdrop of earlier theories of conic sections. That brings me directly to the second and final point.

If we now ask Kant whether reason can adjudicate between Apollonius and his predecessors, perhaps he would be honest enough to say 'no'. This does not mean, however, that no judgement can be made about their relative merits in light of reason's demand for systematic unity. It is just that reason itself would not pass this judgement – for lack of a schematization. But it may well be that this judgement is made by reflective judgement, which both shares (as I claim) reason's preference for systematic unity and has access to the resources needed for understanding and judgement (in particular the schematism of imagination). Reflective judgement would be able to follow the steps that lead demonstratively to Apollonius' results and also those that led to Euclid's results (had his treatise not been lost). Having understood them, given its preference for systematic unity, it feels greater astonishment for the former and thus prefers the former to the latter. But I think, in fact, Kant should have seen that it would ultimately prefer Desargue, Pascal or – anyway – de la Hire to all of them.

## Acknowledgements

I would like to express my thanks to Vincenzo De Risi, Stephen Menn and Roshdi Rashed for kindly taking the time I know they did not have to give me extensive feedback on earlier versions of this paper. I have done my best to take this feedback into account. The remaining infelicities are my fault.

## Notes

1. No doubt his reason for doing so is that the circle is the figure that results from passing a cutting plane through a cone parallel to its base. It should be noted, though, that Kant is departing in this respect from Apollonius – his point of reference for conic sections. Apollonius never speaks of circles in the *Conica*, but always of circumferences of circles, namely as traced on a conic surface by the cutting plane.
2. All translations in this paper, from Kant and other sources, are my own. Quotations from Kant's works are by volume number of the Academy edition, page number and line number. Passages from the first *Critique* indicate page number from the first (A) or second edition (B), as the case may be.
3. The long-winded title may be translated as 'concerning a discovery that is supposed to show that every new critique of pure reason is made dispensable by an earlier one'. The work was brought out by the publishing concern of Friedrich Nicolovius.
4. By 'ordinate' I mean all those lines parallel to one another, meeting the section at two points, bisected by the diameter or axis of the section. By 'abscissa' I mean that part of the diameter or axis of a section lying between the vertex of the section and the point of intersection between the diameter and one of its ordinates. By the 'latus rectum' of the parabola, I mean the length upon which is formed the rectangle whose width is the abscissa and the magnitude of whose area is equal to the square on the ordinate.
5. It is important not to assume that these two planes are orthogonal to each other. For precisely in the case when they are not orthogonal, it will turn out the angle of inclination of the ordinates to the diameter will not be a right angle.
6. This diameter of the circular base is itself the line of intersection between the circular base and the first cutting plane. This line of intersection has to be a diameter of a circle, because the first cutting plane passes through the axis of the cone, which joins the summit of the cone to the centre of the circular base.
7. Proposition 46 says that, in any parabola, any line parallel to the first diameter is a diameter. Proposition 47 says that, in any ellipse or hyperbola, any line passing through the centre is a diameter. Proposition 48 says that, in any double-branched hyperbola, any line that joins a point on one branch to a point on the other branch and that passes through the centre is a diameter.

8.   Perhaps, we may regard this thought as on display in the problems of determination in *Conica*, Book Two. Propositions 44 and 45 in the Arabic translation (they are treated as a single problem in Eutocius' edition) address the problem of finding a diameter of any conic section. The analysis in Proposition 44 turns on the fact that ordinates are defined relative to a certain diameter, as I have indicated. The synthesis then involves constructing parallel lines inside a given conic section, finding their mid-points, joining the mid-points and extending the line joining the mid-points to the section. It follows that this line is a diameter from Book Two Proposition 28 and the relation between ordinates and diameters invoked in the analysis of Proposition 44.

9.   There is a serious and difficult question here about what could have been Eberhard's source. Eberhard explicitly refers to Borelli, the editor of the books of Apollonius that survive only in Arabic (translated into Latin by Abraham des Echelles in 1661). Indeed, Eberhard purports to quote from Borelli's *Admonitio* prefaced to the Apollonius edition (1789, 159). But Eberhard later retracts the attribution and says that the quoted passage is from Claudius Richardus. At least, this is what one learns from the note of the Academy edition to 8.191.27–29 of Kant's works. I have not yet been able to consult the Richardus. It remains a question, in my mind at least, what could be the source of Eberhard's information on Apollonius. There is an obvious question related to this one, namely what is the nature of Kant's access to Apollonius. Vincenzo de Risi suggested to me that Kant knew Apollonius through the Borelli edition – at least partly. The plausibility of the idea rests on the fact that Kant himself mentions the edition in the passage from the Reply to Eberhard at issue, and indeed quotes part of the passage quoted by Eberhard. But given that Eberhard himself retracts the attribution and given that Kant himself did not catch the original mistake of attribution, one may doubt whether Kant had been reading the Borelli edition or its preface.

10.  The relevant treatises are lost, but we learn these things from remarks in Pappus' *Collectio* and Eutocius' commentary on Apollonius. Pappus' report is reprinted from Hultsch's edition as Appendix II and translated into German in Zeuthen (1966). For Eutocius' commentary, see Apollonius (Heiberg 1893, 168–170).

11.  Prior to Apollonius, the property used to characterize both the ellipse and the hyperbola seems to have been one of the two proportions established by Apollonius in Proposition 21 of *Conica*, Book One. Let AΓ be the latus rectum of an ellipse or hyperbola, and let AB be the diameter. Let ZH be an ordinate of AB. Let AH be the abscissa. Then, we have AΓ:AB :: square ZH, ZH: rect.AH, HB. At any rate, Archimedes uses this proportion to characterize conic sections other than the parabola in *Conoids and Spheroids*; he seems to be a witness to the older theory.

12.  Indeed, whereas the equation states both necessary and sufficient conditions to be satisfied by points lying on the curve, Apollonius' geometrical characterization states only the necessary conditions – so too his geometrical characterization of the ellipse and the hyperbola. It would not be hard at all for him to have stated the sufficient conditions as well. But he does not do this. Moreover, he uses the converse of the proposition that gives the geometrical characterization of the parabola (Book One, Proposition) in subsequent propositions of Book One, like the problem of constructing a parabola in a given plane under the conditions spelled out in Proposition 53. Roshdi Rashed points this out in his commentary on the Arabic translation of the *Conica*. In particular, he draws our attention to the sensitivity to this point of later mathematicians writing in Arabic, responding to Apollonius and contributing (or at least aware of) advances in algebra. Thus, Nāṣir al-Dīn al-Ṭūsī (1201–1274) points out in his gloss on the *Conica* the use of the converse of Book One, Proposition 11 in Book One, Proposition 27 and Proposition 53. Notes

Complémentaires 36 and 68 of Rashed's edition of the Arabic translation. See too Rashed (2005).

13. Zeuthen's *Die Lehre von den Kegelschnitten im Altertum* is a classic. But the idea that Apollonius was presenting fundamentally algebraic insights, in geometric guise, is no longer considered respectable. See, for example, Saito (2004, 139–169) and Fried and Unguru (2001, 17–56).

14. Gordon Brittan may well reject my reading of this passage. He takes Kant to be saying that the algebraic mathematicians do not need the relevant constructions in Apollonius to secure the objective reality of their treatment of conic sections. Brittan goes on to argue in light of a passage in the first *Critique* at A716/B744 that, for Kant, algebra is shored up by a different kind of construction, namely symbolic construction and that this is why the geometrical construction is not needed. I take Kant to be saying that it (the geometrical) is indeed needed (for the theory of conic sections), but not necessarily in the actual presentation of the theory. Brittan is no doubt right to say, in effect, that Kant needs to say more about algebra. To be sure, Kant says that, even though the algebraic mathematicians do not bother showing how to generate parabolas from cones, their results are objectively real. But he also says pretty clearly that they do not need to do so, because 'they are always completely conscious ... of the pure, merely schematic construction along with [the relevant definition expressed as an algebraic formula]' (8.192.11–12). By 'pure, schematic construction' Kant means the construction we carry out by 'mere imagination according to a concept a priori' (8.192.26–27) as opposed to a construction we carry out by applying some instrument to some kind of material like paper or copper, if we are preparing copper plates. Apollonius' construction of the cone and the parabola would count, for Kant, as pure, schematic constructions. On balance, the passage here is best read as saying: the algebraic mathematicians do not have to explicitly present Apollonius' construction in their treatises, precisely because they are always conscious of it. See Brittan (1992, 315–339).

15. As indicated in the previous footnote, there is an obvious and immediate question about the source of the objective reality of the algebraic treatment of curves other than conic sections. This will be especially urgent in the case of the curves characterized as 'mechanical' by Descartes. I do not have anything special to say about this question, beyond appealing, as everybody does at this point, to the passage in the first *Critique* at B745 where Kant claims that algebra is based on a 'symbolic' construction in intuition where 'the constitution of the object is completely abstracted'. I am struck by his use of the word 'symbolic'. Where there is a symbol, there must be something symbolized. If algebra has objective reality by virtue of its underlying symbolic construction, perhaps this is by virtue of whatever has been symbolized and the way it has been symbolized. In the case of the equation for the parabola, discussed in our passage from the 'Reply to Eberhard', the thing symbolized is the relevant geometrical construction. Perhaps, in the case of the algebraic treatment of 'mechanical' curves, in Descartes' sense, the thing symbolized is the set of rules needed to mechanically generate the curve. I realize that Kant polemicizes in our passage from the 'Reply to Eberhard' against anybody who could think that knowing how to 'mechanically', i.e. 'graphically', produce a curve really knows anything about the curve in question. But I do not think that this really affects my suggestion, because the sense of 'mechanical' is not the same. In any case, this paper raises another difficult question for Kant: the question that I raise at the end of Section 3.

16. The only hitch of this kind I might conceivably encounter is if I tried to construct something impossible. That can and does happen. Typically, it happens in Euclid in the course of a proof by reduction. But we discover (or rig up) the auxiliary

construction in such a way that the attempt to produce it in the intended way produces a contradiction instead. That is not the fault of space, understood as a reservoir of every that I might be given or need for the purposes of solving constructive problems in geometry.

17.    I try to spell out this idea in detail and as precisely as possible in the chapter on Section 17 of the B-Deduction in my forthcoming book *Kant's Transcendental Deduction: A General Cosmology of Experience*.

18.    Perhaps, this title might be translated into English, thus 'Concerning a Lofty Tone Lately Struck in Philosophy'.

19.    The full title of the book was *Plato's Briefe, nebst einer historischen Einleitung und Anmerkungen, Königsberg bey Friedrich Nikolovius 1795*.

20.    It is amusing to note that Schlosser himself seems to have pre-emptively read Kant as an intellectual drudge, i.e. as someone determined to put everybody to work in the factory of reason. This can be seen in the first footnote to Schlosser's translation of what we usually refer to as Plato's 'seventh letter (Schlosser prints it as the sixth). The note is a comment on the famous passage of this letter where Plato tries to explain to the friends of Dion why it is impossible for him (Plato) or for anybody to write up true philosophy: there is something ultimately ineffable in the objects of philosophical enquiry that cannot be communicated to others in this way (342ff). The passage gives Schlosser an opportunity to reflect on the current state of German philosophy. He happily seizes the opportunity and proceeds to excoriate Kant, without mentioning him by name (the illusions to a 'critique of reason' and to the 'purifying of reason' are among the many tip-offs that Kant is the intended target.) Thus, Schlosser is led, by a chain of reflections that are not always easy to follow, to say that 'a critique that would deny reason [the right to proceed by analogies] would not so much purify as unman it. It even seems to me that a philosophy that would, through such a purification, so sequester itself from reason, would itself run the risk of soon transforming into a mere factory production of forms [*Formgebungsma-nufactur*] ....' (1795, 183). Kant replies to this tit with a tat: 'This dismissive way of writing off the formal aspect of our knowledge (which aspect is indeed the chief concern of philosophy) as pedantry, by calling it a 'factory production of forms', confirms the suspicion of a secret motive: to hang out the philosopher's shingle and thereby banish all philosophy, while giving oneself out as victor by disporting over it [sc. philosophy] in a lofty way' (8.404.3–8).

21.    It is worth pointing out the Plato himself stresses the significance, value and unavoidability of intellectual labour in the Seventh Letter. See especially 340b–c. This passage must have resonated with Kant. Perhaps, that is why Kant tries to claim Plato for himself in a footnote. Plato himself is naturally guilty of believing in some kind of intellectual intuition. But Kant claims that he almost discerned the essential points of the critical philosophy (8.391.30ff – the footnote that continues to the next page).

22.    I am here ringing the changes of Roshdi Rashed's observations on Proposition 17 in his commentary to the Arabic translation of the *Conica* (Apollonius 2010, 314–316).

23.    One thing really should be said here, though briefly. It is obvious that, in the case of mathematical concepts like the one on display in Apollonius III.17, the story about concept formation Kant is understood to be telling in the Jäsche logic is totally inadequate. The idea in Section 6 is that three 'logical acts' of the understanding produce 'concepts according to their form' (9.94.20). The example given in the footnote to this section is that of the concept Tree. We apparently get this concept by comparing, say, the foliage of a fir tree and a linden tree in a meadow, then by reflecting on what they have in common and then by abstracting from this reflection the concept Tree. Even if that is adequate as an account of how we form empirical

concepts like that of Tree, it clearly cannot account for the way Apollonius and his predecessors formed the general concept Conic Section. It may well be that there are different moments in the *Conica* that might be characterized as comparisons, reflections or abstractions. But nothing much will have been gained by using these words unless we also try to understand precisely what Apollonius was doing mathematically in the *Conica*. Whatever he was doing, it was different from what I do when I form my concept Tree.

## References

Apollonius. 2008. *Apollonius de Perge: Les Coniques,* tome 1.1: Livre I. Commentaire historique et mathématique, édition et traduction du texte arabe par Roshdi Rashed. Berlin: Walter de Gruyter.

Apollonius. 2010. *Apollonius de Perge: Les Coniques,* tome 2.1: Livres II et III. Commentaire historique et mathématique, édition et traduction du texte arabe par Roshdi Rashed. Berlin: Walter de Gruyter.

Brittan, Gordon. 1992. "Algebra and Intuition." In Chap. 12 in *Kant's Philosophy of Mathematics: Modern Essays,* edited by Carl Posy, 325–339. Dordrecht: Kluwer.

de la Hire, Philippe. 1685. *Sectiones Conicae*. Paris: Etienne Michallet.

Dijksterhuis, E. J. 1987. *Archimedes*. Translated by C. Dikshoorn. Princeton, NJ: Princeton University Press.

Eberhard, Johann. August. 1789. *Philosophisches Magazin*. Halle: Gebauer.

Fried, Michael, and Sabetai Unguru. 2001. *Apollonius of Perga's Conica: Text, Context, Subtext*. Leiden: Brill.

Friedman, Michael. 1992. *Kant and the Exact Sciences*. Cambridge, MA: Harvard University Press.

Friedman, Michael. 2013. *Kant's Construction of Nature*. Cambridge: Cambridge University Press.

Heath, Thomas. 1981. *A History of Greek Mathematics*. 2 vols. New York: Dover.

Heiberg, J. L., ed. 1891. *Apollonii Pergaei Quae Exstant Cum Commentariis Antiquis*. 2 vols. Leipzig: Teubner.

Heis, Jeremy. 2011. "Ernst Cassirer's Neo-Kantian Philosophy of Geometry." *British Journal for the History of Philosophy* 19 (4): 759–794. doi:10.1080/09608788.2011. 583421

Kant, Immanuel. 1902. *Kant's gesammelte Schriften Herausgegeben von der Königlich Preußischen Akademie der Wissenschaften*.

Kant, Immanuel. 1916. *Kritik der reinen Vernunft*, edited by Benno Erdmann. 6th ed. Berlin: Walter de Gruyter.

Knorr, Wilbur. 1986. *The Ancient Tradition of Geometric Problems*. New York: Dover.

Laywine, Alison. 1998. "Problems and Postulates: Kant on Reason and Understanding." *Journal of the History of Philosophy* 36 (2): 279–311.

Laywine, Alison. 2003. "Kant on Sensibility and the Understanding in the 1770s." *Canadian Journal of Philosophy* 33 (4): 443–483.

Newton, Isaac. 1972. *Philosophiae Naturalis Principia Mathematica*, edited by Alexandre Koyré and I. Bernard Cohen. 3rd ed. Cambridge, MA: Harvard University Press.

Pappus. 1876–1878. *Collectionis Quae Supersunt*, edited by F. Hultsch. 3 vols. Berlin: Weidmann.

Pascal, Blaise. 1964. *Oeuvres Complètes*, edited by Jean Mesnard. 2nd vol. Paris: Desclée de Brouwer.

Plato. 1992. *Opera*, edited by Burnet. 4th vol. Oxford: Clarendon Press.

Rashed, Roshdi. 2005. "Les premières classifications des courbes." *Physis* XLII (1): 1–64.

Saito, Ken. 2004. "Book II of Euclid's *Elements* in Light of the Theory of Conic Sections." In Chap. 7 in *Classics in the History of Greek Mathematics*, edited by Jean Christianidis, 136–169. Dordrecht: Springer Verlag.

Schlosser, J. G. 1795. *Plato's Briefe, Nebst Einer Historischen Einleitung und Anmerkungen*. Königsberg: Friedrich Nikolovius.

Steiner, Jakob. 1881. *Gesammelte Werke*, edited by Karl Weierstrass. Vol. 1 of the *Gesammelte Werke*. Berlin: G. Reimer.

Sutherland, Daniel. 2006. "Kant on Arithmetic, Algebra and the Theory of Proportions." *Journal of the History of Philosophy* 44 (4): 533–558.

Zeuthen, H. G. 1966. *Die Lehre von den Kegelschnitten im Altertum*. Hildesheim: Georg Olms.

# 'With a Philosophical Eye': the role of mathematical beauty in Kant's intellectual development

Courtney David Fugate[a,b]

[a]Civilization Studies, American University of Beirut, Beirut, Lebanon; [b]FCHI, Emory University, Atlanta, GA, USA

This paper shows that Kant's investigation into mathematical purposiveness was central to the development of his understanding of synthetic a priori knowledge. Specifically, it provides a clear historical explanation as to why Kant points to mathematics as an exemplary case of the synthetic a priori, argues that his early analysis of mathematical purposiveness provides a clue to the metaphysical context and motives from which his understanding of synthetic a-priori knowledge emerged, and provides an analysis of the underlying structure of mathematical purposiveness itself, which can be described as unintentional, but also as objective and unlimited.

My aim in this paper was to show how Kant's investigation into mathematical purposiveness, or what in §62 of the *Critique of the Power of Judgment*[1] he also describes as 'objective formal purposiveness', was central to the development of his understanding of synthetic a-priori knowledge. Specifically, I will (i) provide a clear historical explanation as to why Kant points to mathematics not only as a prime, but even as an exemplary case of the synthetic a priori throughout his critical writings, (ii) argue that his early analysis of mathematical purposiveness provides a clue to the metaphysical context and motives from which his understanding of synthetic a-priori knowledge emerged and (iii) provide an analysis of the underlying structure of formal objective purposiveness itself, which can be described as a kind of unintentional but also objective and unlimited purposiveness. Although it is not essential to demonstrating the thesis of this paper, I have also sought to provide a complete account of the sources and content of the mathematical examples Kant uses to illustrate mathematical purposiveness in the texts I will discuss. I have done this for three reasons. First, no such account is to be found in any of the extant commentaries. Secondly, in the text itself, Kant provides only the briefest description of his examples without indicating their original sources. This indicates that he expected his readers to be relatively familiar with the mathematical classics in a way that most contemporary readers will not be. Reaching a perfectly precise understanding of what Kant means by mathematical purposiveness will thus require some historical investigation. Thirdly, knowing the origins of the examples adduced by Kant can provide us with an insight into the kinds of sources he drew upon and the place of mathematical literature in his intellectual development.

The paper is structured as follows. In Section 1, I will examine the several examples and descriptions of mathematical purposiveness Kant provides in the critical period with the aim of determining the essential structure of this kind of purposiveness. In Section 2, I will then recall some key aspects of the historical background that prompted Kant's development of his own concept of mathematical purposiveness. I will subsequently draw on this in Section 3 to interpret the relevant elements in Kant's pre-critical philosophy, bringing this historical excursion to an end in Section 4 with an interpretation of Kant's first explicit discussion of mathematical purposiveness in his book, *The Only Possible Argument in Support of a Demonstration of God's Existence,* which was published in 1763. In Section 5, I will return to the critical writings, and in particular to §62 of the *Critique of the Power of Judgment*, in order to refine my earlier account and to explain some further aspects of these texts, the full import of which will only be evident in view of the investigations of the intervening sections.

## 1.   The Significance and structure of mathematical purposiveness in the critical writings

Kant's first and principal discussion of mathematical purposiveness in his critical writings is found in §62 of the *Critique of the Power of Judgment*, published in 1790. At first blush, the purpose of the discussion appears to be merely to provide a typological introduction to the topic of the Analytic of the Critique of the Teleological Power of Judgment, of which it forms the first section.[2] This part of the work specifically concerns material objective purposiveness, i.e. the purposiveness that is found in the real objects of physical nature. In the previous section, Kant had distinguished *objective* purposiveness, which concerns a purposiveness determined by the *concept* of an end, from *subjective* purposiveness or beauty, which concerns only the subjective effect of the representation of an object on our mental faculties and is not determined by the concept of any end. The apparent aim of §62 is thus to further distinguish between two types of *objective* purposiveness, namely the merely *formal*, which concerns the relative properties of mathematical objects, i.e. the spatio-temporal form of experience, and the *material*, which concerns the actual causal relations of objects of experience to some end.

However, this straightforward interpretation of §62 as well as the tidiness of its typology are both belied by a closer inspection of the text itself. First, throughout the third *Critique*, Kant defines objective purposiveness in general as the purposiveness that 'can be cognized only by means of the relation of the manifold to a determinate end' (*KU* 05:226), where the 'end in general is that the *concept* of which can be regarded as the ground of the possibility of the object itself', from which it follows that 'in order to represent an objective purposiveness in a thing the concept of *what sort of thing it is supposed to be* must come first' (*KU* 05:227). On the strength of this, Kant goes so far as to say that 'to represent a formal *objective* purposiveness without an end, i.e. the mere form of a *perfection* ... is a veritable contradiction' (*KU* 05:228). Yet precisely in this section devoted to the formal *objective* purposiveness of mathematical objects, Kant writes that such 'purposiveness still does not make the concept of the object itself possible' (*KU* 05:362), indeed that it 'is not grounded in a purpose' (*KU* 05:364) and that such objects are purposive 'without an *end* or any other ground having to be the basis of this purposiveness' (*KU* 05:364). So unless more is going on in this passage than first appears, Kant himself is caught in a 'veritable' and entirely unnecessary contradiction.

A second clue pointing in the same direction is found in the length and complexity of the passage itself. As we will see later in this paper, in this passage, Kant does many remarkable things that would not be required simply to establish a typology of purposiveness: he provides three different and rather elaborate examples of mathematical purposiveness; explains that this form of purposiveness generates a unique kind of transcendental illusion – which, as it turns out, is precisely that which led Plato to his dogmatic theory of the ideas; he provides a diagnosis of the 'natural' grounds within reason that generate such an illusion and takes time to explain that even when the illusion is unmasked, such purposiveness still genuinely points up an 'inexplicable' harmony between our mental faculties, which 'enlarges the mind, allowing it, as it were, to suspect something lying beyond those sensible representations' (*KU* 05:365). Thus, the very length of Kant's discussion as well as this last suggestion that mathematical purposiveness, just like aesthetic purposiveness, points us towards the supersensible basis of our mental faculties, again suggests that it has far greater significance for Kant than it would if it were introduced merely for typological reasons.

A third and final clue to the wider significance of mathematical purposiveness for Kant lies in his discussion of it in completely different contexts in both his pre-critical writings and his critical writings after 1790. In Section 4, we will see that the core understanding of mathematical purposiveness in the *Critique of the Power of Judgment* is already present in Kant's pre-critical writings, where he interprets it, however, as indicating that space has the ground of its possibility in a divine being, and thus where it forms an integral part of his early argument for the existence of God. Kant also makes use of the same material in the drafts for his essay on the progress of metaphysics, written in the early to mid-1790s and then again in 'On a Recently Prominent Tone of Superiority in Philosophy', published in 1796. In both cases, Kant uses it to suggest, as he does more briefly in the third *Critique*, that Plato's own theory of the ideas arose from a transcendental illusion made possible by his clear recognition of the synthetic a priori character of mathematical purposiveness. Mathematical purposiveness is thus of such significance that it leads Kant to break his usual account of the history of philosophy according to which the existence of synthetic judgements a priori, particularly in mathematics, had never before been recognized.

On the basis of these three clues, I take it as evident that the superficial features of §62 mask a deeper level at which Kant is engaged with mathematical purposiveness. In the remainder of this section, my goal is simply to clarify the structure of mathematical purposiveness itself by taking a closer look at examples and descriptions of it contained in this passage. This will provide a platform for my investigation into its historical and metaphysical roots, which will occupy me in the middle sections of this paper. I will then return to the other elements of §62 of Kant's third *Critique* in Section 5 of this paper.

### 1.1. The Structure of mathematical purposiveness

Kant begins his discussion in §62 by observing that:

> All geometrical figures that are drawn in accordance with a principle display a manifold and often admired objective purposiveness, namely that of serviceability for the solution of many problems in accordance with a single principle, and indeed each of them in infinitely many different ways. (*KU* 05:362)

What Kant has in mind here can be better determined from the three examples that follow. The first is based on the general problem of constructing a geometrical figure given two constraints, namely 'a triangle from a given baseline and the opposite angle'. As Kant notes, this has an infinite number of possible solutions, 'but the circle comprehends them all as the geometrical locus for all triangles that satisfy this condition' (*KU* 05:362). Kant does not provide any further explanation of the mathematical theorem in question, but it seems clear that he is pointing to a consequence of the Law of Sines, which states that, for any Euclidean $\triangle ABC$,

$$\frac{AC}{\sin \angle ABC} = \frac{CB}{\sin \angle BAC} = \frac{BA}{\sin \angle ABC} = D.$$

Now, given the length of any side and the measure of its opposite angle, say $AC$ and $\angle ABC$, which is the case Kant mentions, it is possible to compute the constant D according to the above formula, and thereby to determine the exact ratio of the length of each remaining side to the sine of its opposite angle. Clearly, there are an infinite number of possible triangles satisfying this condition, just as Kant remarks. Yet, as it turns out, if we construct a circle with diameter $D$ such that its circumference also passes through the endpoints of the given line $AC$, then the set of all points on the circle above this line comprehends all possible solutions. That is to say, if and only if point $B$ lies on this line, will the resulting triangle satisfy the conditions of having the given length $AC$ with the given opposite $\angle ABC$. For instance, if length $AC = 1$ and $\angle ABC = 30°$, then the circle with diameter 2 constructed as in Figure 1 is that which 'comprehends [all solutions] as the geometrical locus for all triangles that satisfy this condition'.[3]

This is indeed a very beautiful theorem. But its particular purposiveness for Kant lies precisely in that such a 'simple figure' (*einfache Figur*) provides the 'basis for the solution of a host of problems, for each of which by itself much preparation would be required, and which as it were (*gleichsam*) arises from this figure itself as one of its many splendid properties' (*KU* 05:362). It is notable that Kant does not simply ascribe purposiveness to the circle as such. Rather the purposiveness Kant has in mind becomes visible only through his presentation of the circle as the solution to a certain kind of problem, which is only fully expressed in the general theorem. More precisely, in this case, the beauty seems to depend upon the combination of several factors: (1) the simplicity of the circle itself; (2) the appearance that the initial problem would require 'much preparation', i.e. purposive design; and (3) the recognition that, despite its simplicity, the circle nevertheless fully and exactly satisfies this purpose

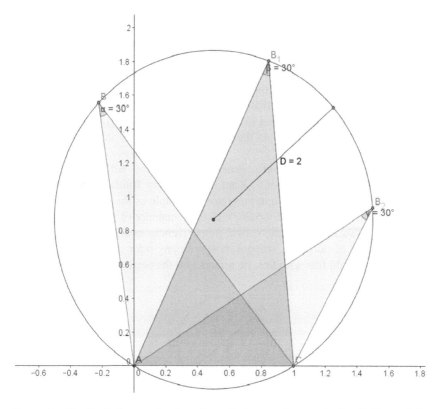

Figure 1. Kant's first example of mathematical purposiveness in the *Critique of the Power of Judgment*.

(providing the general solution), which itself comprises an infinity of other possible purposes (providing particular triangles as solutions).

The second example is likewise based on a problem regarding an infinite set of solutions, namely, that of finding two lines 'that intersect in such a way that the rectangle constructed from the two parts of the one is equal to the rectangle from the two parts of the other' (*KU* 05:362). Although Kant does not mention the fact, this is just the problem whose solution is given in the so-called Intersecting Chords Theorem, which is demonstrated by Euclid in *Elements*, Book II, Proposition 35: 'If in a circle two straight lines cut one another, the rectangle contained by the segments of the one is equal to the rectangle of the segments of the other' (1908, vol. 2, 71). Or for any two chords $AC$ and $BD$ drawn in the same circle as in Figure 2,[4] $(AE){\cdot}(EC) = (BE){\cdot}(ED)$.

As before, Kant locates the purposiveness in question not simply in the circle itself, but in the theorem regarding the circle that provides a complete solution to the initial problem. As Kant explains, 'the solution of this problem looks as if it will be very difficult' and yet all solutions are again provided in an incredibly simple construction, which is yet not specifically designed beforehand to provide any particular solution (*KU* 05:362–363).[5] In summarizing the first two examples, Kant thus remarks that the circle and 'other curves yield in turn other purposive solutions that were *not thought of at all in the rule that constitutes their construction*' (emphasis added).[6] Again, mathematical purposiveness seems to depend upon the same three factors, namely, (1) the simplicity of the circle, which means the simplicity of the 'rule that constitutes its construction', i.e. its concept; (2) the appearance that solving the problem would require much purposive design on our part and (3) the recognition, in a final theorem, that the circle nevertheless comprehends the infinitely many ways in which it can be solved.

The third example is very general, but as we will see, still quite illuminating. 'All conic sections', Kant explains, 'are fruitful in principles for the solution of a host of possible problems, as simple as the definition is which determines their concept' (*KU* 05:363). In the remainder of the paragraph, Kant then extends this idea of mathematical purposiveness as a kind of fitness to an infinite number of purposes, without, however, any prior determination to any specific purpose. In line with this, he remarks on the pleasure he takes in seeing

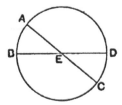

Figure 2. Kant's third example of mathematical purposiveness in the *Critique of the Power of Judgment*.

how 'the ancient geometers investigated the properties of such lines without being distracted by the question of limited minds: for what is this knowledge useful?' They rather 'delighted' directly in the mathematical purposiveness in such objects themselves, which, however, *also* made such objects infinitely useful – and so useful for posterity, e.g. the ellipse for the study of planetary orbits – despite not being constructed for any specific use (*KU* 05:363). Kant's point seems to be that although this kind of purposiveness consists in a mathematical figure's being suitable to the solution of an infinite number of unexpected problems, thereby generating a multitude of important theorems, the concept of the figure does not consist in this and is not specifically determined by this relation. The purposiveness here lies precisely in the intrinsic and infinite fitness of such a simple and undesigned figure to providing all kinds of such solutions, i.e. cognitive purposes.

We are now in a position to approach the two key descriptions of mathematical purposiveness that Kant provides in this section. In the first, Kant states that this purposiveness 'expresses the suitability of the figure for the generation of many shapes aimed at purposes', but 'still does not make the object [i.e. the figure] itself possible, i.e., it is not regarded merely with respect to this use' (*KU* 05:362). In other words, while the figure seems to be derived specifically so as to serve these purposes (because it serves so many, so precisely and so fully as is shown in the general theorem), its concept is in no way founded upon or derived from the concepts of such purposes: 'the many rules ..., which are purposive in many respects, without an *end* or any other ground having to be the basis of this purposiveness' (*KU* 05:364). This explains the strange and seemingly contradictory status of mathematical purposiveness we noted above. All the other examples of perfection, i.e. objective purposiveness, that Kant cites are analytic,[7] and these must have been what Kant had in mind in forming his definition of objective purposiveness earlier in the third *Critique*. In these, the structure of the object is derived, without the need for construction in a priori intuition, purely from the concept of the end it is to serve. For example, the perfection of a square that is judged by its being constructed in accordance with the concept of a square is purely analytic. However, the mathematical purposiveness Kant has in mind here is evidently *synthetic*, i.e. it consists in the suitability to ends that are not contained in the concept of the figure at all, but which can be precisely expressed in the form of a theorem demonstrable in pure intuition. In this, it is more like subjective purposiveness, which is not determined by a concept. And yet, just like other cases of objective purposiveness, mathematical purposiveness is based on the fact that a figure indeed serves purposes, and, moreover, the fact that it does serve these purposes can be demonstrated as necessary a priori, although not from the mere concept of the figure involved, but from a construction made possible by a priori intuition. This peculiarity of mathematical purposiveness seems to explain why Kant describes it neither as purposiveness without a purpose (aesthetic), nor as intentional (analytic objective purposiveness), but rather 'as if' it were intentional, i.e.

determined by a purpose. This yields the second key description of mathematical purposiveness given in the passage, as 'that which is purposive and so constituted as if (*als ob*) it were intentionally arranged for our use, without any regard to our use' (*KU* 05:363). Because what we are dealing with here are mathematical figures and relations, we can say that in such a theorem we represent as necessarily and reciprocally related the individual figure and an infinity of solutions, i.e. possible cognitive ends, although we can clearly see that the concepts of these things are not internally connected in a way that could account for this relation. In *On a Recently Prominent Tone*, published 6 years later, Kant draws all of this together in what is perhaps his most precise description of mathematical purposiveness as

> fitness to resolve a multiplicity of problems, or multiplicity in resolving one and the same problem, from a principle, just as if (*als ob*) the requirements for constructing certain quantitative concepts were laid down in them *on purpose*, although they can be grasped and demonstrated as necessary *a priori*. (*VT* 08:391)

The deeply perplexing character of this purposiveness should be clearly understood before continuing further. First, it is an *objective* feature of geometrical space *consisting precisely in* the unity expressed in a synthetic a priori mathematical theorem that is demonstrable only through a construction in pure intuition. Therefore, it is not purely an expression of a subjective fitness of certain concepts to our mental powers or even to our contingent cognitive aims. Rather, the purposiveness here, being objective, is itself a sort of unity, expressible in a theorem that belongs to the geometrical objects in question. Yet, secondly, this purposiveness, although it is itself a particular unity, also expresses the unity of an infinity of other synthetic but lower-order geometrical propositions, each of which can be provided, at least in principle, with its own a priori demonstration. To this extent, the less general propositions do not seem to be dependent for their truth on the general theorem any more than do the concepts of the figures involved. Thirdly, and most significantly, the higher-order proposition or theorem cannot be derived analytically even with the lower-order or less general propositions being given, because it provides them not with mere distributive unity, e.g. the way blueness in a way ties together all blue books, but rather with genuine *collective* unity, the kind of unity that is conferred by the derivation of many things from one common ground, e.g. the way a species of animal can be shown to share a common trait because of their generation from a common ancestor. This seems to be why Kant describes such purposiveness as sharing features of the intentional; for it is necessarily 'as if' the general theorem here provides the reason for the truths it is seen to unity, since 'purposiveness is thinkable only through relation of the object to an understanding, as its cause' (*VT* 08:391). Yet this thought must remain in the mode of an 'as if' precisely because all such truths can be shown to be necessarily true a priori without the knowledge of this higher-order truth, and so seem, in this respect, to be as independent of the general theorem as are the concepts of the figures involved. As we will see further below, it is Kant's view that the only way we can successfully combine these two seemingly incompatible features of mathematical purposiveness, i.e. the ground-like character of the general theorem and the apparent independence of the truths and concepts for which it seems to provide the ground, is by regarding it as deriving from the deeper foundation of both, namely the essential ground of all mathematical objects as such.[8] In this way, the infinite purposiveness exhibited in certain theorems is seen to be a consequence of the supreme unity of the single common and essential ground of all mathematical truths.

We can now see more clearly why Kant thinks mathematical purposiveness offers us a kind of original and unique *datum*; for not only is it synthetic a priori, but it is at the same time an *infinite purposiveness*. In other words, from neither the concepts involved, nor from the mere fact of its synthetic character, can one explain why there is not only a necessary connection there, but also an *infinitely purposive* one. Mathematical purposiveness thus not only provides an example of synthetic judgements a priori, it provides an example that uniquely calls out for an explanation of the general possibility of such truths. It is this uniqueness, according to Kant, that forces us to look beyond the figures and their concepts, indeed beyond the individual theorems themselves, to the supposedly purposive constitution the space that makes them possible, in order to locate the synthetic source of this necessary purposiveness. That mathematical purposiveness forces us to move in this way from the concepts of mathematical figures to an investigation of the common ground of their possibility is, as we will see, precisely what connects all Kant's discussions of mathematical purposiveness from the pre-critical period onward.

Before moving on, I would like to draw attention to the specific language Kant uses in describing this mathematical purposiveness in these passages, because this will figure in our historical investigation in the next few sections. This language is in fact extremely uncharacteristic of Kant's critical period. Whereas he routinely speaks of intuitions, concepts, appearances, experience and phenomena when discussing what is knowable, Kant describes this mathematical purposiveness as 'a purposiveness in the essence of things' or 'perceived in the essence of things' or, again, as 'purposiveness observed in the essence of things (as appearances)', which seems to 'pertain originally to the essence of things (insofar as their concept can be constructed)' (*KU* 05:364–365). This usage of 'essence', which is underscored by Kant's insistence that this purposiveness is not arbitrary, but rather pertains 'necessarily' to things, is thus as consistent as it is striking. We will see the source of it below.

## 2. Wolff on the beauty of mathematical truths

To gain a deeper insight into what is going on in Kant's account of mathematical purposiveness in the third *Critique*, we must briefly look back at the work of Christian Wolff. It will be helpful for the sake of clarity to anticipate what we will see in this section. The work of Wolff, of course, provides the immediate context of Kant's intellectual development in general. In particular, we will see that Kant develops his own account of mathematical purposiveness in part as a direct response to the explanation of mathematical beauty in Wolff's *Rational Thoughts on the Purposes of Natural Things* (hereafter German Teleology). To show this, however, I will have to first explain some specific aspects of Wolff's metaphysics, because Kant's rejection of Wolff's theory of mathematical beauty is made possible by his rejection of Wolff's theory of essences. So in this section, I will explain this metaphysical background and show how it motivates Wolff's theory of mathematical beauty.

In his German Teleology, Wolff sets out to explain how investigations into the physical constitution of the actual universe can provide a moving insight into the freedom and specific character of its divine creator. He proposes by this to build a 'ladder' from nature to God, which will confirm his separate metaphysical demonstrations of creation's design while also providing the reader with edification for strengthening their moral and religious devotion. In short, the aim of the work is to improve the reader's moral state by allowing them to better contemplate God's perfection as it is mirrored in the physical world. To accomplish this task, Wolff begins by explaining that we should only look for evidence of God's design in what he calls the 'contingent' order of nature. For Wolff, a thing or a proposition is contingent if it is not absolutely necessary, i.e. if its non-existence or falsehood, respectively, do not imply a contradiction. In such a case, whether a thing exists or a proposition is true cannot be determined through the essences of these things alone, and consequently, if they do exist or are true, then there must be some *other* ground or reason for this being the case. Because of this, only those existent or true things in nature that are also contingent in respect to the essences of all things within nature taken as a whole require a ground external to nature to account for their existence or truth. Furthermore, if something is actually the case, then although it remains contingent, it is also necessary in the sense that, given the external ground of its being the case, it could not have failed to obtain. So some contingent things, namely those that exist or are true, are nevertheless necessary, although only hypothetically, not absolutely so.

All of this means that in order to properly gather evidence for God's existence and character a posteriori from physical nature, we must be able to clearly distinguish the contingent in nature, or, what is the same, the hypothetically necessary in nature, from what is absolutely necessary. Now, according to Wolff, the essences of all things, as well as all that can be derived from these essences by the principle of contradiction alone, are absolutely necessary. In his *Philosophia Prima, sive Ontologia* (hereafter Latin Ontology), Wolff offers the following two-step proof of this claim:[9]

§299. The essences of things are necessary. The essences of things consist in the non-repugnance of those things that are in one and the same thing, and yet are not at the same time determined to be in it by a third (§143). Since it is impossible for the same thing to be and simultaneously not to be (§28); it is furthermore not possible for those things that are non-repugnant when taken together, as well as *per se* neither determining of one another, nor simultaneously determined by anything else, to also be mutually repugnant. These are hence necessarily non-repugnant, and consequently the essences of things are necessary.

E.g. The essence of a triangle consists in that it is not repugnant to three lines that they be united in order to comprehend a space in such a way that any two taken together are greater than the third. Is there indeed anyone who would declare that it would actually be impossible to unite three straight lines such that two together would be greater than the third, and so as to enclose a space? No one would seriously assert this except one who would concede that the same thing can both be possible and not possible, that is, one who was devoid of human understanding. [...]

§303. The essences of things are absolutely necessary. They are in fact necessary when considered absolutely, since their necessity is demonstrated by supposing nothing other than a definition. Hence they are absolutely necessary.

The first paragraph claims to establish that essences are necessary; the second claims that this necessity is not hypothetical, but absolute. The first step starts in the second line with the definition of an essence. An essence, according to Wolff, consists of 'those things in a being itself that are not repugnant and do yet not *per se* determine one another' (1736, §143). Because these components of the essence do not mutually determine one another, the essence as a whole must also be the primitive ground of all the being's internal determinations[10] or, what is the same for Wolff, the components making up the being's essence must be the first grounds of its intrinsic possibility. Furthermore, as these things are first, there can be no intrinsic reason why essences are in certain beings (Wolff 1736, §. 156). Other things for which the essence by itself provides a sufficient reason, i.e. what can be deduced from it by the principle of contradiction, are what Wolff terms 'attributes' (1736, §146). Finally, because the components of an essence are all primitive and not mutually determining, the only possible test of the possibility of an arbitrarily assumed essence is the lack of contradiction between its components. As Leibniz wrote some time before: 'All that which implies contradiction is impossible, and all that which implies no contradiction is possible' (Leibniz 1996, 235). Wolff holds this as well.

In the second sentence, Wolff uses this last feature of essences to argue that they are all necessarily possible. Because it is impossible for something to be and not to be (this is just the principle of contradiction), and essences are clearly possible, it is impossible for them to also be impossible, i.e. for the components of an essence to be repugnant or to contradict one another. Now, he argues, if it is impossible for them to be repugnant, it follows that they are necessarily non-repugnant, which is to say that the essence is necessarily possible. The second step, which takes place in §303, simply notices that this proof assumes nothing but the definition of essence, and consequently its conclusion is absolutely necessary, i.e. essences are absolutely necessarily possible. The point here is that if the proof of necessity rests on nothing but the definition of the subject of the proof, then this necessity rests on no further condition. Thus, its necessity is not hypothetical, but absolute.

Moving one step closer to our goal, Wolff understands geometrical or mathematical figures in particular to be what he calls 'imaginary notions' (1736, §110). His reason for saying this lies in his nominalism; as intrinsically abstract entities, mathematical objects are not fully determinate, and are not capable by themselves of existing. For example, triangular objects are indeed really triangular and this fact is indeed informative regarding many other real properties of the triangular object, but when we think of a triangle as itself an object, we represent it as if it were something that could exist apart, although it really cannot. Thus, a triangle thought of as an abstract entity is the thought of a property as if it were a complete thing, and therefore it is something that has its existence only in the imagination. Yet, and this is extremely important, when we imagine a triangle, we are at the same time thinking certain truths that really do belong to the essences of all triangular things. Now, according to Wolff, the method of mathematics is precisely to demonstrate all knowledge from the absolutely first and complete real definitions of quantitative and geometrical objects. It thus begins from definitions composed precisely so as to contain all and only those first conceivable internal determinations of its objects, and it proceeds by demonstrating all further mathematical truths from these alone using the principle of contradiction. Consequently, just as essences in general and all of their logical consequences are absolutely necessary, so are all specifically mathematical essences as well as the propositions that are derived from them. Because all mathematical truths can and must be proven by resolution back to such essences, it follows that all mathematical truths whatsoever are true by the principle of contradiction alone. Their opposites are not only false, they are indeed internally contradictory. In Kant's language, Wolff holds them to be analytic.

On this basis, we can now return to Wolff's German Teleology, where he explains that, for the reasons just cited, examples of mathematical beauty, no matter how striking, must not be mistaken for indications of God's existence or taken as clues to his design. As Wolff explains:

> One who understands mathematics can easily make this matter clearer and more comprehensible. We meet with the most beautiful order in the infinite series of numbers. However, one cannot for that reason conclude that someone has placed numbers in such order. The order is absolutely necessary and cannot be other than it is: it can be comprehended from the nature of number and in the matter itself is found this order's sufficient ground, so that no further ground outside of it need be sought. [...] Take the infinite series that Herr von Leibniz has provided for the circle, which proceeds in the most beautiful order, as Herr von Leibniz himself has shown to be more thoroughly the case than it even first appears. Regardless of this, however, it flows from the necessary foundations of mathematics and requires no further ground, when one desires to comprehend why it is so. (1741, §9)

The infinite series to which Wolff here refers is of course the famous Leibniz–Gregory series:[11]

$$\frac{\pi}{4} = 1 - \frac{1}{3} + \frac{1}{5} - \frac{1}{7} + \frac{1}{9} - \cdots.$$

The immense beauty of the series derives not from the fact that it provides a way to calculate $\pi$, but rather from the fact that it allows us to express $\pi$, which is an irrational number that thus neither repeats nor follows any general pattern simpler than itself, in terms of a perfectly orderly pattern based on the consecutive odd integers. However, the point Wolff wishes to make with the series is that although it clearly exhibits immense and unexpected beauty, just as do many other similar truths, a trained mathematician will nevertheless recognize that it follows entirely from the definitions of the concepts involved and the principle of contradiction, and hence is absolutely necessary. Indeed, because all mathematical truths contain the complete sufficient ground of their own truth, at least in Wolff's view, they do not depend in any way on God's existence or his specific character. Consequently, their truth provides no evidence regarding either. Using Kant's later terminology, Wolff is basically arguing that because all truths about essences, one species of which consists of all mathematical truths, are entirely analytic, they do not require us to infer a third or external thing as ground to explain their specific synthesis.

## 3. Kant's rejection Wolff's doctrine of the necessity of essences

Surprisingly enough, Kant takes his first major step away from Wolff and towards his later critical understanding of mathematical purposiveness already in a few rough handwritten sketches dating from around 1753–1754. In these notes, Kant records his earliest reasons for taking the radical step of rejecting the theory of essences we located in Wolff. Yet Wolff is not specifically mentioned in the sketches; the topic of discussion is rather the deficiencies of Leibniz's *Theodicy*. One will recall that in the *Theodicy*, Leibniz seeks to defend God's goodness against the objection raised by all the evils in the world. In a key passage, Leibniz for instance explains:

> The ancients attributed the cause of evil to matter, which they believed uncreated and independent of God; but we, who derive all being from God, where shall we find the source of evil? The answer is, that it must be sought in the ideal nature of the creature, in so far as this nature is contained in the eternal verities which are in the understanding of God, independently of his will. For we must consider that there is an original imperfection in the creature before sin, because the creature is limited in its essence. (1996, 135)

In other words, Leibniz argues that God is not responsible for evil because its true source is, so to say, in the 'ideal matter' of creation, i.e. in the essences of things themselves. Also, God is supposedly not responsible for these or their consequences, precisely because they are the eternal verities 'which are altogether necessary, so that the opposite implies contradiction' (Leibniz 1996, 74). Among such essential truths are those of 'mathematics, where the opposite of the conclusion can be reduced *ad absurdum*, that is, to contradiction' (Leibniz 1996, 87). God's antecedent will is thus indeed to create a perfect world, but given the limitations inherent in the eternal essences, his consequent will can only be to create the best possible world (Leibniz 1996, 136–137).

After briefly summarizing what he sees as the core elements of Leibniz's book, Kant responds:

> The errors of this theory are indeed too serious for us to be able to accept it. Leibniz presents the rules, which aim at perfection, as conflicting with each other in their application. ... What is it which causes the essential determinations of things to conflict with each other when they are combined together, so that the perfections, each of which on its own would increase God's pleasure, become incompatible with each other? What is the nature of the unfathomable conflict which exists between the will of God, which aims only at the good, and the metaphysical necessity which is not willing to adapt itself to that end in a general harmony which knows no exceptions? ... The whole mistake consists in the fact that Leibniz identifies the scheme of the best world on the one hand with a kind of independence, and on the other hand with dependence on the will of God. All possibility is spread out before God. God beholds it, considers it, and examines it. He is inclined in one direction by the determinations inhering in the possibilities, in accordance with their particular perfections, and he is inclined in another direction according to the effect produced by their combination. (*Refl* 3705)

Thus in Kant's view, Leibniz's system contains an incorrect account of the relation of essences to the divine will. They must not be conceived as independent in respect to it, although this is required in order to understand how they could possibly limit, and in this sense come into conflict with, the full realization of God's antecedent will. Indeed, on the view Kant expresses here, the very division between an antecedent and a consequent will rests on a view of essences as prior and independent conditions on God's creation of the world. But, Kant asks, 'what is one to say [then] of God's infinity and independence'?

In a related note from the same time period, Kant explains that the only scheme suitable to a proper conception of God's independence is one that

> even subjects every possibility to the dominion of an all-sufficient original Being; under this being things have no other properties, not even those which are called essentially necessary, apart from those which harmonise together to give complete expression to His perfection'. (*Refl* 3704)

We should expect in such a scheme that

> the essential and necessary determinations of things, the universal laws which are not placed in relation to each other by any forced union into a harmonious scheme, will adapt themselves as if spontaneously to the attainment of purposes which are perfect. (*Refl* 3704)

Already here, we can see that behind Kant's criticism of Leibniz, there lies a radically new positive account of the teleological structure of the created universe. For as we saw in Wolff, but as is also true of Kant's other major German predecessors, genuine purposiveness is thought to be a feature only of the contingent order of nature, i.e. of the order of nature that is not the consequence merely of the essences of things and thus of what can be specifically and externally arranged to bring about some end. Specifically in Leibniz and Wolff, the conflict between essences made it the case that not all ends that might have pleased God individually could exist together in the same world. Some less than perfect things must exist merely as the means to other greater perfections. The system Kant introduces in these lines, by contrast, holds that essences themselves must, both individually and as a whole, possess an absolute and intrinsic fitness to the realization of every end God could possibly have. Moreover, as nothing will exist in nature only because it must be there to achieve some other end, i.e. nothing will exist merely as a means, all things in creation will be both means and ends, and they will be this in virtue of their *essential* determinations. Put differently, because there is no conceivable limitation that could force God to make anything a means but not also an end, and a creation in which all things are ends is clearly more perfect than one in which only some things are ends and others are mere means, we should expect all things to be both means and ends.

Kant's writings of 1755 amplify this new positive theory of essences by which things manifest an intrinsic, unlimited and reciprocal relation to ends. In the *Natural History and Theory of the Heavens*, Kant explains that we can see just such a relation to ends in the most basic elements of matter and in the necessary laws of physics. As he writes,

> The better we get to know nature, the more will we gain the insight that the universal characteristics of things are not foreign to and separate from each other. We shall be adequately convinced that they have essential affinities through which, *by themselves*, they prepare to support each other in the establishment of perfect constitutions … and that altogether the individual natures of things in the field of eternal truths among themselves already constitute, as it were, a system in which one relates to the other; we shall also become aware that the affinity is a part of them from their common origin out of which they all drew their essential determinations. (*NTH* 01:364)

Thus due to the common origin of their essences in God, all created things will possess '*by themselves*' and essentially, thus intrinsically, an affinity in which each 'supports' every other as an end. By this we will see even that 'the universal development of matter in accordance with mechanical laws has been able to bring about connections that have created benefits for creatures with reason without requiring laws other than their universal determinations' (*NTH* 04:294). 'Nature', in other words, has 'an essential determination to perfection and order' (*NTH* 01:348). Kant recounts this same theory at various points in the *Nova dilucidatio* of the same year, while also providing it with a partial metaphysical foundation. The Principle of Coexistence demonstrated in that work, for instance, aims to prove that real interaction of substances requires precisely the view of essences outlined above. It is this view of essences, as already in themselves attuned to a harmonious and reciprocal relation to all other things, that makes it possible to think of them, once created, as influencing one another through their own forces. 'The same indivisible act', Kant explains,

> which brings substances into existence and sustains them in existence, procures their reciprocal and universal dependence, so that the divine act does not need to be determined, now one way, now another, according to circumstances. There is rather a real reciprocal action between substances. (*PND* 01:415)

The really new and interesting aspect of the *Nova dilucidatio* for our purposes lies in its clear rejection specifically of Wolff's doctrine of the absolute necessity of essences:

> For although essences (which consist in inner possibility) are ordinarily called absolutely necessary, nonetheless, it would be more correct to say that *they belong to things absolutely necessarily*. For the essence of a triangle, which consist in the joining together of three sides, is not necessary. For what person of sound understanding would wish to maintain that it is itself necessary that three sides should always be conceived as joined together? I admit, however, that this is necessary for a triangle. That is to say: if you think of a triangle, then you necessarily think of three sides. And that is the same as saying: if something is, it is. But how does it come about that the concepts of sides, of a space to be enclosed, and so forth, should be available for use by thought; how, in other words, it comes about that there is, in general, something which can be thought, from which then arises, by means of combination, limitation and determination, any concept you please of a thinkable thing – how that should come about is something which cannot be conceived at all, unless it is the case that whatever is real in the concept exists in God, the source of all reality. (*ND* 01:395–396)

In this passage, Kant is clearly and consciously challenging, indeed mocking, Wolff's supposed proof of the necessity of essences in his Latin Ontology, which I quoted in Section 2. To Wolff's rhetorical question implying that no one with understanding would assert that it is possible for a triangle to both be possible and impossible, and thus deny that it is necessarily possible (i.e. its essence is necessary), Kant responds with his own, asking 'what person of sound understanding would wish to maintain' that a triangle is actually necessary? Put simply, Kant is here willing to admit that if a triangle is something possible, then it is necessary that it is possible, just as Wolff maintains, but he denies that it follows from this that a triangle is absolutely necessarily possible. This is because even granting it to be possible, it still makes sense to ask *why* it is possible or to ask what ground makes it possible for us to represent a triangle in the first place. The definition does not explain the possibility of representing a triangle, but rather presupposes it. So the necessity demonstrated by Wolff is only a hypothetical necessity, which presupposes the real possibility of a triangle. But couldn't it have been the case that a triangle was impossible? In somewhat different terms, the real possibility of a triangle, which its definition presupposes, is not itself explained or grounded in the definition alone, and so the second step of Wolff's proof, which claims to establish the absolute necessity of essences falls apart. Underlying this criticism, of course, is Kant's distinction between the logical and real elements of possibility, which will be carried over into his later work. According to this distinction, Wolff had illegitimately reduced all possibility to merely logical possibility, i.e. the non-contradictoriness of a thing's essential components. But as Kant notes here, we can still ask about the possibility of these components themselves. What makes these possible? In other words, from where do we derive the real content of our thoughts and on what ground are we able to ascribe them reality? Whereas the logical possibility of a thing requires no ground external to the essence, this real element of possibility must have a further ground. And, it is precisely by relation to this common foundation of their real possibility that Kant thinks it makes sense to expect the very essences themselves will possess an intrinsic and infinite affinity or unlimited reciprocal purposiveness.

As we have seen, this move is essential to Kant's strategy for seeking a ground of unity even for essences in the wisdom of God. More importantly in the present context, it also provides Kant with a strategy for reinterpreting the significance of mathematical purposiveness in his next major philosophical work.

## 4. 'With a Philosophical Eye': mathematical purposiveness in Kant's *The Only Possible Argument in Support of a Demonstration of God's Existence*

This masterwork of 1763 is really a natural theology, a metaphysics of natural science and a cosmogony all wrapped into one comprehensive system. In the first section, Kant provides the definitive statement of his pre-critical theory of essence and possibility, outlines a demonstration of God's existence based on it, argues that all essences must have their common ground in God, derives the most important of God's attributes from his pure concept and finally argues that based on these attributes we should expect only the greatest possible harmony and unity to issue from essences of things themselves, both individually and in conjunction. In the second section, Kant then tries to show in an a-posteriori way, i.e. from our knowledge of creation itself, that it exhibits precisely the kind of superabundant harmony and unity the a priori arguments of the first section would have us expect. 'Our purpose' in this part, Kant explains,

> will be to see whether the internal possibility of things is itself necessarily related to order and harmony, and whether unity is to be found in the measureless manifold, so that, on this basis, we could establish whether the essences of things themselves indicate an ultimate common ground. (*BDG* 02:92)

The heading of the first section informs us that he will thus seek to demonstrate the existence of God '*a posteriori* from the unity perceived in the essences of things', and in particular, 'by appeal to the properties of space' (*BDG* 02:93).

Kant's teleological theory of essences is now ready to bear fruit in a completely original account of mathematical purposiveness. The key idea, of course, is that because essences as such are not absolutely necessary, but rather possess a common origin, it is possible to interpret the purposiveness specifically of mathematical essences in a teleological fashion as well and to infer from these an underlying divine cause. Kant recognizes, however, that this is not apparent to the working mathematician. The mathematicians, he notes, may find many splendid properties in mathematical figures, such as 'the certainty of their conviction, the exactitude of their execution, and the extensiveness of their application' (*BDG* 02:93). But just like Wolff, they unwittingly assume the real possibility of the concepts of such objects and treat them analytically, or with the principle of contradiction, in order to demonstrate particular mathematical truths.

The philosopher, however, sees these same things from another perspective, i.e. from the perspective of their real possibility: 'Looking at it with a philosophical eye', Kant explains, 'I come to notice that order and harmony, along with such necessary determinations, prevail throughout space, and that concord and unity prevail throughout its immense manifold' (*BDG* 02:93). Thus, when looked at 'with a philosophical eye', there becomes visible a particular feature of mathematical forms, i.e. their unique purposiveness, which provides the philosopher with a new *datum* to be explained.

As a first illustration of such purposiveness, Kant cites the Intersecting Chords Theorem, which, as we saw above, he recycles in the third *Critique*. What is so remarkable about the theorem, he explains here, is that 'I have no reason at all to suppose that such a simple construction (*einfältige Construction*) should conceal something highly complex which is itself subject, in virtue of that very construction, to major rules of order' (*BDG* 02:94). The second illustration, which immediately follows, is an unnamed consequence of *Elements*, Book III, Proposition 36, or the so-called Secant Tangent Theorem. This theorem itself states that if two lines are drawn from the same point outside a circle, one of which intersects the circle and the other of which is a tangent, then the square of the length of the tangent is equal to the product of the parts of the other line that intersects the circle. In Figure 3, this means that $DB^2 = (DC) \cdot (CA)$. Kant does not refer to this theorem directly, but to the following consequence[12]: For any given circle, the parts of all lines intersecting in the manner of DA will have the same product of their parts as does *DA*, namely whatever $DB^2$ is. So for any such line *DA*, $DC = DB^2/CA$, where $DB^2$ is a constant for a given circle and point outside it. That is to say that the lengths of the parts of all such lines will always relate inversely, and the precise quantity of this inverse proportion will depend on $DB^2$. But *DB* and *DA*, or length of the tangent and the length of the whole line, are proportional, as is clear from Figure 3,[13] i.e. if *DB* increases or decreases so will *DA*. Thus, the parts of the line will relate inversely to one another also according to the length of the total line, or as Kant says, such lines 'are always divided into parts which are related to each other in inverse proportion as their wholes' (*BDG* 02:94).

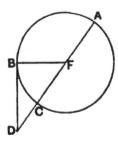

Figure 3.   Kant's first example of mathematical beauty in *The Only Possible Argument*.

'One cannot but be surprised', he then notes,

> in spite of the ease with which these truths are understood, that this figure requires
> so little design (*so wenig Anstalt in der Beschreibung dieser Figur*), and yet that so
> much order and such complete unity in the manifold should arise from it. (*BDG*
> 02:94).[14]

In other words, what is so remarkable here is that such complete purposive unity, expressible in a single general theorem, should arise from a figure that can be described with so little design or purposive arrangement of its own.

A third example Kant silently adopts from Galileo's *Two New Sciences*, Second Day, Theorem 6, Proposition 6: 'If from the highest or lowest point in a vertical circle there be drawn any inclined planes meeting the circumference the times of descent along these chords are each equal to the other' (Galileo 1914, 188–189).[15] In Figure 4, this means that excluding frictional forces, for any chords BA and CA in a circle, a freely rolling body will traverse them in equal times. Galileo provides three proofs of this theorem, but the result can easily be shown to follow from the Pythagorean square rule for the composition of forces, because the impeding force exerted by the plane and the decrease in speed caused thereby, relative to the free fall down *AF*, turn out to be exactly compensated for by the corresponding decrease in the length of the path also relative to the length of *AF*.

After describing this theorem in some detail, Kant recounts how its demonstration caused his students to be as impressed as if they had witnessed a true 'miracle of nature'. And then, he continues to claim that:

> One is, indeed, amazed and rightly astonished to find, in such a seemingly
> straightforward and simple thing as a circle, such wondrous unity of the manifold
> subject to such fruitful rules. Nor is there a miracle of nature which could, by its
> beauty and order, give more cause for amazement .... (*BDG* 02:94)

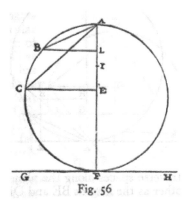

Fig. 56

Figure 4.   Kant's third example of mathematical beauty in *The Only Possible Argument*. *Source*: Original figure from the Italian edition of the *Two New Sciences* as reproduced in Galileo (1914, 189).

Kant's fourth and final example is that the area, $A$, of the annulus between two concentric circles, where the radius of the larger circle is $R$ and the smaller circle is $r$, is $\pi R^2 - \pi r^2$ or $\pi(R^2 - r^2)$ (see Figure 5). But because $\Delta AOB$ is a right triangle, $d^2 = R^2 - r^2$ and so simply $A = \pi d^2$.[16] That is to say, the area of the annulus is equal to that of the circle with radius $d$.

Again, the mathematical purposiveness Kant locates in this particular construction concerns specifically the ease with which it provides the solution to a general problem, which itself comprises an infinite number of special cases. The problem in this instance is that of turning the annulus into a circle. As Kant notes, the annulus and the circle seem to be completely different shapes and so this 'strikes everyone as a difficult undertaking requiring great art for its execution' (*BDG* 02:95). But when the above truth is realized, one 'cannot but be taken aback at the simplicity and ease with which the solution sought is revealed in the nature of the matter itself, requiring almost no effort on my part at all' (*BDG* 02:95). This theorem is striking, in Kant's view, precisely because it means that the annulus is just the same as the circle with radius d, and so turning the annulus into a circle proves amazingly easy to accomplish, despite seeming to require a complex plan. This is a case, Kant explains, 'where what seems possible only as a result of complicated preparation presents itself without artifice (*ohne alle Kunst*), as it were, in the thing itself' (*BDG* 02:95). Like in the first example, Kant interprets the uniqueness of mathematical purposiveness to lie in the fact that figures with little or no design for the purpose or purposes at hand nonetheless fulfil these purposes completely and with exact precision, indeed such that it is expressible in a single general theorem. They serve purposes, necessarily and precisely, without evidently being directed towards them intentionally at all.

Despite the differences in all four examples, they seem to be framed so as to support Kant's specific interpretation of mathematical purposiveness as consisting a feature of theorems that express the necessary harmony between figures that we are nevertheless able to conceive by themselves or as unrelated. In the first example, there is no purely conceptual link between the proportions of the parts of the two intersecting lines and the figure of the circle; in the second,

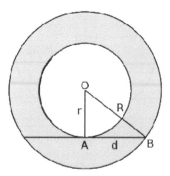

Figure 5.   Kant's fourth example of mathematical beauty in *The Only Possible Argument*.

none between the proportions of the parts of the single line, the lines length and the figure of the circle; in the third, none between the inclines which take equal times to descend and, again, the figure of the circle; and finally in the fourth, none between the shape of the annulus and the shape of the circle. Yet in all cases, in the act of construction, a universal and necessary theorem relating these figures is at once discovered, *as if* they had been arranged for this more general purpose. Thus, as Kant specifically writes in summary,

> the purpose ... has been to draw attention to the existence, in the necessary properties of space, of unity alongside the highest degree of complexity, and of the connection between things *where all seem to have their own independent necessity.* (*BDG* 02:95; emphasis added)

This point is extremely important both for how Kant goes on to use mathematical purposiveness in the following pages of the *Only Possible Argument* and for how it provides a basis for the later discussion in the third *Critique*. For Kant's point is clearly that in each example there are two things that, when considered by themselves or according merely to what is contained in their concepts (e.g., of a circle as 'a space [produced] by moving a straight line around a fixed point'), possess no necessary connection, but which, when actually considered as figures in space itself, prove to possess an infinity of purposive and yet also necessary connections, i.e. to be 'constantly subject to the same law, from which they cannot deviate' (*BDG* 02:94; cf. *BDG* 02:126–127). Using the terminology Kant would only develop later, these theorems are synthetic a priori truths; they go beyond the concepts involved and yet are universal and necessary.[17]

With care we can thus extract from the above examples essentially the same general understanding of mathematical purposiveness that we located in the critical period. Here again three points are repeatedly emphasized: (1) The concept of the figure, e.g. of the circle, is immensely simple and contains no reference to possible ends it might serve. (2) To bring about these consequences would otherwise seem to require incredible preparation and purposive design. (3) Yet an infinity of such consequences can be demonstrated to flow from the figure itself through construction. In both texts, Kant therefore describes what he takes to be an objective form of purposiveness, expressible in a general theorem, that is found really in the mathematical objects under consideration, and specifically thinks of this purposiveness as a kind of unforeseeable (from the concepts involved), but infinite fitness to possible purposes. Because the actual design of something for a determinate purpose is generally called intentional, one could also describe mathematical purposiveness as an unintentional but yet infinite fitness to ends.

Before returning to the third *Critique*, we must examine Kant's assessment of the significance of such mathematical purposiveness in the present context. After noting that common teleological arguments usually infer the existence of a designer from the merely contingent order of nature, he asks:

> Is this harmony any less amazing for being necessary? I would maintain that its necessity makes it all the more amazing.[18] A multiplicity, in which each individual

had its own special and independent necessity, could never possess order, or harmoniousness, nor could there ever be unity in their reciprocal relationships to each other. Will this not lead one, as the harmony in the contingent designs (*Anstalten*)[19] of nature leads one, to the supposition that there is a supreme ground of the very essences of things themselves, for unity in the ground also produces unity in the realm of all its consequences? (*BDG* 02:95–96)

Thus circa 1763, Kant is fully willing to infer the existence of God as the only possible explanation of the source of certain instances of mathematical purposiveness and, on this basis, to regard these instances as particular illustrations of the kind of supreme unity that belongs to the essence of space. This argument depends upon four major points: (1) The connection displayed in certain mathematical theorems goes beyond anything aimed at or captured in the concepts of the figures involved, and thus requires a 'synthetic' basis in an underlying common ground of their possibility. (2) This ground must be constituted so that it can specifically explain the kind of unbounded purposiveness found in the mathematical essences of things. Both of these depend in turn on (3) Kant's having made sense of the thought that essences *could* be dependent on a principle outside of themselves, thus that they are not absolutely necessary. Finally, the inference to the divine being rests on (4) Kant's belief at this time that space is transcendentally real and thus that the explanation of its possibility requires an equally real metaphysical ground.

## 5. With a transcendental eye

We are now prepared to close the circle by returning to the third *Critique*. As we now know, Kant had already rehearsed much of the material for his discussion of mathematical purposiveness in §62 some 30 years earlier during his pre-critical period. We have seen in particular that Kant recycles for this purpose the Intersecting Chords Theorem, and we can now also conjecture with some confidence that his consistent but peculiar reference to 'the essences of things' in §62 derives from the pre-critical context in which his thoughts on this general topic first emerged. We have also seen that, in both the critical and pre-critical treatments, mathematical purposiveness has been held up as an exemplary and even emotionally striking case of necessary synthetic truths, which express an infinite and most precise relation to purposes that are yet not contained analytically in the concepts of the figures concerned, i.e. 'purposive solutions that were not thought of at all in the rule that constitutes their construction' (*KU* 05:363). That is to say, we can now clearly see that the issue of mathematical purposiveness is intimately related to the problem of synthetic a priori truths.

But there is much more than this that both binds together and separates Kant's pre-critical and critical discussions of this topic. In the earlier of the two, we saw that Kant sought to infer the existence of God, as the unifying ground of space, precisely from the synthetic and necessary character of such purposiveness. As one might expect, the critical treatment thus attempts to explain the deception involved in such an inference. But rather than simply recalling his own previous position, Kant hits upon rather curious device of placing his earlier views in the mouth of another philosopher of at least equal renown. 'Plato', Kant explains:

> himself a master of this science [i.e. mathematics], was led by such an original constitution of things, in the discovery of which we can dispense with all experience, and by the mental capacity for drawing the harmony of things out of their supersensible principle (to which pertain the properties of numbers, with which the mind plays in music), to the enthusiasm that led him beyond the concepts of experience to ideas, which seemed to him explicable only by means of an intellectual communion with the origin of all things. ... He thought he could derive that which Anaxagoras inferred from objects of experience from the pure intuition internal to the human mind. For in the necessity of that which is purposive and so constituted as if it were intentionally arranged for our use, but which nevertheless seems to pertain to the essences of things, without any regard to our use, lies the ground of the great admiration of nature .... (*KU* 05:363)

Of course, as an interpretation of Plato, what Kant says here is dubious at best,[20] particularly because the specific understanding of mathematical purposiveness he attributes to Plato is Kant's own original creation. But read as a reflection on Kant's own earlier position, it offers us a unique insight into how he views his own intellectual development. Such 'enthusiasm', Kant tells us, was 'surely excusable' because it was based on an unavoidable deception that has its genuine foundation in our reason. The rules or theorems that exhibit this purposiveness

> are one and all synthetic and do not follow from a *concept* of the object, e.g., from that of a circle, but need the object to be given in intuition. But it thereby comes to seem as if this unity empirically possesses an external ground, distinct from our power of representation, for its rules, and thus as if the correspondence of the object with the need for rules, which is characteristic of the understanding, is in itself contingent, hence possible only by means of an end expressly aimed at it. (*KU* 05:364)

That is to say, the *empirical* independence of intuition from our concepts, and thus from the understanding, unavoidably makes it look, even to the 'philosophical eye' of a Plato, as if the unity and purposiveness found in objects of intuition for our understanding have an external source. Notably, this already implies that Plato (or rather the pre-critical Kant) clearly recognized the synthetic a priori character of mathematical purposiveness. He went astray, however, by thinking that the empirical independence of thought and concept requires the inference of a transcendentally real external ground to account for it.

But as Kant goes on to explain, having the eye of a philosopher is not enough to avoid this error, excusable though it may be; one must look at the matter through 'a critical view of reason', i.e. with the eye of the *transcendental* philosopher, who sees it essentially in light of the question as to how such truths can be possible in the first place. Such a transcendental-philosophical gaze, according to Kant, would see immediately that the a priori character of these truths means that they cannot be about 'a property of the object outside of me, but merely a kind of representation in me, and thus that I *introduce* the *purposiveness* into the figure that I draw *in accord with a concept*' (*KU* 05:365). By this Kant does not mean to say, however, that the purposiveness here is a mere product of the unity of the understanding, somehow inserted transcendentally into the manifold of intuition. The unity of such truths is not found in the concepts or indeed in the understanding at all, even transcendentally. So Kant admits that the admiration of such purposiveness

> cannot be criticized insofar as the compatibility of that form of sensible intuition (which is called space) with the faculty of concepts (understanding) is not only inexplicable for us insofar as it is precisely thus and not otherwise, but also enlarges the mind.... (*KU* 05:365)

The real point is therefore not that such purposiveness is a mere product of the transcendental understanding, but rather that it is not a purposiveness between objects in a transcendentally real space. What it is, much rather, is a purposiveness between two things transcendentally within us, namely, our own a priori form of sensible intuition and our own a priori forms of understanding. The original error that led to the inference of a divine cause was thus based only in part on the failure to recognize the transcendental ideality of space. It was also based on the *correct* recognition of the unique character of the mathematical purposiveness, which, if interpreted correctly, would have led to the recognition of a purposiveness between our own transcendental faculties of intuition and understanding that is genuinely inexplicable, and which ought to be an object not only of admiration and wonder, but also of a whole new science, namely, transcendental philosophy.

In *On a Discovery*, which appeared the same year as the *Critique of the Power of Judgment*, Kant elaborates on this remarkable harmony between sensible intuition and the understanding, ascribing the recognition of it now not only to Plato, but also to Leibniz. Leibniz's pre-established harmony, Kant suggests in this text, was at bottom an attempt to express the harmony

> only of the mental powers in us, sensibility and understanding, each in its own way for the other, just as the [first] *Critique* teaches that for the *a priori* cognition of things they must stand in a reciprocal relationship to one another in the mind. (*ÜE* 08:250).

According to Kant, the *Critique of Pure Reason*

> has definitely shown that without it [i.e. this harmony between sensibility and understanding] no experience is possible ... but could still provide no reason why we have precisely this mode of sensibility and an understanding of such a nature, that by their combination experience becomes possible; nor yet, why as otherwise fully heterogeneous sources of cognition, they always conform so well to the possibility of cognition in general.... (*ÜE* 08:249–250)

As is clear from the continuation of the passage, what Kant says here is supposed to apply most generally to the relation of sensibility and understanding insofar as this is required for the possibility of experience. Thus, it also includes, though it does not explain, the harmony of intuition and understanding that is required for the cognition of mathematical forms in space and time, which are also the forms of possible objects in experience. Although the *Critique* demonstrates that we must presuppose such harmony, it still does not allow us to anticipate the supreme degree of harmony that becomes evident through the kinds of mathematical examples that Kant cites in the illustrations.

We have now completed our analysis of §62 of the third *Critique*, which is Kant's central treatment of mathematical purposiveness. This is not, however, the end of the larger story, either for this concept itself, or for Kant's attempts at explaining through Plato the revolution that took place in his own understanding of mathematical purposiveness as a result of the transcendental turn. In the essay 'On a Recently Prominent Tone', published 6 years later in 1796, Kant greatly expands this use of Plato beyond what can be discussed here, introducing also Pythagoras so as to generalize these observations to both space and time, and so to both geometry and arithmetic. In the same text, Kant also more clearly explains how Plato mistook sensible intuition for an intellectual intuition of things themselves, and reiterates that mathematical purposiveness must remain a matter of amazement even to the transcendental philosopher. The difference, however, is that, unlike the ordinary philosopher, the transcendental one realizes that such purposiveness is a presupposition of the possibility of experience, and so although it cannot be reduced to anything less amazing, any attempt to explain its origin can only lead to ungrounded enthusiasm, 'because its [i.e. reason's] oracle falls silent, once the question has been elevated so high that it now no longer has any meaning' (*VT* 08:393). Finally, as if he were trying to directly confirm my thesis here that the wonders of mathematical purposiveness were central to his discovery of the synthetic a priori, Kant writes:

> In all these inferences, Plato at least proceeds consistently. Before him there undoubtedly hovered, albeit obscurely, the question that has only lately achieved clear expression: "How are synthetic propositions possible *a priori*?". (*VT* 08:391n)

## 6. Conclusion

This paper provides an historical explanation as to why Kant points to mathematics as a prime example of the synthetic a priori throughout his critical writings, while at the same time revealing the specific importance of mathematical purposiveness to the early development of this notion. Indeed, it seems quite evident that there can be no clearer or more striking example of the existence of the synthetic a priori than this form of objective formal purposiveness, because it lies precisely in a unity that *infinitely* exceeds the unity contained merely in the definitions of the mathematical figures themselves. Thus, although there is no way to estimate precisely how significant this might have been for Kant's development, the fact that he found certain theorems concerning mathematical figures and constructions to exhibit such a super-abundance of necessary purposiveness, which he thought could yet not be derived merely from their individual concepts, quite obviously served as a major motivation for, and later as a confirmation of, the thought that many mathematical theorems require a deeper common ground of unity, and thus are not analytic but synthetic in character.

The analysis contained in this paper also brings to light a particularly unique example of the unity of synthetic a priori knowledge, which may have significant implications for the discussion of Kant's views on non-conceptual content. It is unique in this respect, because it provides perhaps the only clear illustration of how this synthetic unity really must be understood as the source of a kind of absolute affinity and harmony of objects in space, such that we should always expect there to be more unity in actual experience than we can anticipate merely through any finite concept or set thereof. It thus indicates that not only are these truths synthetic, but that this synthesis should not be understood even *transcendentally* to be a mere effect of understanding on sensibility, but rather as evincing an inexplicable harmony or fitness between sensibility and understanding that exceeds any finite conceptual determination. Thus, unlike the usual examples of synthetic a priori knowledge, it shows that the unity underlying them is not merely the result of the application of conceptual unity to the manifold of intuition. Indeed, Kant's precise point in the passages we analysed from the third *Critique* is that this unity is not found in the concepts, but only in the combination of concepts with intuitions in us, the possible compatibility of which is nevertheless really an inexplicable cause for wonder. This indicates that the forms of intuition not only serve to realize, in constructions, the conceptual unity found in mathematical concepts, but indeed that the compatibility between intuition and understanding must make an original transcendental contribution to the *unity* found in the resulting constructions. It should be noted that the irreducibility of this unity either to intuition or to concept taken separately also clearly makes it impossible to fully reformulate the resulting truths as definitions, thereby directly providing a refutation of the thought that all synthetic truths could be rendered analytic by such a process. This is because mathematical purposiveness is seen by Kant as pertaining to the essential ground of mathematical objects in a way that *infinitely* exceeds not only what could be captured in any finite set of concepts, but even what could be demonstrated in any finite set of constructions, and thus captured in any finite definition.[21]

## Acknowledgements

I thank the editors of this special issue, the members of the Fox Center at Emory University and ENAKS, as well as Oliver Thorndike, Dilek Huseyinzadegan, Rudolf Makkreel and Mark Risjord for providing helpful critical comments on earlier drafts of this paper.

## Notes

1. It must be kept in mind that in the *KU*, Kant does not want to apply the term 'beauty' in a strict sense to mathematical objects or propositions. However, because the rather technical considerations motivating this typology do not figure in Kant's earlier work, he does speak there of the *'Schönheit'* of mathematical figures (e.g. *BDG*

02:94–95). I will use the term 'mathematical beauty' in this less strict sense, except in the context of the *KU*, where I will employ the more precise locution 'formal objective purposiveness'. As we will see, this difference in terminology does not arise from a change in Kant's view of mathematical purposiveness, but from a refinement in his understanding of other forms of beauty. All translations from the Critique of the Power of Judgment are from Kant (2002) unless otherwise noted.

2. This is how commentators uniformly treat the passage, when they discuss it at all. For some recent examples, see Nuzzo (2005, 330–332) and Wicks (2007, 189–193).

3. This diagram was drawn using the software GeoGebra 4.4.41.0.

4. This diagram is from Euclid (1908, vol. 2, 71).

5. It is not entirely clear whether the sentence 'The other curves … construction.' (*'Die andern krummen Linien … gedacht war.'*) introduces another example or is rather the conclusion to the second. Here I take the latter to be the case.

6. Cf. Nuzzo 2005, 332.

7. They are analytic in the sense that the perfection of, e.g. a circle, consists in its possessing precisely and only the features contained analytically in the general concept of a circle. Likewise, the perfection of something produced intentionally is judged purely by a logical comparison of it with the concept of what was to be produced. So, that a figure is or is not perfect in this sense is merely an analytic judgement. Kant speaks of such analytic perfection, e.g. at B113–B116.

8. It is very likely that what Kant has in mind here is something like a purely mathematical analogue of a higher-order physical law, such as de Maupertuis' Principle of Least Action. What both de Maupertuis and the young Kant found so remarkable about such a principle was that it is unified, in a single universal law and in one principle, so many diverse individual laws, which until that point had been demonstrated from entirely independent grounds.

9. Wolff offers a similar proof in his German Metaphysics (cf. Wolff 1747, §38), but as we will see below, it is this passage from the Latin Ontology in particular that Kant has in mind when later criticizing this doctrine.

10. An internal determination not grounded in any of the remaining components of the essence would have itself to be a component of the essence for this very reason.

11. Leibniz first announced his discovery of this series in the paper '*De vera proportione circuli ad quadratum circumspriptum in numeris rationalibus expressa,*' which appeared in the *Acta Eruditorum* 15 September 1713. See Leibniz (1858, 118–122). Wolff later discusses the series in a letter to Leibniz, mentioning how some involved in the controversy over the invention of calculus were stating that the series had previously been discovered by the English mathematician James Gregory (1638–1675) (Leibniz and Wolff 1860, 152–153). The series is also discussed in Wolff (1738, 182–183).

12. This has not been previously noted by commentators. Walford and Meerbote seem to suggest Kant is referring to the Secant Tangent Theorem itself, because their explanatory note gives only reference to it (Kant 1992, 431, n. 42). Kreimendahl and Oberhausen, 173, n. 117, explicitly follow Walford and Meerbote, and do not provide any further explanation. But the theorem Kant describes is clearly not the Secant Tangent Theorem, although it is related to it as described.

13. The diagram is from Euclid (1908, vol. 2, 74).

14. I have altered the Cambridge Translation here, which by using 'straightforward' obscures the significance of the term Kant uses, namely '*Anstalten,*' and masks its relation to the later occurrence of this term in the same passage, as we will see below. This term occurs throughout the *BDG*, and always means a kind of purposive arrangement or design. The Cambridge Translation translates this word in other contexts variously as 'design' and 'provisions.' See *BDG* 02:127–137 passim.

15.  Again, no other commentator has noted the source of this example, which in this case is historically quite significant, because it suggests that Kant was familiar with Galileo's book. Indeed, I think that if Kant had simply learned it as a standard part of physics, it would be quite a coincidence for him to choose it as one of his examples. Most likely his choice is motivated by a desire to demonstrate his knowledge of Galileo and his sophistication at not needing to mention him by name. Walford and Meerbote provide a modern demonstration in Kant (1992, 431, n. 43). Kant (2011, 173, n. 118) again follows Kant (1992).

16.  No explanation of this is given in either Kant (1992) or Kant (2011). Of course, the mathematics in this case is elementary and of no particular historical significance. Yet the point Kant makes with it is important for our purposes. The diagram given here is in the public domain and copied from Wikipedia.

17.  This does not conflict with the fact that this section is supposed to provide an a-posteriori demonstration of God's existence, because Kant is using the term here in its traditional meaning of moving from the effects to the causes, not from causes to effects, which would be a priori in this way of speaking. This is essentially different from Kant's later use of the term to mean what is cognized independently of experience.

18.  Kant seems to be echoing the language of de Maupertuis, who he cites in the next section:

     If it is true that the laws of motion and rest are indispensable consequences of the nature of body, that proves all the more the perfection of the supreme being: It shows that everything is ordered such that blind and necessary mathematics executes what is prescribed by the most enlightened and free intelligence. (1751, 65–66)

     I think it is reasonable to assume that de Maupertuis' teleological interpretation of the Principle of Least Action, as well as his related criticisms of all supposed teleological proofs of God's existence from the contingent design of nature, had as much influence on Kant's views on mathematical purposiveness as did Hutcheson's writings. However, as discussion of this would take us rather deeply into the field of physics, it must be omitted here.

19.  I have again changed the translation of '*Anstalten*' from 'provisions' as in the Cambridge Translation, to 'design,' so as to keep the link with the use of the term earlier in this passage.

20.  Fistioc (2002) seems to overlook this point, taking Kant's remarks as genuinely about the historical Plato. See esp. 36–43.

21.  Thus, although it might seem as if we could render any given mathematical truth analytic by redefining its subject so as to include its predicate (e.g. render '$2 = 1 + 1$' analytic by defining '2' as meaning '$1 + 1$'), Kant thinks it is true of most if not all mathematical concepts that the number of such truths would in fact prove to be inexhaustible. So the process of redefinition could never be completed. One familiar with the wealth of purposive relations in mathematics would thus at once realize mathematical truths to be non-analytic. Of course, the synthetic character of such truths is not dependent upon the failure of such a process, although its failure does seem to indicate that there is something in these truths that *essentially* exceeds what can be captured in an analytic judgement.

# References

Unless otherwise noted, all translations of Kant's writings are from the Cambridge Translation of the Works of Immanuel Kant. In referring to Kant's writings in German, I have used the following abbreviations: AA=Kant 1902, *BDG=Der einzig mögliche Beweisgrund zu einer Demonstration des Daseins Gottes* (AA 02), *KU=Kritik der Urtheilskraft* (AA 05), *PND=Principiorum primorum cognitionis metaphysica nova dilucidatio* (AA 01), *Refl=Reflexionen zur Metaphysik* (AA 17), *ÜE=Über eine Entdeckung* (AA 08), *VT=Von einem neuerdings erhobenen vornehmen Ton in der Philosophie* (AA 08).

de Maupertuis, Par M. 1751. *Essai de Cosmologie* [Essay on Cosmology]. Published in German the same year as *Versuch einer Cosmologie* Berlin: C.G. Nicolai.

Euclid. 1908. *The Thirteen Books of Euclid's Elements*. Translated with introduction and commentary by T.L. Heath. 3 vols. Cambridge: Cambridge University Press.

Fistioc, Mihaela C. 2002. *The Beautiful Shape of the Good: Platonic and Pythagorean Themes in Kant's Critique of the Power of Judgment*. London: Routledge.

Galileo, Galilei. 1914. *Dialogue Concerning Two New Sciences*. Translated by Henry Crew and Alfonso de Salvio. New York: Macmillan.

Giordanetti, Piero. 2008. "Objective Zweckmässigkeit, objective und formale Zweckmässigkeit, relative Zweckmäwwigkeit [Objective Purposiveness, Objective and Formal Purposiveness, Relative Purposiveness]." In *Kritik der Urteilskraft*, edited by Otfried Höffe, 61–63. Berlin: Akademie Verlag. Chapter 12.

Kant, Immanuel. 1902. *Kants gesammelte Schriften* [Kant's Collected Writings]. 29 vols. Issued by the Prussischen Akademie der Wissenschaften (vols. 1–22), the deutchen Akademie der Wissenschaften (vol. 23), and the Akademie der Wissenschaften zu Göttingen (vols. 24–29) Berlin: de Gruyter.

Kant, Immanuel. 1992. *Theoretical Philosophy, 1755–1770*. Translated and edited by David Walford in collaboration with Ralf Meerbote. Cambridge: Cambridge University Press.

Kant, Immanuel. 2002. *Critique of the power of judgment*. Edited by Paul Guyer and translated by Paul Guyer and Eric Matthews. Cambridge: Cambridge University Press.

Kant, Immanuel. 2011. *Der einzig mögliche Beweisgrund zu einer Demonstration des Daseins Gottes* [The Only Possible Argument for a Demonstration of God's Existence]. Philosophische Bibliothek, vol. 631. Edited by Lothar Kreimendahl and Michael Oberhausen. Hamburg: Felix Meiner Verlag.

Leibniz, G. W. 1858. *Leibnizens mathematische Schriften* [Leibniz's Mathematical Writings]. Part 2, vol. 1. Edited by C. I. Gerhardt. Halle: Schmidt.

Leibniz, G. W. 1996. *Theodicy*. Edited with an introduction by Austin Farrer and translated by E.M. Huggard. La Salle, IL: Open Court.

Leibniz, G. W., and Christian Wolff. 1860. *Breifwechsel zwischen Leibniz und Christian Wolff* [Correspondence between Leibniz and Christian Wolff]. Edited by C. I. Gerhardt. Halle: Schmidt.

Nuzzo, Angelica. 2005. *Kant and the Unity of Reason*. West Lafayette, IN: Purdue University Press.

Wicks, Robert. 2007. *Kant on Judgment*. London: Routledge.

Wolff, Christian. 1736. *Philosophia prima, sive Ontologia, methodo scientifica pertractata* [First Philosophy, or Ontology, treated with the Scientific Method]. New edition Frankfurt: Libraria Rengeriana.

Wolff, Christian. 1738. *Der Anfangs-Gründe aller Mathematischen Wissenschaften* [The First Principles of all Mathematical Sciences]. Part 1. Fourth improved and expanded edition. Frankfurt and Leipzig: Rengerischen Buchhandlung.

Wolff, Christian. 1741. *Vernünfftige Gedancken von den Absichten der natürlichen Dinge* [Rational Thoughts on the Purposes of Natural Things]. 4th ed. Halle: Rengerischen Buchhandlung.

Wolff, Christian. 1747. *Vernünfftige Gedancken von Gott, der Welt und der Seele der Menschen, auch allen Dinging überhaupt* [Rational Thoughts on God, the World and the Soul of the Human Being, and all Things in General]. New and enlarged edition. Halle: Rengerischen Buchandlung.

# Index

Note: Page numbers in **bold** represent figures
    Page numbers followed by 'n' refer to notes

*a posteriori* concepts 100
*a priori* concepts 39, 62–3, 80, 81n, 119
*a priori* intuition 119, 208, 247
*a priori* knowledge 63; synthetic 241
abstract concepts 132
abstract reasoning 142–4
abstract representation 152
acceleration: instantaneous 169, 178
acquisition: conditions of concepts 2;
    of geometrical concepts 2, 62–86;
    knowledge 62, 68
acute-angle cone 212
Adickes, E. 108n
*Admonitio* (Borelli) 236n
aesthetic purposiveness 243
aestheticism 34, 38–41, 46, 50–2n, 54n,
    59n, 203
algebra 142–54, 206, 215; geometrical 215
algebraic equations 206, 215–16
algebraic reasoning 146
algebraic techniques 206
algebraic treatment 205–6, 215–17
algorithms 154
Allison, H. 48n, 52n
*Analytic of Concepts* (Kant) 122
analytic geometry 205–7
anatomy 122
ancient geometers 247
angle: vertical 212–13

Apollonius of Perga 206–17, 221–5,
    228–35, 235–8n; *Conica* 206–15,
    221–5, 228–9, 233, 235–6n, 238n;
    theory of conic sections 4, 201–40
*Apollonius of Perga's Conica* (Fried and
    Unguru) 206
apprehension: instantaneous 187, 192–4
Arabic language 236–7n
Arabic numerals 154–6
arbitrariness 152–3, 159
arbitrary combination 140–67
arbitrary synthesis 69, 72, 137n, 158, 164n
Archimedes 236n
Aristaeus 212–14, 231–2
Aristotle 13, 121, 137n, 205
arithmetic 152, 155; modern formula 156
arithmetical concepts 157, 160
arithmetical identities 154
arithmetical magnitudes 148
arithmetical thought 155
astronomy: observational 226–7
atomic containment 35
Augustine 65
axial triangle 209–14
axiomatic concepts (*Grundbegriffe*) 104,
    110n
axiomatic truth 94
axioms 89, 98–107, 110n, 149; genuine
    94; geometrical 87, 89; of geometry 89;

of intuition 11, 18, 21, 27n, 197;
mathematical 106; of parallels (Euclid)
89; syntheticity of geometricals 89

beauty 242, 266n; mathematical 250, **258,
259, 260**, 267n
beauty of mathematical truths (Wolff)
249–53
Beck, L.W. 135n, 137n
*Berliner Monatsschrift* 219
biangles 135–6n
biological taxonomy 15
blind cognition 142
Borelli, G.A. 105, 236n; *Admonitio* 236n
boundary-determination 114–16
Brittan, G. 237n
Buroker, J.V. 48n

Callanan, J.J. 2, 62–86
Carson, E. 43, 48n, 54n, 56n, 151–3, 158;
and Shabel, L. 1–5
Cartesian coordinates 44
Cassirer, E. 83n, 227
categories 3; definitions 113–39; pure
126–8; schematized 126–32
characterization of space (Kant) 15
circle 87–90, 102–6, 108n, 115–16,
136n, 244–6, 258–62, 267n; Euclid's
definition 93–4, 95, 97; real definitions
3, 92–107, 107–9n, 118, 129, 134n,
138–9n
circle chords 223
closed orbits 172–3
cognition 120–2, 125–6, 136n,
151–2, 204–5, 220–1, 264; blind
142; discursive 67; human 122;
mathematical 4, 12, 33, 79, 150, 168,
197–8; philosophical 150, 168; rational
3, 91, 102, 107n; symbolic 142, 162n;
two stems of 1, 6–32
cognition theory: Kant 169; Wolff 148
cognitive activity 40
cognitive capacity 6, 17, 25
cognitive contribution 38

cognitive faculties 89, 212
cognitive resources 201–2, 228–32
combinatorial characteristic 145
composition of magnitudes 180
composition of motion 168–200
concepts 2–3, 113–15; *a posteriori*
100; *a priori* 39, 62–3, 80, 81n,
119; abstract 132; acquisition
62–4, 72–6; acquisition conditions
2; arithmetical 157, 160; axiomatic
104, 110n; construction of 90–1;
derived (*Lehrbegriffe*) 99, 104, 110n;
elementary 122; empirical 67, 71–2,
77, 114, 130, 135n, 137n; exhaustive
114; fundamental 121–4; geometrical
2, 62, 67, 80–1, 83–4n, 88, 105,
117; hypothetical 99; mathematical
71–2, 81–2n, 90–3, 97–101, 107n,
151–3, 157–61, 163n, 230–1, 268n;
metaphysical 146; philosophical 113,
157; physical 4; pure 122, 133, 138n;
schema (Kant) 92; sensible 92; species
124; subordinate 126; technical 135n
conceptual representation 2, 33–5, 39, 48n
conceptualism 7; non- 7
cone: right-angle 212
conic sections 4, 100, 201–40; Apollonius'
theory 4, 201
conic surfaces 208, 213
*Conica* (Apollonius) 206–15, 221–5,
228–9, 233, 235–6n, 238n
conjugate hyperbolas 214, 224
constructability: of motion 179; of rest
179; of speed 179
construction 3, 7–8, 22–4, 170; of concepts
90–1; mathematical 3, 7–8, 22–4, 170;
of motion 168–200
construction theory (Kant) 22, 29n
contingent order of nature 261
continuous velocity 175
Cooperation Requirement 6–7, 24–5
Copernican Turn (Kant) 113, 119–21,
130, 133
corollaries 99; practical 99, 102–3

Corr, C. 163n
cosmology 257
Critical Philosophy (Kant) 80
*Critique of the Power of Judgement* (Kant) 203, 241–3, **245**, **246**, 264
*Critique of Pure Reason, The* (Kant) 1–4, 26n, 28–9n, 33–5, 47n, 50n, 52n, 54n, 59n, 67–72, 81–2n, 107n, 134n, 153–5, 158–61, 226–9, 232–4, 235n, 237n; *The Discipline of Pure Reason* 38, 62
curvilinear motions 173

de la Hire, P. 226–7, 235
de Maupertuis, P.M. 267–8n; *Principle of Least Action* 267–8n
definitions 90–1, 95–100, 118, 158; geometrical 3, 88; immediacy condition on real 103–7; of Kant's categories 113–39; logical 126, 139n; nominal 117, 126, 129, 137n; of parallel lines (Euclid and Leibniz) 94–7, *see also* real definitions
demonstrative knowledge 72
derived concepts (*Lehrbegriffe*) 99, 104, 110n
Desargues, G. 226, 235
Descartes, R. 55n, 97, 161n, 206, 237n; *La Géométrie* 206; *Sixth Meditation* 161n
determinate spaces 2, 18, 24, 43
determination: reciprocal 124
Detlefsen, M. 164n
developmental psychology 82n
diagrammatic representations 73
dichotomy 33
*Die Lehre* (Zeuthen) 237n
differences: numerical 15-16 28n; qualitative 14–16, 28n
Dionysius II 219
*Discipline of Pure Reason* (Kant) 38, 62
discontinuous motion 174
*Discourse on Metaphysics* (Leibniz) 97
discrimination 45
discursive cognition 67
discursive representation 33, 37–9, 67

discursive unity 10
distance: spatial 183
double-branched hyperbola 211, 224, 235n
Dunlop, K. 3, 83n, 108n, 135n, 140–67

Eberhard, J.A. 136n, 207–12, 215–16, 236n
eighteenth century 88–9, 94–5, 104–5, 109n, 161n, 163n, 168–9, 176–8, 228
ekthesis 230
*Elementa Matheseos Universae* (Wolff) 148
elementary concepts 122
*Elements* (Euclid) 23, 72–3, 87–9, 149, 190, 221–4, 246, 258
empirical concepts 71–2, 77, 114, 130, 135n, 137n; acquisition 67
empirical intuition 3, 16–18, 70, 124, 132, 141, 160; Kant's theory 6
empirical objects 133, 135n, 160
endpoints 171–4; fixed 176
Engfer, H-J. 141
English language 137n
Engstrom, S. 54n, 135n
epistemological doctrines 1
epistemology 80
equality 87, 162n, 182, 197, 225
equality relation 180–1, 185, 197
equilateral triangles 103–6
*Essay on the Criterion of Truth* (Lambert) 101
essences: necessity of (Wolff) 253–7; new positive theory of 255
essential unity 42–6
Euclid 27n, 30n, 57n, 72–4, 107–9n, 212–14, 231–2, 237n, 267n; axioms of parallels 89; circle definition 93–4, 95, 97; *Elements* 23, 72–3, 87–9, 149, 190, 221–4, 246, 258; geometrical proofs 87; parallel lines definition 94–7, 98–103; third postulate 93, 102
Euclidean geometry 13, 29n
Euclidean theorem 203
Eudemus of Pergamum 224

Eudoxus 190; theory of magnitudes 190; theory of proportions 13

Eutocius of Ascalon 231–3, 236n

Evans, G. 20; primitive theory of space 20

ever-diminishing intervals 178; of space 184; of time 184

ever-diminishing magnitudes 178

ever-diminishing speeds 186

Ewing, A.C. 48n

exhaustive concepts 114

exhaustiveness 114

explication of motion (Kant) 170–1

expression: symbolic 148

extended interval of space 171, 174–9, 188, 191, 195–8

extended interval of time 174–7, 188, 191, 195–8

extended magnitude 176

extensive magnitude 125, 186–94, 197

extramental objects 140

extramental reality 153

Falkenstein, L. 48–9n, 54n

fallibilism 142

falling bodies: law of (Galileo) 205

First Critique (Kant) 113, 132, 135–6n

first space argument 29n

Fistioc, M.C. 268n

fixed endpoints 176

fluxions theory (Newton) 199n

formal objective purposiveness 267n

formal sensibility 228

formalism 140, 153–4, 160

forms: medieval theory of intension and remission 187, 194–6

fourteenth century 173, 187

fourth century 149

Fourth Space Argument 9

France 220; Jacobin Revolution (1789) 220; Paris 187

Frege, G. 15, 114

Fried, M.: and Unguru, S. 206

Friedman, M. 29–30n, 58n, 84n, 87, 92, 170, 185–9, 196–7, 198–9n, 227;

Kant and the Exact Sciences 202; Metaphysical Foundations 188, 200n

Fugate, C. 4, 241–70

functions 94

fundamental concepts of nature 121–4

fundamental truths 151

galaxy orbit 226

Galileo Galilei 188, 195, 204, 212, 218, 259, 268n; law of falling bodies 205; Two New Sciences 259

Gardner, S. 48n

generality 35–6, 76–81, 102, 159

genuine axiom 94

genus-species relationship 146

geometer 22, 29–30n, 73–6, 92, 102–3, 208, 216; ancient 247

geometrical algebra 215

geometrical analysis of situation (Leibniz) 87

geometrical axioms: syntheticity 87

geometrical characterization 215, 236n

geometrical concepts 2, 62, 67, 80–1, 83–4n, 88, 105, 117; acquisition of 2, 62–86

geometrical construction 2–4, 62, 211, 214–17, 237n

geometrical definitions 3, 88

geometrical knowledge 66, 81

geometrical locus 244–5

geometrical objects 248, 252

geometrical proofs (Euclid) 87, 97

geometrical reasoning 161n

geometrical representation 163n

geometrical treatment 216–17

geometrical truths 97, 140

Géométrie, La (Descartes) 206

geometry 43–4, 50n, 56n, 65–7, 81n, 84n, 136n, 142–5, 181–2, 215–17, 225–7, 238n; analytic 205–7; axioms 89; Euclidean 13, 29n; Greek 207; philosophy of (Lambert) 88; projective 207, 226–8; real definitions 87–112

Germany 94–5, 104; Königsberg 37, 94

gravitation: law of (Newton) 204, 218; universal 169

gravity: terrestrial 188

Greek geometry 207

Greek theory of magnitudes (Eudoxus) 190

Gregory, J. 267n

Grünbaum, A. 57n

*Grundbegriffe* (axiomatic concepts) 104, 110n

Guyer, P. 48–9n, 158, 163n, 199n

Haag, J. 27n

Heath, T. 206; *History of Greek Mathematics* 206

Hegel, G.W.F. 56n

Heiberg, J.V. 210, 230

Heis, J. 2–3, 87–112, 227

herky-jerky motion 172–3

Hertz, H. 93

heterodoxy 33

Heterogeneity Thesis 6–8, 12, 15–17, 22–5

hexagons 225

Heytesbury, W. 188

higher-order proposition 248

Hintikka, J. 141, 154

*History of Greek Mathematics* (Heath) 206

holistic containment 35

holistic mereological structure of space 42–6

holistic structure to space 41

homogeneity 159; strict logical 12–21

homogeneous plurality 129

human: cognition 122; ideal 37; intuition 59n; knowledge 54n, 219; sensibility 19, 202, 219

Hume, D. 75, 80, 229

hylomorphic terminology 8

hyperbola 213–14, 221, 230, 236n; conjugate 214, 224; double-branched 211, 224, 235n

hypothetical concept 99

hypothetical motion 177

ideal human 37

idealism 57n

ideality: transcendental 1

identity: conditions 171; numerical 14; relation 185

imagination: productive 6, 12, 17–21, 25, 27–8n; visual 79

immediacy condition on real definitions 103–7

immunization: philosophical 219

*Inaugural Dissertation* (Kant) 82–3n

indexicals: spatial 19; temporal 19

infinitesimal 176–8

inner intuition 12

*Inquiry Concerning the Distinctness of the Principles of Natural Theology and Morality* (Kant) 64–7, 72, 77, 140

instantaneous acceleration 169, 178

instantaneous apprehension 187, 192–4; intensive magnitude 192–4

instantaneous motion 170, 173, 177–9, 182–9, 194, 199–200n; medieval philosophy 187–8

instantaneous quality 188

instantaneous rest 177–9, 185–6, 189, 200n

instantaneous speed 169–70, 173, 176–8, 184–9, 194–7, 199n

instantaneous velocity 168–9, 174, 186–9, 194–6, 199n; in Newtonian physics 196

instants of time 174–8

intellectual development (Kant): role of mathematical beauty 4, 241–70

intellectual intuition 238n, 265

intellectual representation 11

intellectual snobism 220

intensive magnitude 178, 186–96; instantaneous apprehension 192–4

Intersecting Chords Theorem 246, 258, 262

interval of motion 179

interval of space 171–3, 176–91, 195; ever-diminishing 184; extended 171, 174–9, 185, 188, 191, 195–8

interval of time 173–90, 195; ever-diminishing 184; extended 174–7, 185, 188, 191, 195–8
intervals: spatial 185
intuition 38–42, 48–52n, 54n, 59n, 73–9, 81n, 84n, 90–3, 120–33, 208–9, 263–6; ; a priori 119, 208, 247; empirical 3, 6, 16–18, 70, 124, 132, 141, 160; functional conception (Kant) 40–2; human 59n; inner 12; intellectual 238n, 265; Kantian 1; outer 13, 16–18, 24 pure 21–2, 27–8n, 35, 53n, 89, 140–1, 158–60, 247–8, 263; sensible 6–9, 18, 26n, 45, 57n, 59n, 79, 155, 264–5; singularity 46–7; spatial 21
intuition-dependent 7
intuitive knowledge 72–3
intuitive origin of spatial representation 33–61
intuitive representation 2, 33–5, 41–2, 45, 48n
intuitive thought 194
inverse-square law 205

Jacobin Revolution (France) (1789) 220
Jäsche, G.B. 134n
judgment theory (Kant) 21
judgments: mathematical 2, 71, 87; teleological 4
justification conditions 2

Kant and the Exact Sciences (Friedman) 202
Kästner, A.G. 50n, 76, 106
Keill, J. 57n
Knorr, W. 224
knowledge: a priori 24, 63; acquisition 62, 68; demonstrative 72; geometrical 66, 81; human 54n, 219; intuitive 72–3; justification 62; mathematical 21–3, 63, 71, 81n, 89–91, 106; philosophical 21; rational 90–2, 106; scientific 150; synthetic 78
knowledge theory (Kant) 77
Königsberg (Germany) 37, 94, 207

Körner, S. 160
Kreimendahl, L.B.: and Oberhausen, M. 267n

Lambert, J.H. 3, 87–95, 98–107, 107n, 227; Essay on the Criterion of Truth 101; Neues Organon 227; philosophy of geometry 88; real definitions 98–103; theory of parallel lines 98
Land, T. 1–4, 6–32
language: natural 66–7, 77, 82n, 164n
Latin 13
Latin Ontology (Wolff) 256, 267n
law of falling bodies (Galileo) 205
law of gravitation (Newton) 204, 218
Law of Sines 244
Laywine, A. 4, 201–40
Lehrbegriffe (derived concepts) 99, 104, 110n
Leibniz, G.W. von 3, 14, 55–7n, 72, 83n, 92–100, 108–9n, 134n, 142–50, 161n, 163n, 251–4, 264, 267n; definitions of parallel lines 94–7; Discourse of Metaphysics 97; geometrical analysis of situation 87; Mediations on Knowledge Truth and Ideas 142; Nouveaux Essais 72, 83n, 95–7, 108n, 143–5, 162n; Theodicy 253
Leibnizian rationalism 76
Leibnizian tradition 150
line segments 90, 181, 209, 217, 223
linear speed 173
lines: perpendicular 96, 105; straight 94–6, 105, 108n, 111n, 129, 135n, 159, 208–9, 224, 251; vertical 156, see also parallel lines
Locke, J. 57–8n, 83n, 144, 161–2n, 229
locus: geometrical 244–5
locus problem: three and four-line 224
Loemker, L. 148–9
logical definitions 126, 139n
logical homogeneity: strict 12–21
Longuenesse, B. 53n
lower-order proposition 248

McLear, C. 9–10
macroscopic properties 135n
magnitude 1, 12–16, 170; composition 180; concept 2, 13–15, 18; ever-diminishing 178; extended 176; extensive 125, 186–94, 197; Greek theory of (Eudoxus) 190; intensive 178, 186–96; mathematical 198; mathematically homogeneous 195; scalar 199n; spatial 177, 181
Maimon, S. 88
Manders, K. 84n
manifoldness 20, 29n, 205
material sciences 113
mathematical beauty: role in Kant's intellectual development 4, 241–70
mathematical physics 43–4, 192, 198
mathematical proof 92, 96–8, 101–5
mathematical proofs theory (Kant) 105
mathematical purposiveness 4, 241–9, **245, 246**, 253, 257, 260–2, 265
mathematization 169, 181, 188
mechanical laws of motion 182, 196
*Mediations on Knowledge Truth and Ideas* (Leibniz) 142
medieval philosophy 187–8; instantaneous motion 187–8
medieval theory: of intension and remission of forms 187, 194–6; of motion 173, 189
Meerbote, F.: and Walford, D. 267–8n
mental activity 42–3, 46
mereological account 8–10; simple 9–17, 22, 25
mesoscopic properties 135n
metaphysical 4, 36, 41, 65, 138n, 160, 162n, 187–8, 196, 250
metaphysical deduction 38, 51n
metaphysical doctrines 1, 145
metaphysical exposition of space concept 2, 9, 12–13, 34, 47n
*Metaphysical Foundations* (Friedman) 188, 200n
*Metaphysical Foundations of Natural Science* (MFNS) 3–4, 168–70, 181–2, 190, 227

metaphysical truths 140, 145
metaphysics 66–8, 134n, 140–2, 145–8, 156–8, 202, 257; traditional 202–3; transcendental 119
*Metaphysik Vigilantius* (Kant) 76
methodology: mathematical 89, 103, 141
Milky Way 226
modal phenomenology 2, 75
modalities: sensory 68
modality 138n
modern formula arithmetic 156
moral devotion 250
motion: composition of 168–200; constructability of 179; construction of 168–200; curvilinear 173; discontinuous 174; explication of 170–1; herky-jerky 172–3; hypothetical 177; instantaneous 170, 173, 177–9, 182–9, 194, 199–200n; interval of 179; mechanical laws of 182, 196; medieval theory 173, 189; net 172; phoronomical conceptions of motion (Kant) 170–8; phoronomical treatment of 178; rectilinear 171, 173; relative 170; and rest at points in space 174–8; uniform 173–5, 195
movement: uniform 187
moving vector 183

*Natural History and Theory of the Heavens* (Kant) 255
natural language 66–7, 77, 82n, 164n
natural philosophy 173, 178, 196, 202–3; Newtonian 168–70, 188, 197
natural science 52–3n, 169, 179, 189, 192, 196, 203, 257
natural teleology 203
natural theology 140, 257
natural thought 194, 205
natural world 4
nebulous stars 226
necessity of essences (Wolff) 253–7
*Negative Magnitudes* (Kant) 49–50n
net motion 172

net speed 172
*Neues Organon* (Lambert) 227
neuroscience 82n
new positive theory of essences 255
Newton, I. 178, 188, 202–5, 218, 228, 231;
    law of gravitation 204, 218; *Principia*
    205, 228; theory of fluxions 199n
Newtonian mathematical physics 4
Newtonian natural philosophy 168–70,
    188, 197
Newtonian physics 174, 196; instantaneous
    velocity 196
Newtonian science 189, 198n
Newtonians: reification of space 55n
Nicolovius, F. 235n
nineteenth century 89, 163n, 176–7
nominal definitions 117, 126, 129, 137n
nominalism 252
non-intersecting lines 99
nonconceptualism 7
*Nouveaux Essais* (Leibniz) 72, 83n, 95–7,
    108n, 143–5, 162n
*Nova dilucidatio* 255–6
numbers: polygonal 146–7
numerical differences 15–16, 28n
numerical identity 14
numerical symbols 144
Nunez, T. 3, 113–39

Oberhausen, M.: and Kreimendahl, L.B.
    267n
object-giving representation 34, 40, 46
objective formal purposiveness 241, 265
objective purposiveness 241–4, 247
objective reality 141, 208–11, 215–17,
    237n
objects: empirical 133, 135n, 160;
    extramental 140; geometrical 248,
    252; mathematical 100, 118, 158–60,
    229, 243, 266, 266n; quantitative 252;
    spatial 116, 130
observational astronomy 226–7
Oldenburg, H. 142–5
*On a Discovery* (Kant) 264

*On a Recently Prominent Tone* (Kant) 248
*Only Possible Argument in Support of a
    Demonstration of God's Existence, The*
    (Kant) 242, 257–62
orbit: closed 172–3; galaxy 226; planetary
    247
Oresme, N. 187
oscillations 172–3
outer intuition 13, 16–18, 24
Oxford (UK) 187

Pappus of Alexandria 149, 224, 236n
parabola 208–11, 235n
parallel lines 87–9, 95–106, 156, 209–10,
    225
parallel lines definitions 94–7; Euclid
    94–7, 98–103; Leibniz 94–7
parallel lines theory (Lambert) 98
parallelism 108n, 143
parallels theory 95–6
Paris (France) 187
Parsons, C. 53–4n, 154, 163n
Pascal, B. 225, 235
Paton, H.J. 48n
perpendicular lines 96, 105
phenomenology 75, 227; modal 2, 75
*Philosophia Prima sive Ontologia* (Wolff)
    250
philosophical cognition 150, 168
philosophical concepts 113, 157
philosophical eye 257–63
philosophical gaze: transcendental- 263
philosophical immunization 219
philosophical knowledge 21
philosophical reasoning 157
philosophical treatise on mathematics
    201
*Philosophisches Magazin* 207
philosophy: of geometry (Lambert) 88;
    Newtonian natural 168–70, 188, 197;
    theoretical 202, 203; traditional 119,
    122; transcendental 113, 132
phoronomical conceptions of motion and
    speed (Kant) 170–8

phoronomical treatment of motion 178
Phoronomy (Kant) 4, 168–200; explication
    of motion 170–1
physical concepts 4
physical sciences 227
physics 120; mathematical 43–4, 192,
    198; Newtonian 174, 196; Newtonian
    mathematical 4
Pippin, R. 48n
planetary orbits 247
planimetric figures 204
Plato 219–20, 237n, 243, 263–5, 268n;
    mathematical purposiveness 243
plurality 129, 169; homogeneous 129
plurality of spaces 42
political snobism 220
polygonal numbers 146–7
Porphyrian tree 15, 117, 124, 146
practical corollaries 99, 102–3
practical sphere 231
primitive theory of space (Evans) 20
*Principia* (Newton) 205, 228
*Principle of Least Action* (de Maupertuis)
    267–8n
Prize Essay (Kant, 1763) 140–67
productive imagination 6, 12, 17–21, 25,
    27–8n
productive synthesis of imagination 1
projected unity 227
projective geometry 207, 226–8
*Prolegomena to Any Future Metaphysics*
    (Kant) 70, 81n, 129, 204–7, 221,
    227–8, 233
proof: mathematical 92, 96–8, 101–5
proportion: theory of (Eudoxus) 13
pure categories 126–8
pure concepts 122, 130–3, 138n
pure intuition 21–2, 27–8n, 35, 53n, 89,
    140–1, 158–60, 247–8, 263
pure reason 201–3
pure science 169
pure sensibility 121
pure synthesis 2
purification 238n

purposiveness 245–9, 264; aesthetic 243;
    formal objective 267n; mathematical 4,
    241–9, **245**, **246**, 253, 257, 260–2, 265;
    mathematical (Plato) 243; objective
    241–4, 247; objective formal 241, 265;
    subjective 242; unlimited 241
Pythagoras 265; square rule 259

quadrilaterals 225
qualitative differences 14–16, 28n
qualitatively identical 14–17
quality: instantaneous 188
*quantitas* 13
quantitative objects 252
*quantum* 13

Rashed, R. 236n, 238n
rational cognition 3, 91, 102, 107n
rational knowledge 90–2, 106
*Rational Thoughts on the Purposes of
    Natural Things* (Wolff) 250
rationalism 142–3; Leibnizian 76
real definitions 3, 93–107, 107–9n, 118, 129,
    134n, 138–9n; of circle (Wolff and Kant)
    92–4; in geometry 87–112; immediacy
    condition 103–7; Lambert 98–103
real definitions theory (Kant) 3, 90, 99
reality: extramental 153; objective 141,
    208–11, 215–17, 237n
reason: pure 201–3
reasoning: abstract 142–4; algebraic 146;
    geometrical 161n
Rechter, O. 154–5, 163–4n
reciprocal determination 124
rectangle 213, 221–3, 246
rectilinear motion 173; uniform 171
reflective judgement 4, 203
Reichenbach, H. 89
Reinhold, K.L. 90
relative motion 170
relativity 170
religious devotion 250
representation: conceptual 2, 33–5, 39,
    48n; diagrammatic 73; discursive 33,

37–9, 67; geometrical 163n; intellectual 11; intuitive 2, 33–5, 41–2, 45, 48n; intuitive origin of spatial 33–62; object-giving 34, 40, 46; sensible 11–12, 17, 45; singular 8, 46; spatial 1, 6–32, 33–61, 73; spatio-temporal 46, 52n, 80; symbolic 149; systematic symbolic 149

rest: constructability of 179; instantaneous 177–9, 185–6, 189, 200n

rest and motion at points of space 174–8

Richardus, C. 236n

right angles 184, 210–14, 223

right-angle cone 212

Risi, V. de 236n

Robinson, A. 199n

Rohault, J. 57n

Rosenberg, J.F. 48n

Rosenkoetter, T. 54n, 134n

Saccheri, G. 105

scalar magnitude 199n

schema: concepts 92

schemata 159

schematism 4, 159, 229–35

schematization 130, 233–4

schematized categories 126–32

Schlosser, J.G. 219–20, 225, 238n

Schultz, J. 96, 106

Schwaiger, C. 162n

science: Newtonian 189, 198n; pure 169

science theory (Wolff) 98

scientific knowledge 150

Secant Tangent Theorem 258

Sellars, W. 26n, 29n

sensations: visual 193

sensibility 1–3, 6–8, 18–19, 26n, 28–9n, 40–7, 47n, 54n, 59n, 120–3, 140–2, 219–21, 228–9, 233–4, 264–6; formal 228; human 19, 202, 219; pure 121

sensible concepts 92

sensible intuitions 6–9, 18, 26n, 45, 57n, 59n, 79, 155, 264–5

sensible representation 11–12, 17, 45

sensory modalities 68

seventeenth century 142, 173–8, 186, 197

Shabel, L. 22, 27n, 48n, 83–4n, 87, 148, 159, 163n; and Carson, E. 1–5

signs: mathematical 156, 164n; use 3

Simple Mereological Account 9–17, 22, 25

singular representation 8, 46

singularity 35–8; of intuition 46–7

situation: geometrical analysis of (Leibniz) 87

*Sixth Meditation* (Descartes) 161n

Smith, K. 48–9n

Smyth, D. 2–4, 27n, 33–61

snobism: intellectual 220; political 220

solar system 226

solitary space 36

sortal-identity 55n, 57–8n

space: characterization of 15; ever-diminishing intervals 184; extended interval of 171, 174–9, 188, 191, 195–8; holistic mereological structure 42–6; holistic structure 41; interval of 171–3, 174–91, 195–8; Newtonians' reification of 55n; primitive theory of (Evans) 20; rest and motion at points of 174–8; solitary 36; transcendental exposition of 136n; unitary 34–6

space concept: metaphysical exposition of 2, 9, 12–13, 34, 47n

spaces: determinate 2, 18, 24, 43; plurality of 42

spatial composition 182

spatial distance 183

spatial exhibition 131

spatial indexicals 19

spatial interval 185

spatial intuitions 21

spatial magnitude 177, 181

spatial objects 116, 130

spatial representation 1, 6–32, 73; intuitive origin 33–61

spatial vectors 182–3

spatio-temporal representation 46, 52n, 80

species concepts 124

speed: constructability of 179; definition (Kant) 170–1; instantaneous 169–70, 173, 176–8, 184–9, 194–7, 199n; linear 173; net 172; phoronomical conceptions of (Kant) 170–8
spontaneity 2, 8, 11–12, 17–25, 27–9n
spontaneity-dependent 6–7
stars: nebulous 226
Steiner, J. 221–3
straight line segment 213
straight lines 94–6, 105, 108n, 111n, 129, 135n, 159, 208–9, 224, 251
Strawson, P.F. 48n
strict logical homogeneity 12–21
subjective purposiveness 242
subordinate concepts 126
Sutherland, D. 3, 13–14, 82n, 87, 148, 163n, 168–200
Swineshead, R. 188
syllogisms 102, 106, 149
symbolic art 145–8, 162n
symbolic cognition 142, 162n
symbolic expression 148
symbolic representation 149; systematic 149
symbolism 144–5; mathematical 151
symbolization 146–8, 161n; systematic 150
symbols: mathematical 3, 141; numerical 144
synthesis 148–50, 168; arbitrary 69, 72, 137n, 158, 164n; pure 2
synthetic a priori knowledge 241
synthetic a priori truths 89
synthetic apriority 1, 4, 78
synthetic knowledge 78
synthetic method 120–1
synthetic truths 262, 266
synthetic unity 10, 18, 24, 26n, 266
syntheticity of geometrical axioms 87
System of Principles 197
systematic symbolic representation 149
systematic symbolization 150
systematic unity 4, 122, 205–7, 212–17, 220, 224–34

tautology 137n
technical concepts 135n
teleological judgement 4
teleological theory (Kant) 257
teleology 203
temporal indexicals 19
temporal relations 19
terrestrial gravity 188
thematic unity 224
Theodicy (Leibniz) 253
theoretical philosophy 202, 203
theoretical sphere 231
Third Postulate (Euclid) 93
third space arguments 9
thought: intuitive 14; mathematical 151, 157; natural 194, 205
time: ever-diminishing intervals of 184; extended interval of 174–7, 188, 191, 195–8; instants of 174–8; interval of 173–90, 191, 195–8
Tolley, C. 29n
Tonelli, G. 83n, 162n, 164n
traditional metaphysics 202–3
traditional philosophy 119, 122
Transcendental Aesthetic (Kant) 78, 120–1
Transcendental Analytic (Kant) 78, 122–3, 126
Transcendental Deduction (Kant) 11
transcendental exposition of space 136n
transcendental eye 262–5
Transcendental Idealism (Kant) 130
transcendental ideality 1
Transcendental Logic (Kant) 122
transcendental metaphysics 119
transcendental philosophy 113, 132
transcendental-philosophical gaze 263
trees: Porphyrian 15, 117, 124, 146
triangles 91–2, 108n, 138n, 159, 209–10, 225, 244–5, 251–2, 260; axial 209, 212–14; equilateral 103–6
triangularity 66
truths: beauty of (Wolff) 249–53; fundamental 151; geometrical 97, 140; mathematical 1, 249–53, 268n;

metaphysical 140, 145; synthetic 262, 266; synthetic *a priori* 89
turn-around point 175–7, 199n
*Two New Sciences* (Galileo) 259
two stems of cognition 1, 6–32
two stems doctrine 7–8, 23
typology 242–3, 266n

Unguru, S.: and Fried, M. 206
unification 37, 130, 233
uniform change in velocity 175
uniform motion 173–5, 195
uniform movement 187
uniform rectilinear motion 171
uniformity 171–2
unitary space 34–6
unity: discursive 10; essential 42–6; projected 227; of reason 226; synthetic 10, 18, 24, 26n, 266; systematic 4, 122, 205–7, 212–17, 220, 224–34; thematic 224
universal gravitation 169
*Universal Theory and Natural History of the Heavens* (Kant) 226
universality 229
unlimited purposiveness 241

Vaihinger, H. 48–9n
vector: composition 181–3; moving 183; representation 184; spatial 182–3

velocity 169, 174, 185, 188, 199n; continuous 175; instantaneous 168–9, 174, 186–9, 194–6, 199n; uniform change in 175
vertical angle 212–13
vertical lines 156
vibrations 172–3
visual imagination 79
visual sensations 193

Walford, D.: and Meerbote, F. 267–8n
Webb, J. 110n
Wikipedia 268n
Wolff, C. 76, 83n, 87–95, 98–100, 105–6, 134n, 141–50, 161n, 267n; background and Kant's Prize Essay (1763) 140–67; beauty of mathematical truths 249–53; doctrine of necessity of essences 253–7; *Elementa Matheseos Universae* 148; *Philosophia Prima sive Ontologia* 250; philosophy 140; *Rational Thoughts on the Purposes of Natural Things* 250; real definition of circle 92–4; symbolic art 145–8, 162n; theory of cognition 148; theory of science 98
Wolff, M. 26n
Wolffianism 153
Wood, A. 199n

Zeuthen, H.G. 215, 237n; *Die Lehre* 237n